The Essential Amory Lovins

The Essential
AMORY LOVINS
Selected Writings

AMORY B. LOVINS

Edited by **CAMERON M. BURNS**

publishing for a sustainable future

London • New York

First published 2011
by Earthscan
2 Park Square, Milton Park, Abingdon, Oxon OX14 4RN

Simultaneously published in the USA and Canada
by Earthscan
711 Third Avenue, New York, NY 10017

Earthscan is an imprint of the Taylor & Francis Group, an informa business

British Library Cataloguing in Publication Data
A catalogue record for this book is available from the British Library

Library of Congress Cataloging in Publication Data
Lovins, Amory B., 1947-
  The essential Amory Lovins : selected writings / Amory B. Lovins ; edited by
Cameron M. Burns.
     p. cm.
  Includes bibliographical references and index.
  1. Environmental policy. 2. Energy policy. I. Burns, Cameron. II. Title.
GE170.L68 2011
333.72--dc22
                             2011000744

ISBN: 978-1-84971-226-2

Typeset in Sabon and Frutiger
by MapSet Ltd, Gateshead, UK

MIX
Paper from
responsible sources
FSC
www.fsc.org FSC® C004839   Printed and bound in Great Britain by the MPG Books Group

Contents

Section 5—Our Human Environment: Building Better, Building Smarter

Section 6—Energy Security and the Military: Blood and Treasure (and Opportunities)

Section 7—Business and Climate: Making Sense, Making Cash, Making Good

Section 8—Miscellany: A Poem, a Letter, and an Opinion

Section 9—Final Thoughts: Finding the Path

Preface

At 63, composing my life so far has included many adventures, but none so much fun as assembling, feeding, and exploring a mind that's a bit like one imagines the attic of a great natural history museum—full of strange stuffed creatures, mysterious artifacts, and wildly eclectic learnings. It's a great place to go wondering in the bewilderness.

As I stroll through this uncatalogued collection, many long-forgotten objects show up. But now order is starting to emerge from chaos. Through the vision and hard work of my Australian-American longtime colleague, friend, editor, and fellow mountaineer Cam Burns, some of the more readable papers he's selected from this diverse collection, drawn from 42 years and about 60 countries, are going on display. You hold the catalogue.

You may wonder where all these odd bits came from. My grand adventure began in a tiny house beneath great maple trees on the end of a quiet little side-street outside Washington, DC. My sister Julie (now a computer linguist) and I played with the ants on the peonies and walked to school along a forested creek. Beneath our family's living quarters, our father, a polymath engineer, designed and built unique scientific instruments like the world's largest optical microscope. His workshop became my playground, setting me on a path to experimental physics. I would disassemble anything to see how it worked. At three or four years of age I liked taking old watches to bits and putting them back together. About half of them resumed working.

Most importantly, our parents had filled the house with all sorts of books, but no television. (Had God meant us to watch TV, She'd have given us square eyeballs.) Ignoring that period's educational dogma of recognizing whole words at once—which doesn't work, so many American contemporaries still can't sound out a word they don't know—our mother, a social-service administrator and editor, held me on her lap, read me poetry, taught me phonics, and had me reading before age four. A sickly child, I was nearly as much at home as at school, so I'd read everything in the house, including the whole *Encyclopaedia Britannica*, by about age seven. Fortunately I recall little of it, but random fragments do pop up at the strangest times.

Our parents also had the wise habit of inviting friends and visitors to dinner, and including us in the adult conversation on all the issues of the day. This immersed us in the life of the mind and the needs of the world. Our musical studies and our father's photography, which I later took up, added culture. Both sides of the family had a heritage of science, art, music, and languages. All four grandparents were immigrants from Ukraine. Our paternal grandfather had fled in the night at 17, just ahead of the secret police, from the inn his parents kept at Tarashcha, where all the merchants stopped between Kiev and Odessa, so he was fluent in a dozen languages. (I briefly knew him when I was a young boy and he was an old man, and he tried to teach me a little of his pre-Revolutionary Russian. He said with a wink that in any language I need know only two phrases—"You are a pretty girl" and "I love you"—so he taught me those.)

One of my piano teachers was the sister-in-law of the great physicist Leo Szilard. Near the end of his life, when I was about nine, we met and I got to ask him why the sky is blue. He gave a good physics answer like "Because the sunset is red," but I gave him to understand I was seeking a more teleological "why" than that. He thought awhile and replied with a twinkle, "My boy, I'm afraid I can't tell you that—not because you're not old enough, but because *I'm* not old enough." His humility encouraged me, as Confucius urged, to keep pursuing knowledge as if chasing after someone with whom one can never quite catch up. Later came other kinds of learning, from Buddhism and the Tao.

By my teens, I'd discovered a subversive secret that I'll share with you now in case you might not have discovered it for yourself: There are few if any disciplines that an intelligent and motivated person cannot learn as much about in six months as most (not all!) people in the field know. This realization opens the doors, nay the floodgates, to the unity of all knowledge—what Harvard biologist E. O. Wilson calls "consilience." It also grants a perpetual license to traverse the entire range of learning, roaming hither and yon without the least inhibition about invading someone's turf or trampling someone's petunias. You can then learn anything you wish, without permission or formality. I've taken liberal advantage of that license ever since, decades before there was an internet. This nomadic hunter–gatherer style of learning complements my distaste for curricula and syllabi. Their compartmentalization of knowledge implies that some things can be learned well and others defectively, all without consequences, because the bits are all separate and unrelated. They're not.

So with this somewhat autodidact background, I set about to learn what seemed interesting and necessary. This naturally led to an unusual academic career: American grammar schools; half the Harvard undergraduate course; two years an Advanced Student at Magdalen College, Oxford; two years a Junior Research Fellow of Merton College, Oxford. I then shifted from academic life to working round the world from a London base for the greatest conservationist of the 20th century, the late David R. Brower, who paid me a little (about enough to cover the telephone bill) to do whatever I thought needed doing. I learned decades later that my parents had worried I was abandoning a promising academic career, but they were kind enough not to say so: their philosophy of child-rearing was to love your children, support their dreams, and get out of their way.

What I thought needed doing in 1971 was mostly linked to energy—clearly about to become a major and persistent global problem. (The Arab oil embargo broke on the world like a thunderclap two years later.) Energy seemed a sort of master key: solving that problem would solve many other problems too, or at least reveal how to solve them.

So after a decade in London, flying back annually to continue a 15-year habit of guiding 100 days a year in the mountains (chiefly in New Hampshire), I moved back to America in 1981 with my first wife L. Hunter Lovins—a lawyer, sociologist, political scientist, forester, and rodeo-riding cowboy. Our goal was to site and build a nest, and to found and lead a small public charity, Rocky Mountain Institute, to make our work more effective. Hunter left 20 years later. RMI continues to flourish in two Colorado sites. It now has about 80 people, gratifying accomplishments, and the vision of a world thriving, verdant, and secure, for all, for ever—pursued by driving the efficient and restorative use of resources.

Such an ambitious goal needs broad foundations. Fortunately, my school studies (commuting on foot to extra classes at nearby Amherst College) had emphasized classics, music, mathematics, and science. At Harvard, alongside senior and graduate research courses in the physical sciences, I enthusiastically added linguistics, humanities, some law, a little social science, and stirred vigorously. "How is all that related?", my faculty advisers kept asking. Well, it's not related, and I like it that way, said I: this is a great *Univers*ity, is it not? Surely the world needs broadly educated people who can make new kinds of connections and thereby solve the problems created by narrow specialists.

Alas, it was all too irregular, so I fled regimentation by migrating to Oxford. I lived two years under the most beautiful tower in England, writing in 1968 my first professional paper on climate change, then two years in the blessed life of a don, before finding it was a few decades too soon to be allowed to do an Oxford doctorate in energy. Meanwhile, though, Dave Brower had engaged my photographer chum Philip Howell Evans and me to create a photographic book about our beloved mountains in north Wales. This made it natural to resign the Fellowship, work for Dave, and gradually make a different path. What a different, stimulating, and wide-ranging learning journey it has become!

In our line of work at RMI, it's normal to pick up a couple of new disciplines a year, and you never know what they'll be: perhaps mining engineering and naval architecture last year, refugee camp design and automotive engineering this year, military doctrine and energy anthropology next year. After a couple of decades of this, everything reminds you of something (as the linguists dryly say about one's first fifty languages). It therefore seemed perfectly natural, in writing a 2002 book on how to get off oil for fun and profit, to blend Thucydides's *Peloponnesian War* and Aristotle's *Nichomachean Ethics* with mid-19th-century economic history and early-21st-century engineering. The only boundaries of the mind are the ones we build. No boundaries mean easier movement. This collection, winnowed by Cam's keen eye from a much larger and more disorderly attic-full, hints at the untrammeled limitless landscape of new solutions awaiting your discovery.

That requires both learning and unlearning. My other great mentor, inventor Edwin Land, said: "People who seem to have had a new idea have often just stopped having an old idea." To which he added, among other things, "Invention is a sudden cessation of stupidity;" "A failure is a circumstance not yet fully turned to your advantage;" and "Don't undertake a project unless it is manifestly important and nearly impossible." Wise teachings all.

The world faces many great challenges, but they look less daunting when their intricate linkages are understood. Yes, the cause of today's problems is often prior solutions, but the cause of solutions can also be prior solutions—once those connections are fully turned to our advantage. This is manifestly important, nearly impossible, and now coming within our grasp.

Brains, as Gifford and Libba Pinchot note, are evenly distributed, one per person. Therefore most of the world's brains are in the South, half are in the heads of women, and most are in the heads of poor people. As an emerging global nervous system and millions of civil society organizations start to knit together that collective intelligence—the most powerful thing we know in the Universe—innovation and collaboration are starting to overcome stagnation and conflict. The search for intelligent life on Earth continues, but we are all striving to become much higher primates, and some promising specimens are turning up just in time. Join the fun!

As my wife Judy (another fine art landscape photographer) and I sit high in the Rocky Mountains in our passive-solar banana farm watching the koi swim, the turtles bask, and the papayas ripen while it blizzards outside, we have much to be thankful for. High on the list are the many wonderful partners in discovery who help keep us ever curious, ever learning, ever joyful. We persevere in the spirit that Gōtō Roshi, a master in Japan, encapsulated for theologian Huston Smith: "What is Zen? Simple, simple, so simple. Infinite gratitude toward all things past; infinite service to all things present; infinite responsibility to all things future."

Thank you for joining us on this exciting journey in applied hope.

Amory B. Lovins
Old Snowmass, Colorado, USA
28 February 2011

Editor's Introduction

This collection of essays, articles, white papers, poems, and letters offers a glimpse into the development of one of the world's great thinkers. That development is no simple linear (or even curved) trajectory. Rather, it's a web of circular paths, weaving in new ideas and data as both society and the observer evolve. Yet it started with two strong threads that still run throughout the tapestry.

As a youngster, Amory was introduced to the natural environment by Peter Johnson, a school chum in Amherst, Massachusetts, who got him "into" the outdoors—mostly hillwalking. This enthusiasm for the natural world would later manifest itself physically in Amory's work as a camp counselor, his passion for scrambling up Welsh mountains during his Oxford years, and a general joy of wild places, and cerebrally in his deep appreciation for our planet and the things it gives us.

Amory's technical comprehension and numerical skills were born of illness. As a kid, he was sick a lot of the time, with various respiratory afflictions, pneumonia, and croup. But his mother taught him to read at a very early age, and he disappeared under the covers with books on everything from mathematics to forestry. (At age ten, his parents learned that he lacked gamma globulin, an important component of the immune system, so his doctor started injecting it, and he's been healthy ever since. Coincidentally, his engineer father had earlier helped commercialize the analytic technology, serum electrophoresis, that revealed the defect.) His technical expertise was well known to family and friends by the time he was in middle school, and some of his high school papers (none included here) are more complex than the energy and financial commentaries he'd write 30 years later—indeed, at about 15 he unwittingly reinvented the integral calculus.

The two threads coevolved. By the early 1970s, when Amory was in his mid-20s, he was using his keen analytical skills to sum up humanity's approach to resources, and how contorted it was (and remains). Yet he was as likely to write a moving essay on the nature of a mountain-climbing trip as a heavily numerical financial or physics analysis of a proposed nuclear power plant.

Energy issues, especially, offered a sort of "grand synthesis" that Amory found many years ago. Most fans of his energy ideas might automatically point to "Energy Strategy: The Road Not Taken?" as a watershed moment. It is, but it is much more than that. In the article, published in 1976, Amory asked a profound and daunting but remarkably simple question: What are we really doing with energy and why? That simple recasting of the question, as would become obvious through Amory's work in subsequent years, is applicable not just to energy but to *everything*—water, cars, food, data centers, communities, trees, soil, buildings, industry, even your everyday life. That one can ask this question in so many arenas shows the universality of digging for our ultimate objectives to see if they might be achieved in a better, more sensible way. This approach has distanced Amory from other commentators on these subjects for more than three decades.

"Energy Strategy: The Road Not Taken?" is a good starting point in understanding Amory's thinking and how he turned energy (and resource) thinking on its head, but the evolution actually began much earlier. During his Oxford years, he got involved in a fight over metal resources in the UK—specifically, copper mining in north Wales. In his writings about the issue, done collaboratively with Friends of the Earth staff in London, it's clear to see that Amory is quickly drifting away from environmental battle per se to environmental common sense (for example, in 1971 he wrote at one point that "scrap brass is a much better ore than Welsh rock"). In the process, Amory helped turn "environmentalism" on its own head, creating a new sort of enviro-*rationalism* that had arguably never existed in the form that Amory has spent his life developing—that is, with the analytical and scientific rigor that might, for example, show the enormous potential and cost-effectiveness of brass recycling, or of not needing brass at all.

This collection is not exhaustive; instead, it is representative, a sampling that will provoke your own thinking. By including certain pieces like some pretty old poetry and a few letters, we hope to entertain seasoned Amory fans, while the material on the environment and energy will hopefully find play with younger readers. By necessity, it has been organized in topical areas (with the exception of "Miscellany")—a mere chronological organization would never have worked, as Amory has circled back on every topic he's ever tackled, sometimes decades later, to keep expanding and updating the facets of the discussion and refining his conclusions. And, you'll note, all topics eventually get drawn into a greater whole, which is exactly how things work on Planet Earth.

We are not here on this Earth for ourselves. We are here for those who came before, and those who will follow—especially them. Amory asks us to think about what we're doing, why, and what the outcome of our choices might be. Read just one of these pieces and you'll find yourself immersed in that great question.

This book wouldn't have been possible without the help of numerous colleagues—mostly, of course, the staff of the public charity Rocky Mountain Institute, where Amory has been practicing whole-system solutions for 29 years, now with about 80 collaborators. Several current and former RMI staff members helped pull together the collection of thoughts, essays, analyses, and insights you now hold in your hands, among them L. Hunter Lovins, Rick Heede, Aaron

Westgate, Maria Stamas, Noah Buhayar, Betsy Ronan Herzog, and Megan Shean. Jonathan Sinclair Wilson at Earthscan and his colleagues Michael Fell, Nicki Dennis, and Anna Rice all deserve thanks for getting this book going in the first place. Thank you all.

Cameron M. Burns
Old Snowmass, Colorado, USA
October 2010

Section 1

Into the Wildness: Mountain Climbing, North Wales, and Poetry

In his teens, Amory became enamored with the outdoors, especially mountainous regions—first in his "native" New England and later (while at Oxford) in Eryri, the mountainous northern part of Gwynedd in North Wales. There he learned of Rio Tinto Finance and Exploration, Inc.'s plan to mine copper in that great "wildness," as Amory called it in the exhibit format book *Eryri: The Mountains of Longing*, commissioned by David Brower to bring attention to the looming threat. As Sir Charles Evans wrote in *Eryri*'s forward:

> *Mr Lovins's sensitive and appreciative description of Snowdonia is evocative, and his frank declaration of opinion about the dangers which threaten the Park's very existence demands thought. He writes both of the Park's natural beauty as it is now and of the movements, as he sees them, which put the Park in jeopardy. His descriptions of what we as a nation possess, and his exposition of how that possession might be lost, must command the attention of all who are concerned to preserve for future generations what is left of the Snowdonia we know.*

Amory and his Friends of the Earth colleagues in London also penned several influential articles for magazines and newspapers about Rio Tinto's plans. As a result of the book, the articles, and a BBC program on which Amory and Philip Howell Evans (his co-photographer on the book) collaborated, the mining proposal was abandoned, Rio Tinto went elsewhere, and the copper is still there. Years later, Tom Burke, who as head of Friends of the Earth in the UK had helped lead the campaign against the mining, became senior environmental adviser to the chairman of Rio Tinto. More recently, Amory and his staff at Rocky Mountain

Institute in Colorado have advised Rio Tinto on the design of various mining projects round the world. Strangely, yet wonderfully, the story has come full circle, and today Rio Tinto is not just a valued industrial partner but a leader in "greening" the global mining industry.

Rather than include essays or articles regarding that old fight over mining, we have instead chosen several essays indicative of Amory's love for mountain areas. Included here are "New England Wanderer," reprinted from *Oxford Mountaineering* (1969), "Longing," from *Eryri: The Mountains of Longing* (1971), and a poem, "To a Poet Met in Cwm Dyli" (1971, also from *Eryri*), that combines a nice (and true) story—of a wandering poet searching for words to take to the next Eisteddfod (a bardic festival)—with haiku stanzaic form and Welsh *cynghanedd* ("binding"—Welsh poetry often echoes consonants rather than rhyming vowels).

Chapter 1

New England Wanderer

1969

Out of dreamless depths as the last stars fall westward. The nylon net-hammock sways gently under the tarp in the first breeze of the world. Too early yet. Doze. No, the birds are awake, promising divinity to any who will join them. Up, then, extrude from the warm cocoon into pine-scented air. Release the hammock prusik and stand in the dawn. Unslip the hitched cord that dangles boots out of reach of porcupines and raccoons. Now over to the stream for a wakening splash. Chittering squirrel bounds away. A leaf slips lazily down the air to ripple the pool, a surge of brown and silver slants up from shadow. Breakfast? Hm, browsing between those rocks. Work round quietly, out of sight. The hands into the water so gently. Brr. Now cup them into a little cave. Patience. Ah, just put your nose in here. No, over this way. Be curious, trout. Come explore this nice cranny. That's right. Now.

Too big for the pan, but easily remedied, thus. Shake glistening dew off the maple billets. Just tap with the axe, there and there, to release the seasoning strains. Explode into dry sticks. Now the knife for a bit of kindling, and some birchbark under, and the sweet smoke, and breakfast, with wild raspberries after. Scour the pan with sand. Then a soft hiss of white oilstone over axe, and stir hatfuls of water into the ashes. Five minutes to pack. Tidy up, and hoist ten kilos to the back, and off into the fourth day of a fortnight's nomadism.

That spur looks good, and up the arête and along the ridge. The right general direction. Warm as oak and maple and ironwood give way to beech and birch: strange silent pillars of pearl-grey, white, reddish-brown, and soft pale gold, smooth and glassy, holding up a roof freshly dipped in molten greenness. All is frozen in time; no leaf flutters. Then the chill of stiff spruces, already getting small and stunted, and there an ancient bonsai cherry-tree only knee-high. The granite starts to show through, the greens and yellows of lichens tinted by sky and water into a misty grey that must be the color of time. Easy slabs, steepening into broken faces. Softly now, nobody to help here. Up the back of that chimney seems simplest. It lies well under the hand. Belay to the pack over a spike, then jam round a short overhang like a granite cornice, and mantelshelf onto the ridge. A light wind with a good taste in it. On the other side a dark green sea whose waves are tipped with stone. Six ranges, there, before the haze of distance intervenes. Thirty or forty miles. Hovering just below, five square miles of cirque, nearly a mile deep. No roads, no sounds, no smoke, no movement, only this.

A deep fissure with snow, and a rock chair on the leeward side close by, so lunch of cakes, cheese, chocolate, raspberry ice. The smell of sun on the rock. Go away, eagle, I'm not for dinner. Time to triangulate. Yes, right. Glacier Cream against the noonday sun, and scramble up the ridge to the peak. A step too exposed, a sling for aid. Good. Under the top stone of the cairn, a plastic bag with a note to Wally, whoever he may be, dated... what day is this? Dated two days back, telling Wally to rendezvous as soon as convenient by following this spur and that stream and the arête to a blasted dwarf spruce, where he will find further directions, and please bring some fresh blueberries for muffins. Well, if they're over there, not to pick the same place. Perhaps better to traverse the next peak to that broad col, then... a noise? Nothing, but move round that buttress to be sure. My God, where did that thunderhead come from so fast? Flash, twelve seconds, moving this way, time to be somewhere else in a hurry. Careless, careless: that golden eagle should have told me, swooping down where there should have been an updraft.

Down that scree chute on the cirque side, brutal or not, use the rope. No, wet and unstable lower down, not healthy to stay in. Must find a trail. If I were a trail, I'd slab along that slope to the left, where the tongue of spruce licks upward. No? Downhill a bit. Confound these roots. Ah, a blaze, and an old one cut with skill. Probably by a timber cruiser coming high for the view. A trail: hardly overgrown down here considering its age, perhaps a game trail still used. Self-belay across the bridge, no time to make a safer one. One log crumbles to dust, the other sags but holds. Ominous mutters now, and the big spruces poised against a black sky. Time to build a nest.

A dry inconspicuous spot. Anorak and overtrousers. Tarp, rope, pack, all nylon, on the ground as insulators. Squat on them, feet close together. Suddenly a piece of hell boils over the ridge and hangs above, bellowing. On the ridge a quarter-mile above me, rocks are shattered and smoking now, boulders blowing away. Here for a few seconds the air turns to water. Then marching columns of rain with hail between, and the wind. If this can be called wind. Lightning is down into timber now. A solid blue bar etches itself with a stunning crash ten meters away. A stabbing splinter torn from a star. Well, if it's there and I'm here, I must be all right. That was a tree. Bits of it are starting to come down now on the leeward side. There was another tree between. A deep hollow sound and the smells of ozone and charring hang in the electric air. Rain washes the world and the storm stalks away. Mumbling, then silence. Strange that it could ever be quiet again. The trees drip-dry in the sun.

The decision is made for me: the ridge will be in verglas, so this valley for the night, then across and onto the next ridge. Down the trail, cool now, to a gorge. Abseil onto a ledge. Half a paper-birch. The other half, uprooted and broken minutes ago, on the bank above. Teatime. Water from the stream, and a few tender birch leaves and shoots in a birch-bark kettle, and a bit of green birchwood for the fire. A green taste to make the sap rise, and the sound of the waterfalls with it.

Into the heart of the cirque, a temple with mountain-laurel for incense. Up a hill on the far side, the soft carpet nearly dry underfoot, and on to the foot of a spur from the next range. A highland meadow, a doe gazes curiously and vanishes. Shuffle loudly to scare the snakes. Slithers in tall grass. Still a sound? Yes, a stream there, falling off the side of the spur. A spring at the top. Someone seems to be playing a flute. A clear ledge for the fire, and fine hardwood, and a breeze to keep the bugs away; tempting, yes, an early halt. Hang the sleeping-bag on the rowan-tree to fluff. Logbook, fifteen-odd map miles. Lie and listen to the leaves and the sky flowing over. Fire. Cocoa, freeze-dried stew, more berries bursting sweetly in the mouth. How about a few snares to get some fresh breakfast? No, too full of stew to bother, and the animals have enough troubles without me. Dusk. Two good trees, ridge-rope, tarp, hammock. Hang up the boots and axe and pack, bank the fire, climb into the hammock. Wriggle back into the bag and take up the hammock slack. Not to think about the next range. It will still be there in the morning.

Chapter 2

Longing

1971

Editor's note: *This essay is excerpted from* Eryri: The Mountains of Longing *(McCall Publishing, New York / Allen & Unwin, London) and is reprinted with permission of Friends of the Earth. This essay does not reflect current Friends of the Earth campaigns.*

The snow-squall is nearly over now. Just a little northerly shower, it was, the sort that on the ridge might bring hail and the strange lightning that flickers without thunder. But now the last flakes are falling from thinning cloud, and great sullen shafts of sun are scanning the cornices down the valley by Foel Goch. It will be a good crisp afternoon. I know just the place to show you what this low November sun can do. We'll take a stroll onto the Glyder ridge. It should be out of the cloud by the time we get that high.

See there, while you lace your boots, look across at the little crag near the lip of Cwm Llugwy: if the wind catches the fine spindrift just right and whirls it into that warm patch of sunlight by the boulders—if we're lucky—there'll be a rainbow. They often roost just by those rocks. But only for a moment. Did you see it? Already the flash of color has blinked and vanished. Today it will be even harder than usual to guess where the sunbolts will strike next.

Business now, the silent checklist before we start out. Anorak, mittens, compass, plastic-covered map sections, photographic gear, band to keep spectacles from blowing away. Spare map, compass, food, and socks. Whistle, light rope and cord, foil blanket, first-aid kit, goggles, torch and spare parts, sheath-knife, cagoule, overmitts, extra film, extra extra film, mountain-rescue flares. Altogether less than five pounds in the pack, enough to make it ride well, and occasionally valuable beyond price. Of course one hopes not, but there may come one of those rare moments of professionalism when everything must be done right.

Could you please toss me that down parka? I want to stuff it into the pack too. It must be about freezing and ten knots, so on top it will be about twenty

Fahrenheit and thirty knots, too chilly for just anorak and windproof trousers if we want to linger. No crampons today; the snow cannot possibly have consolidated enough since last night, and has no crust beneath. Snowshoes would be more useful if we had any. And we had better put on our waterproof gaiters to seal the space from boots to knees, for the ridge looks plastered up there, floundery and heavily drifted.

No, I won't forget the ice-axe. How can I forget part of my hand? Even when there is no snow lying I tend to carry the axe in Eryri, despite the curious looks it attracts. It is indispensable in gales, for crossing streams in spate, for probing swamps and snow-bridges, for balance and braking on very steep wet grass or scree, as a camera monopod, and on one occasion to discourage a ram that wanted to butt me off the north face of Moel Cynghorion. He had been hand-reared as a lamb, most likely, for lack of a foster ewe, and came to lose all fear of men. Whatever the explanation, I was glad of an axe to brandish. It would have been a long way down.

We could, I suppose, put a few pounds of extra food and gear into the pack and stay the night by Castell y Gwynt; but I don't think tonight is a good time to bivouac. Something odd is going on. Look. The high clouds can't agree with the medium clouds which way to go, and those lens-shaped ones past Pen yr Ole Wen should arouse suspicion. It will be good to get up high so we can look to the north and see what is happening. But at least the cloud is still rising and we have a few hours before anything breaks.

Right, we're off. Let me get the gate. Watch out for the cars along here: they get ten points for each pedestrian. Here, I'll show you the map. See that farm on it? And the long gentle spur running up from behind it at an almost constant grade, all the way to the broad col in the Glyder ridge? There lies our route. Simple topography on the way up, though there are many little branch spurs that diverge as you come down. The spur is broad enough that we can usually drop down onto its east side to get in the lee. There are several steps on the way up, where cold blue pools lie quietly under the wind, but most of the spur is dry and rises just slowly enough to need no effort—if you know how to walk uphill.

That cluster of low grey stone buildings across the cattle-grid and up the little road is our farm, Gwern gof Isaf, the Lower Alders-by-the-Cave. We are in the land, remember, where Gwilym, the son of Meri-Anni and Jo Bara (Joe Bread, Joe the Baker) can be called Wilmeriannijobara. By the standard, the tenant farmer here gets off lightly with being called Williams Isa', to distinguish him from his brother at the next farm, Gwern gof Uchaf.

Mr Williams Isa' is one of the few farmers who have official permission for extensive tenting on their land; hence the spatter of mountain tents and the ranks of the Army encampment in the front field. It is a cheerful sight, and probably pays about as well as the thousand-odd sheep. But if we call Mr Williams Isa' a tenant farmer, it gives the wrong impression of his permanence: his family has held this farm for four centuries and more. Here he is now, working in the complex of mysterious sheds and pens behind the farmhouse, a big weathered man in rubber boots and green oilskins.

He gives us a comprehensive glance from the boots up, judging whether we'll be all right. Aye, he remembers me from that time I found his ewe cragbound on the headwall last year. I ask him in the language of the country whether we may please go over his land onto the Glyder. A slow smile comes. He must be happy about something. He would be too polite to smile at my accent. "Of course you may. The place for crossing the mountain wall is up by there, and mind the ice at the stream. The first in three months, you are, to ask my permission. You are welcome."

Perhaps we wouldn't be so welcome if we were the two-hundredth party that day to ask his leave; but if things ever got to that state, he'll know what to do. At the moment, he just wants, very reasonably, to know who is tramping through his back-yard, and whether his visitors are likely to kill his cow with a dropped plastic bag or lame his sheep with a discarded tin. I'm not so sure I would be so patient in his place. About the tenth time a hiker broke down a wall so my weaning lambs rejoined their mothers, or left a gate open so my rams got onto the mountain too early (thus bringing the lambs in the deadly winter weather before the proper time), or brought a pet poodle into my *ffridd* in April and panicked the ewes into miscarriages, I think I might do more than just remonstrate. Action first, maybe, and questions afterwards.

Already we are above the bottom-level land and threading our way through the steeper middle part of the farm, the maze of walled meadows that segregate the various kinds of sheep, each in its season, in the intricate pattern that is the essence of hill-farming. And just here is the mountain wall, solid and very old, with a few long stones set into it crossways to serve as a stile. Its roughness, the integrity of the stone, feels good in the hand. Here is about the lowest the snow lies in late autumn and spring, and the highest any substantial trees would grow if the sheep let them. Above here the unfenced sheepwalk rises uninterrupted to the ridge. The grass is coarse and thin, so poor that an acre will barely support a single sheep, and only if she is hardy. Feel the blades shattering, brittle with frost, beneath our feet; we are carving tracks that will not heal until spring. Let's walk in single file, to trample as little as possible of our host's livelihood. The actual saving of grass is slight, but the idea is not.

You are ten thousand years too late to see the last of the glaciers gouging the cwms out of the north side of the Glyder ridge, but you can easily imagine that the ice down there melted just a few minutes ago, the marks are so new and clear. Our spur, Braich-y-ddeugwm, the Arm of the Two Cwms, gives us a steadily expanding view down into two perfect amphitheatres. That swirling pattern of crags on the walls is like the inside of a clay vessel fresh from the potter's wheel.

Today you should be especially happy that we are on this route, for it is one of the few spurs in Eryri, and the only one from the Glyder to the Ogwen Valley, that is safe whatever the snow conditions might be. The snow is now loose and indecisive, somewhere between when its fresh branching rays interlock it cohesively and when the grains get welded into a solid block. On a steeper and narrower route than this spur we might be avalanched.

Across the valley you can see another of these gentle spurs, dropping south-easterly from Pen yr Ole Wen, south-westernmost outpost of the Carneddau. Nearly every novice hiker in Eryri has fallen afoul of guidebooks that recommend the direct west-face or southwest-buttress approach. Anyone who can read a contour map, however, knows better. To reach the southeast spur you have to be scrupulous about permissions. But the spur is safe and very beautiful. It is surprisingly out of the wind, it rarely ices, and it never avalanches. There are several other spurs in Eryri with the same agreeable properties. I collect them, and the knowledge has served me well. It is a good sort of knowledge to gain in advance.

As you quietly walk up Braich-y-ddeugwm, clearing the stagnant air from the lower storeys of your lungs, be sure you look back. See there, to the north across Nant y Benglog, the old walls of Llugwy. They are fragile tracings on the snow. They recede over the rising moor to the gigantic swelling domes of the Carneddau, the loneliest and wildest of the northern ranges. The great mass of Carnedd Llywelyn bulges high in the distance, usually in cloud. But there it is now; see it while you can. In the wind-blasted mountains radiating from Llywelyn you can find great hidden cliffs, gales and mists that seem almost continuous, and (if you are lucky) a herd of shaggy little ponies, free and swift and elusive. From here the immense landscape of the Carneddau is framed for you on the east by the end of the Glyder range, dropping toward Capel Curig, and on the west by one of the most beautiful mountains in the world.

Tryfan, it is called: a paradigmatic mountain, a mountain you will see in dreams. I can close my eyes and see it at will, every notch and cusp of its silhouette imprinted on my retina from hours of marveling at it. "This is a rugged, stupendous, and steep high rock," wrote William Williams of Llandegai in 1798, and that is probably what the name means. Tryfan is a thin triangular knife-blade of volcanic rock, set on edge on a slanted sea of grass and boulders. It soars to the three-peaked edge, each summit buttressed by a pillar and the highest summit crowned by two natural obelisks. From the central terrace to the jagged summit-ridge, the walls of the upper East Face sweep aloft, cleanly and very strongly, for nine hundred feet.

In the history of mountaineering, Tryfan is as significant as almost any mountain you can name, both as a birthplace and as a symbol. When you enter the Ogwen Valley by road from either direction, you round a corner and see Tryfan, and it moves you. It is always the same, yet always different, for more than any other mountain in Eryri it takes on the color and mood of the light, becoming hundreds of different mountains in one majestic form.

It can also be a menacing mountain; its beauty blinds people to its steepness and iciness. I treat it with great respect, for although it is neither technically difficult (by the North Ridge and other scrambling routes) nor objectively dangerous, it is a place where the careless often come to grief, always in the same rather unimaginative ways and always at the worst possible times and places. The same is true of Bristly Ridge, a steep pinnacled arête connecting Bwlch Tryfan, the south col of Tryfan, with Glyder Fach: it is too scrambly and exposed for most novices to feel happy on it, and in the wet or ice it can be insidiously perilous, committing you before you realize it. Reflect on this, and on the silk of the ice on the sharp ridge to the west, as your boots print blue shadows on the snow.

You cannot walk smoothly up Braich-y-ddeugwm on a fair winter afternoon, for the spell of Tryfan keeps breaking your rhythm. No greater mountain has greater character. From this height Tryfan is in its true proportions, not foreshortened. The light is just right now to skim across the upper buttresses and throw them into the greatest relief. The architecture is perfect Gothic, too perfect for man. There is still enough heat from the sun to send showers of powder-snow down from the tops of those gullies, but it is cold enough that the rock pillars themselves glint with metallic verglas. Now, as often, Tryfan is more pleasant to look at than to be at.

Look across at the rock in the sky, and over the col—a glimpse of far peaks, very snowy and framed by luminous mists. Some secret alchemy of light and air gives this place a strange magic. The light is a palpable fluid pervading all things, each partaking according to its quality. For a time, you are granted here a special virtue of seeing, and images bear a vividness of their own. As simple a sight as three reeds thrusting out of the snow can live in the eye with aching clarity.

Already we are up to the final rise of the spur. There are choices here, but this time we shall traverse on the right, to lose as little as possible of the presence of Tryfan. We cannot turn our backs on a queen. Softly now, feel your boots sinking in securely to make the sound steps; Cwm Tryfan gapes below on your right. It is only a minute from here to the col. Ah, we are in luck: the wind has scoured the snow from the frozen lakes, so we are in no danger of walking over them by mistake.

Suddenly, from one step to the next, we are out of shelter and into the hiss of white wind, and it has a snarl in it. It brings fast clouds of purple and rich black, riding in a deep and endless Himalayan sky. Here on Bwlch Caseg Fraith, as on all the high cols of Eryri, the wind is funneled into a greater hurry and an uninformative direction. On the saddle, north by east; on the spur, northwest by west; in the sky, north by west; everywhere, keen and searching.

Hop the gullies, miss the little frozen pools, mind the windslab. Let's hurry across the col, chased by the wind, weaving between the seams of the ground, to that far tongue of golden rock set in wind-carved teeth of snow. From the stone sprouts a garden of fresh green icicles, each glimmering as a separate star. Come down here a few yards, and we'll be in the lee for lunch and the southern view.

A few yards down—indeed, yes, you are right to gasp. Never have I found it so clear, even in deep winter. It is never like this except just before or after a storm. The direct brilliance of the sun skimming across the Tremadoc estuary is so close that we cannot look there. Nant Gwynant is that pool of intense boiling radiance, pure and without color. The southeast sprawl of snow-capped ranges I have not seen before. They stretch all the way to South Wales. There is no gentle land in sight, at least not until the Border. Ah, look there on the horizon: the Malverns, it must be; there is nothing else there. Details are sharp at eighty miles. Truly for once we have done something right today. My camera sings its own tune, grasping images that leap out of the snow and printing them on film as they print on my brain. Twenty years hence I am sure these things will be clear in memory.

Something is happening to the weather. Sprindrift swishes under the swift low-running sky. A few nibbles of cheese and chocolate and some coffee from the thermos will be all the lunch we have time for, and then we'll have just over an hour before sunset—long enough to spend a few minutes prowling on Glyder Fach, if we don't insist on sunlight to come down by. It looks like a good gamble. In this light there must be something up there. It depends on what the cloud does. Seek and ye shall find.

Lean into the wind now, and draw up the hood against the sting of spicules. We shall take the direct line to the north rim overlooking Bristly Ridge, whose edge looks viciously iced. We are lucky to have had such an easy ascent. Today we seek mere access, not the technical difficulties that would be easier to find. If we had a rudder and centerboard we could tack right up the summit cone, such is the wind here. In our ears beats the pulsing roar of all high places, the sound of restless sky. Shout or you won't be heard. Watch out for the crevices between the big boulders under the snow, perfect leg-traps if you step in the wrong place. Oof. Don't step where I just stepped. What abominable snow. You may find a modified breast-stroke helpful in the drifts.

Today we must have done two things right, for the Grey King is smiling, and as we break onto the upper ridge he lifts the curtain of cloud from it to expose the towers of the Snowdon group shining to the south. We are looking directly onto the triple summit of the hooded north face of Lliwedd, nearly a thousand feet high, where Arthur's knights sleep sealed in a cave until their King shall return. Ice glitters on the battlements that cross to the steep pyramid of Yr Wyddfa, The Tumulus, the summit and temple of all Wales. From the other side of Yr Wyddfa, cupping a wild roaring hollow, Crib y Ddysgl rears back toward us to link with the extremely narrow ridge Crib Goch, the Red Crest. It parallels Lliwedd across the mile-wide gap of Cwm Dyli, the Valley of Rushing Waters. The cwm is full of summery golden haze, lapping gently at the brim, and the globular sun trembles in it like a ripe apricot; but not with warmth. The backlit ice is sheathing Crib Goch, and a gleaming halo surrounds the silhouette of the underlying pinnacles. Today it would be a beautiful and terrible place to be, sitting astride the knife-edge with each leg hanging down over an abyss. Crib Goch used to be considered, in summer, the easiest sort of rock-climb, and now is often called only a scramble, but today it would be exacting and ruthless. It looks stark.

And then the sun drops from under a long red cloud, and the light comes onto the Glyder.

O, the light. It is steady now. It pours molten across the ridge, level with the flowing sky, and it lives in each snow-flake and frost-feather. It makes the ice smoke and the rocks rise up with the sound of trumpets. The mountains burn, and yet they are not consumed. The wind dies and there is nothing else in the world, only this.

Only this, and the miracle. These torn stones on which we stand, this blinding fire and the snow on which it casts our shadows high and thin, this freezing air that sears our throats—of these things are we made, and in the grace of an instant and a place we dissolve into them again, single in exaltation. There is no we and no time, nothing but the blazing silent earth. There are no words.

Shivering. That is my body and my blood pumping. I am here, drunk with light. It is intensely cold. We must awake from this wonder, a little, very slowly. Our bodies drag us back. We must put on our down parkas. They rustle loudly and we are clumsy. Still the spell lingers on us and we cannot speak. Mute dwarves, we are, lost in a forest of rock. A huge horizontal slab supported only at one end and the middle, a spray of radiating columns like a frozen fountain of stone, a bouquet of jagged needles sawing at the clouds, and the lunar plateau near Castell y Gwynt, the Castle of the Winds: eerie monoliths, silent in worship of the setting sun. Around them all sprays the liquid light, and they are within it and of it, world without end. Dark and supple flow the valleys to the north, for night has fallen in the shadow of the Glyder, and a twist of pale pewter river winks out of the cold fields.

Forty miles beyond, standing out in the Irish Sea, a high black wall seethes, sucking all the light out of that side of the sky. It is a southwest gale, getting visibly taller and nearer. It will be here in force, say force ten, within two hours. The wind has returned to moan in the towers of slate, but already it is backing, so our time of grace is over. We must not linger. The coming of the stars must not find us here. All that matters now is the navigation. This would be no time to get entangled with the crags to the north, or the boulder-fields to the south, or the buttress low on Braich-y-ddeugwm: not at night, with a storm coming.

The shadows now fall long and lonely. The snow has lost its sparkle and become heavy. Slow light settles softer on the earth, all glory faded, and into the sky comes the taste of iron, chill against the teeth.

See the last glitter: a thin scarlet line on the farthest rim of the ocean in the south. In Eryri the ocean is never far away: here a man can fall under the twin dooms of mountains and sea, he can follow the two passions of solitude, he can be drawn by both the vertical and the horizontal line. Our line, now, is diagonal, and we have several miles to stumble down it, still stunned a little, under the light of a gaunt moon.

Chapter 3

To a Poet Met in Cwm Dyli

1971

Grey goblet of clouds
not yet drunk by sun: gulls stitch
the sky together.

Under the mountain
a wanderer searches for
substance of his craft:

not metals or gems
or woods or furs but only
words to weave and bind,

words to cast and carve.
He tilts up flat stones to look
underneath for blind

burrowing earth-words.
He parts each stiff clump of grass
to see if quiet

winds, in carding it,
have left any words there. He
seeks words of white flame

to kindle his tongue,
and supple green fern-words, and
words of still water

to flow into the
shapes of vessels. The sun and
the moon bring him words.

The clouds melt and drip
words on him: cold gulls cry words
into the morning.

Section 2

Resources and Energy: Efficiency, Policy, Potential

In the late 1960s, energy policy caught Amory's attention, and he started putting considerable thought into the topic (and has ever since). In early 1971, two years after becoming the youngest Oxford don in quite a few centuries, he approached the authorities and told them he'd like to do his doctorate in energy policy. At the time, it wasn't an academic subject there (or anywhere else) as it is now, so Amory was asked to pick a "real" subject. Instead, he resigned his academic fellowship and went off to do energy policy on his own. The result was an appreciable output of ideas, essays, and analyses on both energy and resources. Until the early 1970s, energy policy received minimal attention from most people, even those in the energy business. As Nobel physics laureate Hannes Alfvén noted in the forward to Amory's 1975 book *World Energy Strategies*, the oil crisis of 1973 changed all that, and "as a consequence, the energy debate has become a mixture of high-class technical–scientific–social arguments on the one hand and a sales pitch of political and economic character on the other. It is not the actual difficulty of the subject alone that has made and still makes the matter so confusing." Alfvén noted that the energy "problem" was divided into two realms:

> the choices among various energy systems and how much energy we think we need. ... From Lovins's analysis, one rightly gets the impression that we can picture reasonable solutions to the first group of problems, but that the second demands a much more profound analysis of what kind of society we really want to have.

While much of Amory's attention before 1976 looked at energy demand and supply growth and conversion, by that year he'd hit on the best questions one could ever ask about energy—the simpler, more fundamental queries of: what do we need it for, and how much, of what kind, at what scale, from what source, can meet each end-use need in the cheapest way? He didn't have the answer entirely, but simply posing these new questions was enough. Published in *Foreign Affairs* in October 1976 under the title "Energy Strategy: The Road Not Taken?", Amory's reframing opened the floodgates on energy policy by not only suggesting that most thinking on the subject until that point had been backwards but by offering a common-sense approach to energy. In December 1976, Amory was called to testify before a joint hearing of two US Senate subcommittees about what he'd written. The official record from those hearings—which included news articles, editorials, published and unpublished reactions, and about three dozen of what he called "tedious responses to fatuous critiques"—eventually totaled more than 2,000 pages. As Hugh Nash later observed in *The Energy Controversy*, an anthology of those exchanges, "the article was catapulted into prominence by people who loathed it." And yet it was widely adored, too, becoming the most-reprinted article in the history of *Foreign Affairs*. Today, some of the points Amory makes seem common sense, almost old-fashioned—a mere three and a half decades later. Yet it's that turning our set thinking on its head from which we can still extract outstanding value today.

The other papers in this section are an excerpt from a short popular book commissioned by Maurice Strong (chair of the 1972 UN Stockholm environment conference) and the late Dame Barbara Ward; a prescient 1976 paper about challenges to the human prospect, originally commissioned by the Stockholm-based International Federation of Institutes for Advanced Study; a puckish 1977 paper invited by *George Washington Law Review* (Amory's readings in the law began at Harvard and continued at Oxford); and a white paper on advanced energy efficiency commissioned by Nobel physicist Steven Chu, later US Secretary of Energy, for an energy study he chaired for the InterAcademy Council—a consortium of about 90 National Academies of Science. Dr Chu asked Amory: "What have we learned about energy efficiency in the past 30 years that nobody has bothered to write down before?" That question proved unusually fruitful.

Chapter 4

Only One Earth

1972

Editor's note: *By 1968, Friends of the Earth had begun focusing a lot of its efforts on the upcoming UN Conference on the Human Environment, which then took place in June 1972. Amory's main contribution to that effort was the book* The Stockholm Conference: Only One Earth, *which he wrote with research support from Michael Denny and Graham Searle (the first FoE UK director), and from which this chapter is excerpted. The book was designed "to amplify and complement" an international television production about the Conference. "Because we think these [environmental] problems are important not just for the politicians and the scientists but for everyone," Amory wrote, "we have tried to keep this book accurate but popular, invaluable but inexpensive."*

The Earth is very old: more than 4,000,000,000 years old, an age we can write but not imagine. We do not know just how the Earth was made. Somehow it gathered into a spinning ball. Its quivering crust slowly began to cool and take shape, a bleak and violent place. Winds of poison blew across its empty seas. But when more than 1,000,000,000 years had passed and the searing fire-tides had ebbed, the mixing of dead atoms in some strange swamp or ocean made by chance a new kind of matter: groups of atoms that could help others like themselves to form, that could break apart other groups and take their energy, that could absorb sunlight and store its energy. When the workings of time and chance brought these new kinds of matter together and linked them to each other in a great cycle driven by the sun, a new kind of chemistry was born, with such power that it swept the whole world, and gentled it, and changed it from brown to green.

For the past 3,000,000,000 years, sunlight has fed the growth of this new chemistry. Very slowly, larger and more complex groups of atoms have built themselves by trial and error; different patterns of atoms have come together to make cells, and cells to make tissues, and tissues to make organs, and organs to make redwoods and bees and sharks and hawks and men: things so wonderful that we know almost nothing of how they are put together.

Of course, most of the ways in which atoms have happened to arrange themselves have not been good for much. The plants and animals that live today are made of the select few patterns of atoms that did work well enough to survive; and for each living thing that has survived, there have been thousands of "mistakes" that are no longer here. By now there has been so much chance mixing of atoms that most of the ways in which groups of (say) a few hundred of them can be put together have already been tried somewhere; and most of the patterns that are not here probably are not here for a good reason—they did not work well enough for plants and animals to make more like them. Yet at the same time, this chance trying of many ways of putting atoms together is still going on: life is never still, always trying to become something else, something more efficient and stable and strong. Life changes to meet its needs: As Robinson Jeffers wrote:

> What but the wolf's tooth whittled so fine
> The fleet limbs of the antelope?
> What but fear winged the birds, and hunger
> Jewelled with such eyes the great goshawk's head?

In the midst of this ceaseless change, life holds to one central truth: that all the matter and energy needed for life moves in great closed circles from which nothing escapes and to which only the driving fire of the sun is added. Life devours itself: everything that eats is itself eaten; everything that can be eaten is eaten; every chemical that is made by life can be broken down by life; all the sunlight that can be used is used. Of all that there is on Earth, nothing is taken away by life and nothing is added by life—but nearly everything is used by life, used and reused in thousands of complex ways, moved through vast chains of plants and animals and back again to the beginning. Any break in these chains can spoil the whole. The web of life has so many threads that a few can be broken without making it all unravel (and if this were not so, life could not have survived the normal accidents of weather and time), but still the snapping of each thread makes the whole web shudder, and weakens it. Thus in the complex world of living things everything depends on everything else, all life is the same life, every effect is a cause, nothing can happen by itself. You can never do just one thing: the effects of what you do in the world will always spread out like ripples in a pond, and will make faraway and long-delayed changes you have never thought about. Yet you can never do nothing either, for you too are part of the web. Doing just as you have always done is itself doing something, and this too has effects.

The Earth is round. We have known this for hundreds of years, but few people even today see what it means: that everywhere on Earth, linked by cause and effect, is in a sense the same place, and that there is only so much earth and sky and water: so much and no more. We do not have unlimited amounts of anything—of land, of wind, of rain, of food, of sunlight, of away to throw things; for the Earth is round, and roundness means limits. Nor do we have immeasurable amounts of anything—for the amounts have been measured and, lately, found wanting. For the first time in his short history, man is now facing the limits of the Earth that he likes to call his.

Man does not like to think his history is short, but so it is—so short that it is the merest instant in the Earth's history. To see this, to put man's life in context with the Earth's, imagine the whole history of the Earth compressed into the six-day week of the Biblical Creation—a scale that makes eight thousand years pass in a single second. The first day and a half of this week is too early for life, which does not appear until about Tuesday noon. During the rest of Tuesday, and also Wednesday, Thursday, Friday, and well into Saturday, life expands and transforms the planet: life becomes more diverse, more stable, more beautiful; life makes a home for itself and adapts itself to live there. At four in the afternoon on Saturday, the age of reptiles comes onstage; at nine in the evening it goes offstage, but pelicans and redwoods are already here, lifeforms now threatened by man's wish to have the whole world to himself.

Man does not appear on the Earth until three minutes before Saturday midnight. A second before midnight, man the hunter becomes man the farmer, and wandering tribesmen become villagers. Two-fifths of a second before midnight, Tutenkhamon rules Egypt. A third of a second before midnight, Kong Fuzi and Gautama Buddha walk the Earth. A fortieth of a second before midnight, the Industrial Revolution begins. It is midnight now, and some people are saying we can go on at the rate that has worked for this fortieth of a second, because we know all the answers. Do we really know that much?

Man likes to think that somehow he has stopped being a part of nature—that he controls nature and can do what he likes to nature without putting himself in danger. This is not true. Nature laughs last. It is one of nature's rules that those who won't play by the rules won't play at all. Nature has often invoked this rule: every living thing that has tried to use more energy or matter than the world can provide is now dead. All that now lives has grown in such a way that it lives on the Earth's terms—otherwise it would not be living. Man must do this also, or he will meet the same fate as any other "mistake" of evolution; for there is no reason why everything should come right in the end. The Earth doesn't care. It was here long before we were, and will be here even longer after we leave—and perhaps it will be more comfortable without us. We have forgotten how to be good guests, how to walk lightly on the Earth as its other creatures do.

Chapter 5

Long-Term Constraints on Human Activity

1976

Editor's note: *This piece, penned in 1976, when Amory was a consultant physicist representing the US Friends of the Earth in London, discusses global ("outer") limits on resources. He suggests that, because humankind knows so little about these resources and the ways in which they interact, we should work to understand them before short-term goals preclude "important long-term options." This piece first appeared in* Environmental Conservation, *Vol. 3, No. 1, Spring 1976.*

Introduction

Mankind is constrained by problems of food, land, water, climatic change, energy, hazardous substances, non-fuel minerals, human stress and social tension, ecological stability, management, and global organization. In the past decade, speculation about how these and other constraints might impose "outer limits" on human activity has become bewilderingly diverse, reflecting both the diversity of the world scientific community and the blurring of distinctions between technical, ethical, and ideological conclusions. Governments, unused to assuming the worst in cases of scientific uncertainty (especially on a time-scale of decades or centuries), have tended to discount heavily the rising concern behind these speculations, and hence to foreclose important long-term options by imprudent short-term pragmatism.

Yet this very emphasis on attacking acute and immediate problems has focused attention not merely on *how Man might* approach some outer limits, but also on *why Man cannot* approach some others. As we identify biophysical or other constraints whose avoidance or evasion will require social and institutional change, we begin to appreciate the difficulty of that change through our everyday efforts to overcome not ultimate biophysical constraints but more proximate managerial constraints. In short, even as we become aware of outer limits, inner limits arise to crowd them from our thoughts.

This dichotomy, or rather continuum, of limits can also be viewed in another way. The amount of most resources theoretically available under perfect management is extremely large; this is true not only of food but also of energy and most raw materials. However, the gap between theoretical and practical availability also tends to be extremely large, because of various constraints—geopolitical, social, technical, biological, economic, and so on. The rate, difficulty, and side-effects of overcoming these constraints will vary widely from case to case, and will be assessed differently by different people. (This is partly because perceptions vary from sophisticated to nil in the sphere of purely social or political constraints, which are of course no less real or effective than "hard" biophysical constraints.) In many cases and places, the constraints (of whatever nature) which stand between us and the more remote outer limits are so hard to overcome in a timely fashion that these outer limits seem to be of purely academic interest. Thus while we must bear in mind their long-term implications, our immediate task is to assess the nature and permeability of the far more-obvious inner limits. But in order not to do this under false assumptions, we may have to reason backwards (Lovins, 1975b, 1975c) from outer limits, however remote they may seem on the short timescale of much political action. Thus we may have to ask: What options must we retain for the long term, and what must we not do now if we are to retain those options? Otherwise we may "jump from the frying-pan into the fire."

This long-range-planning approach is commonly rejected in favor of reliance upon short-term technological fixes. One must therefore ask whether technology is inherently addictive, forcing society into further crises demanding further technological fixes—and so on, until the habit becomes socially unsustainable. This might occur if, for example, fixes are prescribed for biophysical symptoms rather than for underlying social disorders—a common result of misdefining some state of affairs as a "problem" which must *ex hypothesi* have a "solution." If technology evades, obscures, or defers social problems, rather than resolving them, then its ever-increasing use must eventually lead to social disequilibria without technical solution. Likewise, if technical change entails social change, then the pace of required social change must soon become excessive.

Many thoughtful analysts are uneasy about these trends and about the rising dependence of all societies on rapidly devised and deployed technologies whose complex side-effects are unpredictable, unpredicted, or (most often) simply ignored. Yet this malaise seldom crystallizes into explicit consideration of the option of technological restraint: of trying to root out the underlying social problem rather than attacking its symptoms with technical diligence and therapeutic zeal. This option—assessing the wisdom of intervening at all with our technical tools—is generally submerged by more "practical" assessments of the nature and hazards of ad hoc solutions to what "problem-solvers" perceive as their immediate problems. Thus the possibility of social adjustment, for example of stimulated changes in social norms and goals, is generally dismissed by technologists who are unable to distinguish "can" from "should"—dismissed even as they strive to suppress the symptoms or signals which might bring about that adjustment, and even as they cheerfully assume that the equally difficult social adjustments which technical innovation requires will indeed occur on schedule. Social delays and disruption then tend to increase, and as delays become longer (compared with accelerating rates of change), instabilities also increase.

Pragmatists may say that these somewhat theoretical concepts are irrelevant to their needs: that problems must be sorted into a hierarchy of decreasing imminence and addressed as they arise. Such a hierarchy does indeed exist; but it is a dangerously incomplete tool of decisionmaking—a tool whose use by those obsessed with the short term and with the boundless ingenuity of their successors has led to many of our present difficulties. Inner limits (Brown, 1974b; Congressional Research Service, 1975) are in general most imminent in food supply, then in energy supply, and lastly in mineral supply; but the imminence of food–population collisions in some regions (Brown, 1972, 1974a; Borgström, 1973) does not make other kinds of limits irrelevant anywhere. On the contrary it means that, while preparing as best we can for those collisions which it is too late to avoid, we should be addressing the problems that the survivors will have to cope with next—problems that we may have time to solve if we start now. In other words, we must break the cycle of short-term planning that has got us into our present acute difficulties (e.g. Ehrlich, 1974).

Progressive social thinkers who are faced with long-term environmental problems often argue that they cannot afford to worry about the remote and abstract when surrounded by the immediate and concrete—that potential climatic or genetic instabilities are of academic interest in a world full of actual anarchy, war, famine, disease, injustice, and ignorance. This thesis has an element of truth, but a greater element of folly. The problems which overwhelm us today are often precisely those which, through a similar lack of vision and persistence, we failed to solve decades ago—problems which have built-in perceptual and responsive delays of many years. These delays ensure that most conventional institutions can only defer or disguise problems, not solve them.

We have thus to conclude that many of the major problems which we face today, and which so exercise our harried policymakers, are *ipso facto* insoluble; for it takes time to solve problems, and all we have time for now is makeshift remedies. The major problems of the next few years are likewise often insoluble, even though they are the only ones which our politicians are likely to perceive. The major problems of the late 1980s might just, and those of the 1990s may well, be soluble if attacked now, but not if deferred. Many modern institutions are trying to devise ad-hocracy to cope with immediate difficulties: few are trying to solve soluble problems, or even to find which they are. This paper accordingly seeks to identify and explore certain long-term problems that must be addressed promptly if we are to live to enjoy some of the later and more interesting limits to human activity.

It is hard enough to apply synthetic rather than analytic discipline to history; it is far harder to do the same for events which will not occur for decades hence—particularly where present knowledge is not sufficient to *prove* that they will occur at all. Yet it is far more important. The kernel of the incredible tangle of human problems is the principle of interrelatedness, expressed in so many different ways: "Everything is connected," "You can never do just one thing," or "Only One Earth." That this brief survey of present and emerging problems uses for didactic convenience a form of analytic framework does not mean that the reader, absorbing a wide range of seemingly disparate material, can safely do the same. Those who are the prisoners of their categories will not foresee the relationships between climatic change and food production, nor between energy conversion and climatic change, nor between food production and energy conversion—not until, that is, these and a myriad similar relationships, which work inexorably whether they are perceived or not, have drawn themselves unmistakably to the attention of everyone, the sower and the reaper of the future alike.

It would be improper for a state-of-the-world survey of this kind to rely upon or to endorse such common analytic categories as "resources," "pollution," or "population." The first has an arrogantly anthropocentric ring, denying both economic and moral standing to that which Man does not endow with utility; the second suggests an interest more in the event than in its origin or effect; the third implies that a single variable can be validly dissociated from an enormously intricate and nonlinear complex whose control, however it may be exerted, will hardly act through just one variable at a time. Thus if such simplistic categories creep in, it must be with an apology. Let us now embark on our brief and selective review of some major biophysical constraints—a survey that is intended not as an exhaustive catalogue, which would be tedious if it were indeed possible, but rather to flag certain critical problems (especially those demanding action far in advance), and to convey something of the intensely integrative character of the needed perceptions and remedies.

Food

Recent reports (Brown, 1972, 1974a; Borgström, 1973) on the race between agriculture and population are ominous, especially for countries whose food prospects have long been unpleasant because of the terrible momentum inherent in a skewed age-structure. A typical poor country that achieves replacement reproduction by AD 2000 will stabilize shortly after the middle of the next century—at about 2.5 times its present population. If all countries achieve replacement reproduction by the year 2000, and if mortality does not greatly increase—both being assumptions that many experts consider too sanguine—

then world population, now about 4 thousand millions, will stabilize nearly a century hence at 8–9 thousand millions or slightly more (Frejka, 1973).

Population growth is the driving force in the classic Malthusian confrontation with food supplies; but a new competitor, already half as important and becoming steadily more so, has emerged (Brown, 1972, 1974a, 1974b; Borgström, 1973). This is rising affluence, which, expressed as demand for animal proteins, is now a major claimant on scarce supplies of both grains and feedstock proteins. The latter are, on the whole, exported from poor to rich countries, where they are converted into 0.1–0.2 times as much animal protein of similar or lesser food-value but higher price. Grains—comprising more than half the direct food-supply of the world and part of the indirect remainder—are consumed directly at a rate of about 0.5 kg *per caput* per day in most poor countries and less in some rich countries—but at over twice that rate in the latter countries' meat and beverages. Rising demand for animal protein is now colliding with three main limits on its production:

1   Marine protein production, having risen by a remarkable ~5 percent per year during 1950–1970, is now very close to its sustainable limit. This is generally thought to be perhaps 100–120 million metric tons per year (the total catch is currently about 70), but only with far more vigilant control of over-fishing and marine pollution.
2   Beef production, already limited by the fertility of cattle, has now encountered (and in many areas exceeded) the sustainable limits of grazing. Further increases thus generally depend on intensive feedlot agriculture—at an energy cost of the order of 6–12 kg of coal equivalent per kg of beef protein (Slesser, 1973; Leach, 1975).
3   The increase in soybean production in the past few decades has in most cases been due to increased planting; no significant intensification of areal yields is in sight, and demand continues to rise much faster than supply.

Costly, unexpectedly slow, and ecologically risky intensification, rather than areal expansion, has accounted for most increases in world crop-yields since 1950—for about four-fifths of those since 1970 (Brown, 1974a). Readily irrigable land and the water needed to irrigate it have latterly become extremely scarce and now seldom occur in the same region: accordingly, further intensification tends to be subject to diminishing returns. The long-term fertility of established farmlands is jeopardized not only by the poorly understood chemical, ecological, and mechanical stresses of intensification, but also by indirect population pressures. These pressures, surveyed by Eckholm (1976), include for example overgrazing (which helps to drive the 2–10 km/yr advance of the southern Sahara), deforestation of uplands (this may increase lowland flooding), increased farming of fragile uplands (this may silt lowland irrigation systems), urbanization of prime farmland, and the direct and indirect side-effects of mining.

Such forces are rapidly changing the world food market, in the manner of the world energy market, from the buyer's to the seller's province. A global politic of food (especially protein) scarcity is now emerging, with disquieting implications both for hungry people and for world political stability. This new political fabric will surely reflect:

- Competition for exportable grain between countries of widely varying wealth.
- A growing world tendency for main world suppliers of agricultural commodities to withhold supplies from hungry customers for domestic political reasons, regardless of the morality of interdependence.
- Greatly increased volatility of prices in all countries because supply deficits in the past three years have virtually eliminated the two main buffers: the post-1960 US land-bank is now almost empty, and world grain stocks have recently been at their lowest levels in modern history. "Domino" response of prices to relatively small perturbations will therefore become more common.
- The increasing monopoly power of midwestern North America, which controls about two-thirds of the world's exportable cereals and a very high percentage[1] of the world's exportable soybeans.
- The resulting vulnerability of world food supplies to weather in a single region—a region that is subject to a regular 20-years drought cycle which is due to recur about now.
- As scarcities increase, a tendency to view food as a strategic commodity to be controlled in accordance with political goals.

These problems (Brown, 1972, 1974a, 1974b; Borgström, 1973) are not merely imminent but already upon us: we have no safety margins left. Another 1972 international crop failure or 1932 US Dust Bowl could now, with our scant reserves, trigger major regional disasters.

Land and Water

Salinization, erosion, laterization, overuse, paving, and other degradation are tending to decrease irreversibly the world's net stock of arable land: Man may already have changed a tenth of all ice-free land to desert, and reduced total global vegetation by a third (Wilson & Matthews, 1970). This continuing, indeed accelerating, erosion of the Earth's carrying capacity (Eckholm, 1976) is by any measure one of the most critical environmental problems today, and will be with us for a very long time to come.

Of all potentially arable land (about 24 percent of the total area of ice-free land), only about 44 percent is now cultivated. This does not mean, however, that 56 percent awaits the touch of the plow to spring to fertility; what it really means is that almost 56 percent is waterless, degraded, or despoiled. The near-saturation of much potentially arable land—83 percent of which is already cultivated in Asia, 88 percent in Europe, 64 percent in the USSR, 51 percent in North America—gives little cause for delight in a few lower figures (such as 22 percent for Africa, where many soils are lateritic and most are very poor). It is in general not possible to sustain a high net productivity in the humid tropics.

Massive irrigation was once seen as a short-term panacea—short-term because, without the most unrelenting and sophisticated control (which few societies have been able to sustain for very long) of drainage and of watershed land use, irrigation can harm fertility by salinization or waterlogging of the soil. Many experts now believe, however, that irrigation in most areas will be too little and too late. The growth rate of irrigated areas of the world—nearly 3 percent per year during 1950–1970—will probably average only about 1 percent per year over the next few decades: a mere 2–2.5 million sq km (2 percent of all ice-free land) is now irrigated, at great cost and with many unwelcome side-effects. Properly irrigating the main arid and semi-arid regions of the world would require the total continental runoff—a physical impossibility—and water projects so large as to risk substantial changes in regional or global climate. Nor is it clear where such prodigious amounts of water could be found. Present rapid depletion of groundwater resources throughout the world will soon lead to widespread local shortages. The only other source of irrigation water seems to be desalination; this is costly, extremely energy-intensive (Lovins, 1975a), and severely restricted in its possible rate of deployment. Moreover, even when irrigation water is somehow obtained, controlling soil salinity and hydrology and preventing siltation of water systems requires perpetual centralized management that is more exacting than most societies seem able or willing to provide (Eckholm, 1976).

Increasingly stringent water constraints suggest that agricultural innovation is less needed for intensively fertilized and irrigated land than for brackish water and for dry-farming in marginal lands with extreme climatic conditions and fluctuations. This adaptive approach is relatively new, and will be slow to produce results analogous to those of "green revolution" research. Its social side-effects are hard to assess; its ecological effects, especially in such fragile systems as the marginal steppe, savannah, and tundra, are likely to be even more disruptive than are present enterprises.

All areas desperately need cheap, simple, and foolproof technologies for conserving clean water, removing pathogens, and recycling dissolved nutrients. Water quality in many poor countries falls far short of World Health Organization safety standards, posing immediate epidemiological hazards and greatly complicating the formidable problems of dense human settlements. In the longer run, the absence of recovery technologies may impose an early outer limit on intensive agriculture through the escape into surface- and ground-waters of nitrogen, phosphorus, pesticides, and other agricultural additives.

The long-term effects of these substances on soil microbiota are conjectural. Industrial nitrogen fixation is now of very roughly the same amount annually as is computed from natural microbiological processes and is projected to increase by at least an order of magnitude by the end of this century; yet it is already viewed as a potentially severe hazard to water quality through both eutrophication and human toxicity. Phosphorus use is increasing about 2.7 times as fast as human population; its fate in the biosphere, and its long-term availability as a mineral resource, are largely unknown. Perturbations to the scarcely known rates, routes, and reservoirs of natural nutrient cycles through the introduction of such proposed genetic syntheses as nitrogen-fixing bacteria symbiotic with non-leguminous plants might be large, and probably cannot be assessed in advance. Nor are they necessarily ecological only: increased nitrification leads (after some delay) to increased denitrification and hence to release of by-product N_2O (in a proportion dependent on soil conditions); photochemical products of this substance can in turn enhance ozone-destroying stratospheric reactions. Preliminary estimates (Schneider, 1976) suggest that such an effect may be large enough to impose an important constraint on artificial nitrification by fertilizers or otherwise.

Climatic Change

That study of such a basic problem should be effectively starting only now bespeaks our ignorance of many basic details of interactions between geophysical processes and the biosphere. This essential subject has evaded scrutiny through a combination of disciplinary barriers, lack of resources (especially in tropical countries), and concentration on managed (i.e. simple) ecosystems. This ignorance of the impact of climatic change on life, and conversely of human activities on climate, means that intensive and expensive effort is now necessary on a world scale if we are to study climatic outer limits predictively rather than empirically. Such work seems especially urgent now that we know that several major human influences on global circulation (heat, carbon dioxide, and particulates) may tend to act in the same direction in certain respects (Bryson, 1972, 1974, 1975; Kellogg & Schneider, 1974; Schneider & Dennett, 1975), that Man may already be affecting global climate, particularly in the monsoon belt (Bryson, 1974; cf. Mitchell, 1975), and that critical determinants of global climate and solar flux, such as the Arctic pack-ice and the ozone shield, are more sensitive to perturbations than had been thought (e.g. Schneider & Dennett, 1975; Schneider, 1976).

There is already reason to view with grave concern major alterations of water flows, of evaporation patterns, and of surface optical properties (as in deforestation) (Wilson & Matthews, 1971). Moreover, unsuccessful attempts to resolve fully the well-known controversy over stratospheric chemistry and supersonic civil aviation have recently disclosed a host of even more disquieting possibilities—perhaps only the first of many—ranging from "greenhouse" action by halocarbons (Ramanathan, 1975) to coupled climatic and photochemical effects, including stratospheric temperatures affecting reaction rates and thence ozone chemistry (and perhaps climate). The above-mentioned possible interaction between denitrification and the integrity of the ozone shield illustrates the sort of unpleasant surprise that can remain invisible so long as soil chemists do not talk to stratospheric photochemists; and though the biological significance of ozone depletion is obvious, its climatic significance is speculative.

Attempts to develop sound theories of climate and of its variations (Schneider & Dennett, 1975) have tended to concentrate on global changes of state persisting for decades or centuries. Such changes could be induced, for example, by the melting of the thin and short-lived Arctic pack-ice—a probably irreversible change that could be fairly easily accomplished (Lovins, 1973; Manabe & Wetherald, 1975; Schneider & Dennett, 1975; Schneider, 1976) and would have profound implications for human life. Likewise, recent geological, palaeoecological, geophysical, and archaeological evidence shows that rapid areal extensions of the Antarctic continental ice-sheet at intervals of the order of ten thousand years can raise sea levels 10–20 m and reduce average global surface temperature 5–6C° in less than a hundred years (Hermann Flohn, pers. comm., 1973). Such a surge may occur at an unknown (and currently unforeseeable) time in the next five thousand years or so, and constitutes a small risk whose serious consequences need study.

Yet the precariousness of the world's food supply (Eckholm, 1976; Schneider, 1976) requires that more immediate attention be directed towards short-term regional instabilities in climate—periods of persistent aberrations from the mean on a timescale of seasons or years, bringing repeated droughts—or, in some areas, grossly excessive rainfall. The environmental manager, who must plan on the basis of the extreme rather than the mean, must know whether bad years tend to coincide in different regions by mere coincidence or by causality ("teleconnection"); hence, how often will disastrous years such as 1972 recur? This is a fairly new subject of intensive study, both empirical and theoretical, and is fully as difficult as it is important.

Ironically, although it is short-term regional instabilities that pose the major immediate threat to world food production and to such activities as the navigation of Arctic waters, it is often long-term global instabilities that these and other human enterprises might most readily induce. For example, although the biological effect of Arctic oil-spills has long been a subject of concern, the possibility that a large spill might have major effects on global climate has only recently been recognized (Campbell & Martin, 1973; Aagaard & Coachman, 1975) and is now receiving (as it deserves) earnest attention. The suggested sequence is fairly simple: oil can allegedly emulsify into small persistent droplets which could be distributed by currents along the underside of the Arctic pack-ice; the ice melts seasonally on top and freezes on the bottom, so bringing the oil to the surface in a few years; the oil would then reduce the albedo (reflectivity) of the ice during the spring period of intense insolation (illumination and heating by the sun), thus encouraging an irreversible melting which could drastically alter zonal circulation throughout at least the Northern Hemisphere (Wilson & Matthews, 1971). Unfortunately, the Arctic pack-ice is already subject to a number of present or planned attacks (such as high-latitude heating from industrial energy conversion, or altered salinity through the diversion of Arctic rivers) which would tend to act in the same direction as the oil (Wilson & Matthews, 1971; Lovins, 1973; Manabe & Wetherald, 1975; Schneider, 1976), just as most identified attacks on the ozone shield probably tend to deplete rather than to replenish it.

Energy

The climatic effects of Man's energy conversion are already observable locally and regionally, and could become globally important during the first half of the next century (Mitchell, 1975). Until a detailed and exact understanding of climatic change—if attainable—has refined calculations, present schemes for evading resource scarcities through large injections of energy capital are premature: regardless of the technology used, the climatic effects of the inevitable heat release may be intolerable (Lovins, 1973). Other outer limits on energy conversion may arise through:

- The economic impact of shortages, of rising production costs, and of sectoral capital intensity that has recently increased by about an order of magnitude (Lovins, 1975a, 1975b).
- Sociopolitical instabilities (such as alienation, high-technology violence, and centrifugal political stresses) resulting both from deployment of centralized, complex, high-technology (hence vulnerable and unforgiving) energy systems and from gross inequities in the distribution of fuel resources and of energy conversion.
- The tendency of krypton-85 emissions from nuclear facilities to increase the electrical conductivity of the troposphere (Boeck et al., 1975).
- The biological side-effects of producing such fuels as offshore oil, surface-mined coal, and transuranic isotopes.

Some energy resources may be simply too dangerous to use (Edsall, 1974, 1975). Some potentially intractable hazards, such as actinide aerosols (Lovins & Patterson, 1975), submicron particulates and sulfate particles, and liquefied natural-gas spills (Lovins, 1975a), are only just starting to receive serious attention. Although less harmful energy technologies—especially for living on energy income rather than on energy capital—are technically feasible and can be economically attractive, the formidable rate-and-magnitude problems of change in a massive energy economy (Lovins, 1975a, 1975c), as in other major sectors, ensure that no voluntary change will be rapid in rich countries, and hence that the oil/gas economy will persist long enough to cause serious political and fiscal dislocations.

The sociopolitical, environmental, and economic problems of energy conversion, and the difficulty of rapidly deploying complex new energy technologies with their enormous requirements of capital (Lovins, 1975b), materials, and skilled labor, will together enforce a slowing of rates of growth in energy conversion, and now cast serious doubt on our ability to deploy, for example, widespread desalination, synthesis of nitrogen fertilizer, mining of very dilute mineral deposits, and other highly energy-intensive technologies (Lovins, 1975a). Indeed, energy conversion technologies are themselves so energy-intensive that the ability of some under any circumstances, and of many during periods of rapid growth (Price, 1975), to yield net energy is in serious doubt (Lovins, 1975a, 1975b).

It is hard to see how the meager energy rations of poor countries (many of them already dependent on petroleum-based agriculture) can be much increased so long as the rich countries fail to realize that they themselves will have to conserve energy whether they like to or not. Waste will be greatly reduced by disruptive compulsory changes in lifestyles brought about by shortages—if not by a deliberate strategy of increased efficiency (Over & Sjoerdsma, 1974; Armstrong & Harman, 1975; Ross & Williams, 1975; Schipper, 1975) and of more careful distinction between demand and need (Lovins, 1975a, 1975b, 1975c). Such a strategy cannot be devised until a wide range of energy futures and implications of choice has been explored in public, for people cannot choose options that they do not perceive. Research to this end is only now beginning in a few countries (Energy Policy Project of the Ford Foundation, 1974; Chapman, 1975; Lovins, 1975c; Sørensen, 1975).

There seems no question that future energy decisions to be made by both producers and consumers will have profound, and so far unassessed, effects in practically every sector—especially in foreign policy and trade, agriculture, architecture, transport, economic planning, settlement patterns, and social organization. This pervasive influence of energy, however, is not confined to those countries which are now using it most liberally and wastefully: developing countries, hardest hit by scarcity and high prices, will have to make special efforts to find, and to encourage reluctant industrial countries to help them to find, locally appropriate energy technologies and development paths that are not energy-intensive (or at least that are not intensive users of energy capital), as in the long run no other approach can succeed either for them or for anyone else. Nor are energy problems in rich countries likely to be properly addressed until it is widely appreciated, firstly, that most of the important energy issues are not technical and economic but rather social and ethical, and secondly, that increases in energy supply tend to be slow, costly, risky, and temporary, whereas decreases in energy demand tend to be comparatively quick, cheap, safe, and permanent.

A possible outer limit on energy conversion through nuclear fission—a young technology whose rapid proliferation (and potential for nuclear violence and coercion by criminal lunatics) raises difficult ethical questions of transcendent importance (Edsall, 1974, 1975; Lovins, 1975c)—is posed by its production of large quantities of extremely toxic radioisotopes, some with half-lives of the order of thousands to millions of years. Plutonium-239, a biochemically active 24,400-years hard alpha emitter that is toxic in submicrogram quantities and can form respirable aerosols (Lovins & Patterson, 1975), is a prominent example. Such substances require infallible and perpetual isolation from the biosphere, and it is hard to imagine how this can be done. Plans to manufacture large amounts of transuranic isotopes must therefore be urgently re-examined. (Even the mobilization of thorium-230 in the mining of its parent, uranium, can result in a semipermanent release of very uncertain but potentially large amounts of radon-222 and its highly active daughters (Comey, 1975).)[2] More generally, the inherent risks of fission technology are so great that operating safety, waste containment, and the security of strategic materials may be ultimately limited not by cost, ingenuity, or diligence, but by the impact of human fallibility and malice on highly engineered systems (Lovins, 1975c). A rapidly growing body of competent technical opinion suggests that this is the case, and that deployment of the technology should therefore be suspended until enough infallible people have been found and trained to operate it within a sufficiently placid social context for the very long periods required.

Hazardous Substances

Many classes of substances can now be identified which may be too hazardous for Man to trust himself to take care of (examples include some classes of persistent organic carcinogens). Some other persistent substances can also be identified which, though not a priori candidates for producing outer limits by jeopardizing human health or the stability of life-support systems over large areas, nonetheless appear to carry enough risk to require special scrutiny. Identifying such a risk is a technical problem; deciding whether to incur it in the pursuit of some benefit is a political and ethical problem (Green, 1975).

Identifying and regulating toxic, mutagenic, teratogenic, carcinogenic, or otherwise hazardous substances well in advance of their distribution—which, despite the best efforts at containment, may be worldwide and permanent—grows steadily more urgent as Man manufactures each year more than a thousand substances that did not exist before. Our experience of these new substances is generally one of discovering unforeseen risks. For example, though the persistence and toxicity of many halogenated aromatic compounds are well known, some research workers are only now beginning to suspect that low-molecular-weight halogenated hydrocarbons (such as aerosol-propellant chlorofluorocarbons, dry-cleaning fluids, and some other solvents) may inhibit fermentation. The possible impact of some of these halocarbons on stratospheric chemistry is another recent and unwelcome surprise (cf. Report IMOS, 1975).

Likewise, a synthetic analogue of insect juvenile hormone is said to have been found recently to have substantial and perhaps irreversible biological effects on people in certain circumstances—in quantities of one or two *molecules*.[3] Very many surprises of this sort suggest certain conclusions:

- No system is truly closed.
- Toxicity may be hard to determine, even with years of tests (especially for genetic effects, as the introduction time of new substances is far shorter than the generation time of Man).

- The primary (especially the long-term subclinical) effects of man-made substances are virtually as unknown as the rates, routes, and reservoirs of their natural distribution, the ways in which they may be concentrated or altered in natural systems, and the ways in which they may interact with one another.
- We shall never be able to predict the effects of all random combinations (synergistic or antagonistic) of man-made substances *in vivo*.
- We must therefore face ethical questions about the risks and benefits of new substances, the rate (if any) at which future risks should be discounted, and the burden of proving risk.
- Monitoring of the entire biosphere—including the human genetic pool—is an important, though belated, line of defense in detecting risks which we have failed to detect through earlier screening. We must recognize, however, the defects of all monitoring systems: notably that, if a persistent substance is slow to propagate through the biosphere, its concentrations in relevant systems may continue to increase long after the case for regulating its distribution has been recognized, accepted, and acted upon. Therefore, admirable and necessary schemes for global monitoring must not be relied upon to relieve us of the responsibility of striving to avoid making mistakes earlier.

Non-Fuel Minerals

Outer limits to the availability of metals and similar materials might be posed by environmental, geopolitical, or economic side-effects of extraction and processing, and perhaps in some cases by the inventories physically present in the lithosphere. Unfortunately, the basic data needed to assess these limits do not now exist. In particular, the detailed lithospheric distribution statistics (especially the degree of continuity of grade-tonnage distributions) are not now known for most metals (Lovering, 1969); hence neither the geopolitical nor the entropic significance of depletion of high-grade sources can now be assessed with sufficient accuracy for sound policy (Lovins, 1973).

Likewise, nobody seems to have used the power of modern metallurgical theory and experience to assess in detail the likely scope, cost, rate, and difficulty of substitution either in major alloy families or in specialized single-metal applications. Despite some early work on the conceptual structure of this problem, no thorough study of material substitutions throughout the range of modern industry has been attempted or is yet proposed.

Technical change in extractive industry will include extension to new areas with unique and largely unknown environmental problems (such as the Arctic, the seabed, and tropical rain forests), and to regimes of polymetallic extraction based on elaborate multi-stream beneficiation and extractive metallurgy. Both the novel criteria (such as mineral grain-size) and the market structures that are characteristic of such future extractive economies deserve study, as the resulting political and market instabilities may be pronounced. Little work has been done, too, on possible ways to buffer against future tensions and instabilities resulting from the exceptionally skewed geographic distribution of certain mineral reserves: it may be possible, by working far in advance, to distribute more equitably the economic and social benefits of such reserves.

Human Stress and Societal Tension

Little is known of human response to stresses as diverse as noise, boredom, changing age-structure, and urban crowding. Indeed, some authorities accept "stress" as a legitimate concept and identify as pathological states—i.e. clinical entities resulting from "stress"—patterns of behavior which others construe as mere adaptation. Adaptation and adaptability are not necessarily good, however, and it is generally accepted that the survival of Man as a species may be less threatened by stress than is the survival of those qualities which distinguish Man as human. There are thus dangers as well as difficulties in devising social mechanisms and structures which are intended, in widely varying societies throughout the world, to counteract societal trends that conflict with biophysical constraints. Such social engineering will require the most thoughtful consideration by a wide range of both scientists and humanists.

On a more pragmatic level, one must recognize that war and similar societal conflicts, aggravated by the continuing and needless proliferation of nuclear fission technology, are still the gravest threat to the survival both of humanity and of humane values. Unevenly distributed resources of many kinds have enormous conflict potential and can be defused only with the greatest difficulty. Poverty and social inequities, too, can be immensely destructive—perhaps more directly than had been imagined. Thus certain direct links have recently been found, for example, between nutrition (of which relatively little is yet known) and the proper functioning of some of the body's mechanisms of immunity. It is possible that many more surprising connections with other forms of stress may in time be found.

Diversity and Resilience of Ecosystems

Important and largely ignored insights are now emerging from work on the relationship of stability or resilience to diversity in natural communities, and on the dynamic behavior of such systems under stress. At the same time, more data are becoming available on the loss of genetic diversity through extinction of populations and species, through modification of habitats, and through decay of variability in gene pools. (The human genetic pool may be altered not only by spontaneous mutation, chemical mutagens, and ionizing radiation, but also by various patterns of population growth or control: the genetic implications of demographic policies demand careful study but have received almost none.) All these discoveries give cause for grave concern at Man's abuse of both marine and terrestrial life-support systems whose stability and productivity are essential to our own (Polunin, 1974). We know very little about the dynamics of creeping and widespread ecological destabilization, nor about—to choose a specific example—the ability of marine and coastal ecosystems to withstand the various "insults" (including ~1 cu km/yr of artificial materials) of which they are the final recipients (Wilson & Matthews, 1970).

Even if understanding of the dynamics of stressed biological systems is much improved, an important gap will remain in the perceptions of decision-makers who, conditioned by everyday experience of purely physical systems, expect linear and reversible behavior—not the nonlinearity, irreversibility, threshold responses, and long delays of which natural systems are capable. Until the mental models or paradigms of linear economics are replaced by appreciation of the intricacies of an animate world—and of the importance of preserving several thousand million years' worth of design experience—research alone cannot have the needed effect on indifference towards ecological and genetic degradation.

Management

More urgent than most biophysical outer limits are the limits of our abilities to manage large interconnected systems, and so to penetrate even the innermost of the inner limits. The task is to build management systems in accordance with the realities of biophysical cause and effect, in order to extend the limits of social capacity to cope with environmental problems. Lack of properly educated people—generalists with the integrative grasp, broad technical base, systems training, and international orientation needed to understand complex transdisciplinary problems—will be the main constraint. At present, nearly every element of the systems of rewards and penalties in both the public and the private sectors discourages the evolution of a class of environmental managers with these skills and qualities.

Meanwhile, the world's economic managers face some extremely difficult tasks for which they are not well prepared. They will have to deal increasingly with disruptive short-term capital flows having destabilizing effects on the world monetary and trade structure; they will have to begin to adopt a holistic view as it becomes clear that there is not enough capital (nor land, energy, time, etc.) to devote separately to each competing problem in each sector; and they will have to devise, in highly industrialized countries, means for an orderly transition to a micro-variable but macro-stable economy of stock (Daly, 1973), with incentives for longevity and low-entropy design of manufactures, for stimulating recycling and repair, and for adopting low-impact lifestyles. Moreover, they will have to simplify economic structures which are now becoming so complex that their transaction costs may exceed their productivity. No adequate theory exists that might guide any of these efforts. Perhaps most difficult will be the need to face the issue of distribution rather than following the "let them eat growth" theory: physical stabilization will entail much moral growth, together with the recycling of such nearly extinct societal values as thrift, neighborliness, craftsmanship, humility, diversity, and simplicity (Illich, 1973).

Economists now in service will find themselves compelled by circumstance to overhaul their tool-chests at short notice. Discounted-cash-flow decision-making will quickly be seen to rely on a pernicious assumption of infinite substitutability of equally worthy investments (for example, forms of resource exploitation). Current discount rates will lose their relevance as they are seen to make the present value of practically any common resource—from whales to newborn children—virtually zero. The discounting of future risks as though they were real costs will seem more and more a way of masking social irresponsibility. Rapidly changing conditions will subordinate marginal-cost analysis to economic sensitivity analysis, and forecasting based on correlation will foreseeably give way to forecasting based on causality. Whole new sciences, such as energy analysis, are starting to emerge (IFIAS, 1975) and may substantially alter the basis of decisionmaking (Slesser, 1975). And as new tools evolve, so will new problems: an obvious category of limits to human activity—one of which, unfortunately, no examples can be given—is limits we haven't thought of yet. Present research and development, being governed overwhelmingly by short-term market incentives and therefore largely responsive rather than anticipatory, are unlikely to identify such limits far enough in advance for convenient avoidance—especially as only a minute fraction (perhaps 2 percent) of all "Research & Development" is devoted to the world's real problems rather than to military, prestige, and luxury or other projects.

As rates of change increase, decisionmakers will become more acutely aware of a central weakness of the conventional economic paradigm: its failure to recognize the effects of possible delays. This leads to an inability to construct the anticipatory or growth-slowing policies that a lagged system requires. One can show, for example, that a feedback system using free, instantly available, and 100 percent efficient technologies to abate perceived pollution in an exponentially growing industrial system will not be able to prevent exponentially rising pollution—simply because of the perceptual delay in recognizing the need to deploy abatement technologies. Yet an economist who perceives the world through a delay-free paradigm will not expect this result and will be unable to cope with it. Not until delays and nonlinearities in the world are systematically recognized in all sectors—in natural processes, in political perception, in social action, and in technical innovation—can the environmental problems of a dynamic world be successfully addressed. Finally, as Eckholm (1976) emphasizes, incisive analysis, well-communicated to a wide range of people, is not enough to solve problems. For example, otherwise sound proposals for land conservation cannot work without a widely shared conservation ethic that impels ordinary people to demand and support government initiatives; nor without offering ways to meet basic human needs while avoiding environmental degradation; nor without reform of land tenure, social inequities, cultural patterns, and economic and political practice so that those people who are to do the conserving have strong personal incentives to forgo apparent short-term benefits.

Global Organization

In agriculture, industry, trade, or technological transfer—indeed, in any sector that one cares to examine—nearly all present development efforts are tending to produce an increasingly interdependent world. In the process, some sustainable social structures and cultural values, traditional and indigenous, are being submerged by others that are less appropriate and perhaps more ephemeral. Integrity is yielding to homogeneity, appropriate independence perhaps to short-sighted interdependence. There is no doubt that this is commonly done with the best intentions; but before it proceeds much further, should we not assess whether it is wise? Its drawback is that an interdependent world is extremely vulnerable to disruption of small parts (Brown, 1972). The possibility that, in building the sort of world which our liberal instincts demand, we may be destroying its natural defenses against instability, should give us reason to pause.[4]

If the world were under perfect management and had ample safety margins to guard against local wars, droughts, and other misfortunes, there would be little cause to worry about present trends—save, of course, concern at the social and humanistic implications of losing cultural diversity. But the world is under manifestly imperfect management, is torn by sectoral interests, and shows every sign of an increasing frequency of local and regional collapses. This being the sad situation, as a prudent contingency plan we should examine the appropriateness of each new link that makes the world more interdependent, and the possibility of sufficiently decoupling the world, under definite political and economic arrangements, to reduce the likelihood that local disasters might propagate more widely. In so doing, we might be laying the groundwork for a sustainable global society; for one need not allow economic decoupling to degenerate into the moral decoupling of which we have seen far too much lately.

It is hard to be specific about the forms that regional instabilities might take. Industrial societies are in many ways as vulnerable to external or internal dislocation as less industrial, more agricultural, societies; even though the latter may be far less able to buy their way out of trouble, their less centralized social organization and their greater adaptability may well serve them better to resist misfortune. But in an interdependent world, local instabilities may rapidly spread. To make this less likely, we should consider the logistic details of possible regional instabilities, the mechanisms of propagation (for all we know, our customary means of intervention might make matters worse), and—most important—the epidemiological implications of regional disaster. In this connection we must study the plausible patterns of predation on dense human monocultures, the analogy between broad-spectrum antibiotics in such monocultures and broad-spectrum pesticides in agricultural monocultures, possible trends and hazards in the selection and mutation of pathogens, and the implications of worldwide air travel. Such matters are not nice to think about; but not thinking about them is one way to ensure that the consequences of regional instabilities will be not merely regional, and will do much harm that could have been prevented.

Examining what degree of regional integrity or global interdependence is appropriate (rather than taking it for granted) will require equal measures of dispassion and compassion. Most important, it will require worldwide reflection on how best to further Man's highest gifts and goals. Such intellectual interdependence—the gathering and blending of Man's best ideas—differs from physical and economic interdependence by assuredly increasing, rather than potentially jeopardizing, our common security in an uncertain world.

Summary

Biophysical and other "outer limits" of food, land, water, climatic change, stratospheric chemistry, energy, hazardous substances, non-fuel minerals, human stress, and social and ecological stability raise fundamental questions about present trends in management methods and in global organization. The diverse outer limits surveyed in this paper reflect complex, poorly perceived, and often unsuspected interconnections between numerous biological and geophysical processes, many of which are obscure or still unknown. Our lack of predictive power, let alone of quantitative understanding, implies a need to treat essential life-support systems with great caution and forbearance, lest we erode safety margins whose importance we do not yet appreciate.

Even those outer limits which now seem remote are relevant to present policy, as their timely avoidance may require us to discard otherwise attractive short-term policies in favor of others that offer less immediate advantage but that retain options which may be needed later. Such alternative policies may have to rely more on social than on technical innovation in order to address underlying disequilibria rather than merely palliating their symptoms. Moreover, some outer limits are sufficiently imminent, or require such long lead-times to avoid, that fundamental changes in policy, in institutions, and in the degree of global interdependence seem necessary if we are to live to enjoy some of the later and more interesting limits to human activity.

Acknowledgments

The author gratefully acknowledges the support and encouragement of the International Federation of Institutes for Advanced Study (Stockholm) and of the United Nations Environment Programme (Nairobi) in preparing earlier drafts of this paper. He is indebted to both organizations for permission to publish this version, which reflects his own views and not necessarily theirs. Mr Erik Eckholm, Dr Paul Fye, Lady Jackson, Professor Donella Meadows, Dr Sam Nilsson, Dr Walter Orr Roberts, Dr Stephen Schneider, Professor Carroll Wilson, and many others kindly offered advice on part or all of the manuscript, which reflects ideas collected from consultations with a wide variety of authorities.

Notes

1 A referee comments, however, on the "recent major entry of Brazil into the soybean market"—*Environmental Conservation* editor note.
2 See also E. A. Martell in *Am. Scient.*, July 1975.
3 On requesting the author to supply reliable references to this hair-raising new conceivable panecodisaster, we received the response that none could as yet be given. We consequently retain it as the *merest suggestion*, while warning that it should be treated with the utmost caution—*Environmental Conservation* editor note.
4 The conflict between liberal instincts and cold assessment of long-term consequences raises unpalatable ethical issues that may be felt perhaps most acutely by those who advised in vain, shortly after World War II, that the nascent United Nations discourage the distribution of antibiotics to countries which were unwilling to reduce their fertility in proportion to their mortality. Such people argued then, and argue today, that it is morally preferable for a small rather than a large number of people to starve—a proposition which is unlikely to commend itself to modern policymakers who prefer to think that nobody needs to starve. Similar issues are raised by "triage"—the practice, in military medicine or foreign aid, of concentrating one's limited resources on those who need and can benefit from help, and neglecting those who will survive without it or die despite it. The difficulty of thinking about such issues is not lessened by most people's unwillingness to discuss them.

References

Aagaard, R. & Coachman, L. K. (1975). Toward an ice-free Arctic Ocean. *EOS, Trans Amer. Geophys. Union*, **56**, pp 484–6.

Armstrong, Joe E. & Harman, Willis W. (1975). *Plausibility of a Restricted Energy Use Scenario*. CSSP 3705-8, Stanford Research Institute, Menlo Park, California: vi + 199 pp.

Boeck, W. L., Shaw, D. T. & Vonnegut, B. (1975). Possible consequences of global dispersion of krypton 85. *Bull Am. Meteorol. Soc.*, **56**, p 527.

Borgström, Georg (1973). *Focal Points: A Global Food Strategy*. Macmillan, New York: xii + 320 pp, illustr.

Brown, Lester (1972). *World Without Borders*. Random House, New York: xviii + 395 pp.

Brown, Lester (with Erik P. Eckholm) (1974a). *By Bread Alone*. Praeger, New York: xi + 272 pp, illustr.

Brown, Lester (1974b). *In the Human Interest: a Strategy to Stabilize World Population*. Norton, New York: 190 pp.

Bryson, Reid A. (1972). Climatic modification by air pollution. Pp. 133–55 with 12 figs and following discussion in Nicholas Polunin (Ed.) *The Environmental Future*. Macmillan, London & Basingstoke, and Barnes & Noble, New York: xiv + 660 pp., illustr.

Bryson, Reid A. (1974). A perspective on climatic change. *Science*, **184**, pp 753–60.

Bryson, Reid A. (1975). The lessons of climatic history. *Environmental Conservation*, 2(3), pp 163–70, 8 figs.

Campbell, W. J. & Martin, S. (1973). Oil and ice in the Arctic Ocean: possible large-scale interactions. *Science*, **181**, pp. 56–8, fig. [For follow-up, see also *Science*, **186**, pp 843–6, 1974.—Ed]

Chapman, Peter F. (1975). *Fuel's Paradise: Energy Options for Britain*. Penguin, Harmondsworth, Middlesex, England: 236 pp.

Comey, David D. (1975). The legacy of uranium tailings. *Bulletin of the Atomic Scientists*, **31**, September, pp 42–5.

Congressional Research Service (1975). *The Development and Allocation of Scarce World Resources*. Senate document 94-45 (May 1975), US Government Printing Office, Washington, D.C.: vii + 399 pp.

Daly, Herman E. (1973). *Toward a Steady-state Economy*. W. H. Freeman, San Francisco: xii + 332 pp.

Eckholm, Erik P. (1976). *Losing Ground: Environmental Stress and World Food Prospects*. W. W. Norton, New York: 223 pp.

Edsall, John T. (1974). Hazards of nuclear fission power and the choice of alternatives. *Environmental Conservation*, **1**(1), pp. 21–30.

Edsall, John T. (1975). Further comments on hazards of nuclear power and the choice of alternatives. *Environmental Conservation*, **2**(3), pp. 205–12.

Ehrlich, Paul R. (1974) Human population and environmental problems. *Environmental Conservation*, **1**(1), pp. 15–20.

Energy Policy Project of the Ford Foundation (1974). *A Time To Choose: America's Energy Future*. Ballinger, Cambridge, Massachusetts: xii + 511 pp.

Frejka, Tomas (1973). *The Future of Population Growth: Alternative Paths to Equilibrium*. A Population Council Book; John Wiley, New York: xix + 268 pp., illustr.

Green, Harold P. (1975). The risk-benefit calculus in safety determinations. *George Washington Law Rev.*, **43**, pp. 791–803.

Illich, Ivan (1973). *Tools for Conviviality*. Calder & Boyars, London: xiii + 110 pp.

International Federation of Institutes for Advanced Study [cited as IFIAS] (1975). Reports of the Guldsmedshyttan and Lidingö Energy Analysis Workshops. IFIAS, Nobelhuset, Sturegatan 14, S-102 46 Stockholm: ix + 103 pp. [Lidingö item only].

Kellogg, William W. & Schneider, Stephen H. (1974). *Science*, **186**, pp. 1163–72.

Leach, Gerald (1975). *Energy and Food Production*. International Institute for Environment and Development, 27 Mortimer St, London W1, England, and 1525 New Hampshire Avenue NW, Washington, D.C. 20036: ii + 151 pp.

Lovering, Thomas S. (1969). Mineral resources from the land. Pp. 109–34. In NAS/NRC, *Resources and Man*. W. H. Freeman, San Francisco: xi + 259.

Lovins, A. B. (1973). *Thermal Limits to World Energy Use*. Typescript under revision, 16 pp.

Lovins, A. B. (1975a). *World Energy Strategies: Facts, Issues, and Options*. Ballinger, Cambridge, Massachusetts: xvi + 131 pp.

Lovins, A. B. (1975b). Testimony to President's Council on Environmental Quality: in *Hearings on the ERDA RD&D Plan*. Washington, D.C., 3 September 1975: xvii + 115 pp.

Lovins, A. B. (1975c) Introduction (pp. xvii–xxxii) and Part One: Nuclear power: technical bases for ethical concern (pp. 1–104) in A. B. Lovins & J. H. Price, *Non-Nuclear Futures: The Case for an Ethical Energy Strategy*. Ballinger, Cambridge, Massachusetts: xxii + 225 pp., illustr.

Lovins, A. B. (1976). *Openpit Mining* [Earth Island, London.] Available from Friends of the Earth, 529 Commercial Street, San Francisco, California 94111: xx + 115 pp.

Lovins, A. B. & Patterson, W. C. (1975). Plutonium particles: some like them hot. *Nature* (London), **254**, pp. 278–80.

Manabe, S. & Wetherald, R. T. (1975). *J. Atmos. Sci.*, **32**, pp. 3–15 (see especially Fig 4).

Mitchell, J. M. (1975). A reassessment of atmospheric pollution as a cause of long-term changes of global temperature. Pp. 149–73 in S. F. Singer (Ed.) *The Changing Global Environment*. Reidel, Dordrecht: viii + 423 pp., illustr.

Over, J. A. & Sjoerdsma, C. (Eds.) (1974). *Energy Conservation: Ways and Means*. Publication 19, Stichting Toekomstbeeld der Techniek, Prinsessegracht 23, den Haag: 181 pp.

Polunin, Nicholas (1974). Thoughts on some conceivable ecodisasters. *Environmental Conservation*, **1**(3), pp. 177–89.

Price, J. H. (1975). Part Two: Dynamic energy analysis and nuclear power. Pp. 105–223 in A B. Lovins & J. H. Price, *Non-Nuclear Futures: The Case for an Ethical Energy Strategy*. Ballinger, Cambridge, Massachusetts: xxii + 225 pp., illustr.

Ramanathan, V. (1975). Greenhouse effect due to chlorofluorocarbons: climatic implications. *Science*, **190**, pp. 50–2.

Report of the Federal Task Force on Inadvertent Modification of the Stratosphere [cited as Report IMOS] (1975). *Fluorocarbons and the Environment*. Council on Environmental Quality/Federal Council for Science and Technology, Washington, D.C.: vii + 101 pp., illustr.

Ross, Marc H. & Williams, Robert H. (1975). *Assessing the Potential for Fuel Conservation*. Institute for Public Policy Alternatives, State University of New York, Albany: i + 37 pp.

Schipper, Lee (1975). *Energy Conservation: Its Nature, Hidden Benefits and Hidden Barriers*. ERG 75-2, Energy & Resources Group, University of California, Berkeley: v + 79 pp. (mimeogr.).

Schneider, Stephen H. (with Lynne E. Mesirow) (1976). *The Genesis Strategy*. Plenum, New York: xxi + 419 pp., illustr.

Schneider, Stephen H. & Dennett, Roger D. (1975). Climatic barriers to long-term energy growth. *Ambio*, **4**(2), pp. 65–74.

Slesser, Malcolm (1973). *New Scientist*, **60**, p 328.

Slesser, Malcolm (1975). Accounting for energy. *Nature* (London), **254**, pp 170–2.

Sørensen, Bent (1975). Energy and resources. *Science*, **189**, pp 255–60.

Wilson, Carroll L. & Matthews, William H. (Eds) (1970). *Man's Impact on the Global Environment, Report of the Study of Critical Environmental Problems*. MIT Press, Cambridge, Massachusetts: xxii + 319 pp.

Wilson, Carroll L. & Matthews, William H. (Eds) (1971). *Inadvertent Climate Modification, Report of the Study of Man's Impact on Climate*. MIT Press, Cambridge, Massachusetts: xxi + 308 pp.

Chapter 6

Energy Strategy: The Road Not Taken?

1976

Editor's note: *This article is reprinted from Foreign Affairs, October 1976, by permission of the publisher. Copyright 1976 by the Council on Foreign Relations, Inc.*

I

Where are America's formal or de facto energy policies leading us? Where might we choose to go instead? How can we find out?

Addressing these questions can reveal deeper questions—and a few answers—that are easy to grasp, yet rich in insight and in international relevance. This paper will seek to explore such basic concepts in energy strategy by outlining and contrasting two energy paths that the United States might follow over the next 50 years—long enough for the full implications of change to start to emerge. The first path resembles present federal policy and is essentially an extrapolation of the recent past. It relies on rapid expansion of centralized high technologies to increase supplies of energy, especially in the form of electricity.

The second path combines a prompt and serious commitment to efficient use of energy, rapid development of renewable energy sources matched in scale and in energy quality to end-use needs, and special transitional fossil-fuel technologies. This path, a whole greater than the sum of its parts, diverges radically from incremental past practices to pursue long-term goals.

Both paths, as will be argued, present difficult—but very different—problems. The first path is convincingly familiar, but the economic and sociopolitical problems lying ahead loom large, and eventually, perhaps, insuperable. The second path, though it represents a shift in direction, offers many social, economic and geopolitical advantages, including virtual elimination of nuclear proliferation from the world. It is important to recognize that the two paths are mutually exclusive. Because commitments to the first may foreclose the second, we must soon choose one or the other—before failure to stop nuclear proliferation has foreclosed both.[1]

II

Most official proposals for future US energy policy embody the twin goals of sustaining growth in energy consumption (assumed to be closely and causally linked to GNP and to social welfare) and of minimizing oil imports. The usual proposed solution is rapid expansion of three sectors: coal (mainly strip-mined, then made into electricity and synthetic fluid fuels); oil and gas (increasingly from Arctic and offshore wells); and nuclear fission (eventually in fast breeder reactors). All domestic resources, even naval oil reserves, are squeezed hard—in a policy which David Brower calls "Strength Through Exhaustion."

Conservation, usually induced by price rather than by policy, is conceded to be necessary but it is given a priority more rhetorical than real. "Unconventional" energy supply is relegated to a minor role, its significant contribution postponed until past 2000. Emphasis is overwhelmingly on the short term. Long-term sustainability is vaguely assumed to be ensured by some eventual combination of fission breeders, fusion breeders, and solar electricity. Meanwhile, aggressive subsidies and regulations are used to hold down energy prices well below economic and prevailing international levels so that growth will not be seriously constrained.

Even over the next ten years (1976–1985), the supply enterprise typically proposed in such projections is impressive. Oil and gas extraction shift dramatically to offshore and Alaskan sources, with nearly 900 new oil wells offshore of the contiguous 48 states alone. Some 170 new coal mines open, extracting about 200 million tons per year each from eastern underground and strip mines, plus 120 million from western stripping. The nuclear fuel cycle requires over 100 new uranium mines, a new enrichment plant, some 40 fuel fabrication plants, three fuel reprocessing plants. The electrical supply system, more than doubling, draws on some 180 new 800-megawatt coal-fired stations, over one hundred and forty 1000-megawatt nuclear reactors, 60 conventional and over 100 pumped-storage hydroelectric plants, and over 350 gas turbines. Work begins on new industries to make synthetic fuels from coal and oil shale. At peak, just building (not operating) all these new facilities directly requires nearly 100,000 engineers, over 420,000 craftspeople, and over 140,000 laborers. Total indirect labor requirements are twice as great.[2]

This ten-year spurt is only the beginning. The year 2000 finds us with 450 to 800 reactors (including perhaps 80 fast breeders, each loaded with 2.5 metric tons of plutonium), 500 to 800 huge coal-fired power stations, 1,000 to 1,600 new coal mines, and some 15 million electric automobiles. Massive electrification—which, according to one expert, is "the most important

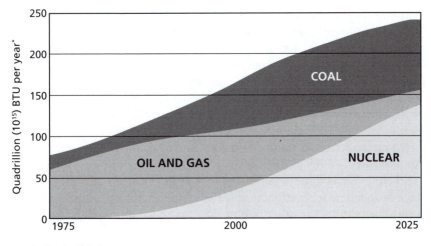

* Or quintillion (10^{18}) joules per year

Figure 6.1 *An illustrative, schematic future for US gross primary energy use*

attempt to modify the infrastructure of industrial society since the railroad"[3]—is largely responsible for the release of waste heat sufficient to warm the entire freshwater runoff of the contiguous 48 states by 34–49F°.[4] Mining coal and uranium, increasingly in the arid West, entails inverting thousands of communities and millions of acres, often with little hope of effective restoration. The commitment to a long-term coal economy many times the scale of today's makes the doubling of atmospheric carbon dioxide concentration early in the next century virtually unavoidable, with the prospect then or soon thereafter of substantial and perhaps irreversible changes in global climate.[5] Only the exact date of such changes is in question.

The main ingredients of such an energy future are roughly sketched in Figure 6.1. For the period up to 2000, this sketch is a composite of recent projections published by the Energy Research and Development Administration (ERDA), the Federal Energy Administration (FEA), the Department of the Interior, Exxon, and the Edison Electric Institute. Minor and relatively constant sources, such as hydroelectricity, are omitted; the nuclear component represents nuclear heat, which is roughly three times the resulting nuclear electric output; fuel imports are aggregated with domestic production. Beyond 2000, the usual cutoff date of present projections, the picture has been extrapolated to the year 2025—exactly how is not important here—in order to show its long-term implications more clearly.[6]

III

The flaws in this type of energy policy have been pointed out by critics in and out of government. For example, despite the intensive electrification—consuming more than half the total fuel input in 2000 and more thereafter—we are still short of gaseous and liquid fuels, acutely so from the 1980s on, because of slow and incomplete substitution of electricity for the two-thirds of fuel use that is now direct. Despite enhanced recovery of resources in the ground, shortages steadily deepen in natural gas—on which plastics and nitrogen fertilizers depend—and, later, in fuel for the transport sector (half our oil now runs cars). Worse, at least half the energy growth never reaches the consumer because it is lost earlier in elaborate conversions in an increasingly inefficient fuel chain dominated by electricity generation (which wastes about two-thirds of the fuel) and coal conversion (which wastes about one-third). Thus in Britain since 1900, primary energy—the input to the fuel chain—has doubled while energy at the point of end use—the car, furnace or machine whose function it fuels—has increased by only a half, or by a third per capita; the other half of the growth went to fuel the fuel industries, which are the largest energy consumers.

Among the most intractable barriers to implementing Figure 6.1 is its capital cost. In the 1960s, the total investment to increase a consumer's delivered energy supplies by the equivalent of one barrel of oil per day (about 67 kilowatts of heat) was a few thousand of today's dollars—of which, in an oil system, the wellhead investment in the Persian Gulf was and still is only a few hundred dollars. (The rest is transport, refining, marketing and distribution.) The capital intensity of much new coal supply is still in this range. But such cheaply won resources can no longer stretch our domestic production of fluid fuels or electricity, and Figure 6.1 relies mainly on these, not on coal burned directly, so it must bear the full burden of increased capital intensity.

That burden is formidable. For the North Sea oil fields coming into production soon, the investment in the whole system is roughly $10,000 to deliver an extra barrel per day (constant 1976 dollars throughout); for US frontier (Arctic and offshore) oil and gas in the 1980s it will be generally in the range from $10,000 to $25,000; for synthetic gaseous and liquid fuels made from coal, from $20,000 to $50,000 per daily barrel.

The scale of these capital costs is generally recognized in the industries concerned. What is less widely appreciated—partly because capital costs of electrical capacity are normally calculated per installed (not delivered) kilowatt and partly because whole-system costs are rarely computed—is that capital cost is many times greater for new systems that make electricity than for those that burn fuels directly. For coal-electric capacity ordered today, a reasonable estimate would be about $150,000 for the delivered equivalent of one barrel of oil per day; for nuclear-electric capacity ordered today, about $200,000–$300,000. Thus, the capital cost per delivered kilowatt of electrical energy emerges as roughly 100 times that of the traditional direct-fuel technologies on which our society has been built.[7]

The capital intensity of coal conversion and, even more, of large electrical stations and distribution networks is so great that many analysts, such as the strategic planners of the Shell Group in London, have concluded that no major country outside the Persian Gulf can afford these centralized high technologies on a truly large scale, large enough to run a country. They are looking, in Monte Canfield's phrase, like future technologies whose time has passed.

Relying heavily on such technologies, President Ford's 1976–1985 energy program turns out to cost over $1 trillion (in 1976 dollars) in initial investment, of which about 70 to 80 percent would be for new rather than replacement plants.[8] The latter figure corresponds to about three-fourths of cumulative net private domestic investment (NPDI) over the decade (assuming that NPDI remains 7 percent of gross national product and that GNP achieves real growth of 3.5 percent per year despite the adverse effects of the energy program on other investments). In contrast, the energy sector has recently required only one-fourth of NPDI. Diverting to the energy sector not only this hefty share of discretionary investment but also about two-thirds of all the rest would deprive other sectors which have their own cost-escalation problems and their own vocal constituencies. A powerful political response could be expected. And this capital burden is not temporary; further up the curves of Figure 6.1 it tends to increase, and much of what might have been thought to be increased national wealth must be plowed back into the care and feeding of the energy system. Such long-lead-time, long-payback-time investments might also be highly inflationary.

Of the $1 trillion-plus just cited, three-fourths would be for electrification. About 18 percent of the total investment could be saved just by reducing the assumed average 1976–1985 electrical growth rate from 6.5 to 5.5 percent per year.[9] Not surprisingly, the combination of disproportionate and rapidly increasing capital intensity, long lead times, and economic responses is already proving awkward to the electric utility industry, despite the protection of a 20 percent taxpayer subsidy on new power stations.[10] "Probably no industry," observes Bankers Trust Company, "has come closer to the edge of financial disaster." Both here and abroad an effective feedback loop is observable: large capital programs→poor cash flow→higher electricity prices→reduced demand growth→worse cash flow→increased bond flotation→increased debt-to-equity ratio, worse coverage, and less attractive bonds→poor bond sales→worse cash flow→higher electricity prices→reduced (even negative) demand growth and political pressure on utility regulators→overcapacity, credit pressure, and higher cost of money→worse cash flow, etc. This "spiral of impossibility," as Mason Willrich has called it, is exacerbated by most utilities' failure to base historic prices on the long-run cost of new supply: thus some must now tell their customers that the current-dollar cost of a kilowatt-hour will treble by 1985, and that two-thirds of that increase will be capital charges for new plants. Moreover, experience abroad suggests that even a national treasury cannot long afford electrification: a New York State-like position is quickly reached, or too little money is left over to finance the energy *uses*, or both.

IV

Summarizing a similar situation in Britain, Walter Patterson concludes: "Official statements identify an anticipated 'energy gap' which can be filled only with nuclear electricity; the data do not support any such conclusion, either as regards the 'gap' or as regards the capability of filling it with nuclear electricity." We have sketched one form of the latter argument; let us now consider the former.

Despite the steeply rising capital intensity of new energy supply, forecasts of energy demand made as recently as 1972 by such bodies as the Federal Power Commission and the Department of the Interior wholly ignored both price elasticity of demand and energy conservation. The Chase Manhattan Bank in 1973 saw virtually no scope for conservation save by minor curtailments: the efficiency with which energy produced economic outputs was assumed to be optimal already. In 1976, some analysts still predict economic calamity if the United States does not continue to consume twice the combined energy total for Africa, the rest of North and South America, and Asia except Japan. But what have more careful studies taught us about the scope for doing better with the energy we have? Since we can't keep the bathtub filled because the hot water keeps running out, do we really (as Malcolm MacEwen asks) need a bigger water heater, or could we do better with a cheap, low-technology plug?

There are two ways, divided by a somewhat fuzzy line, to do more with less energy. First, we can plug leaks and use thriftier technologies to produce exactly the same output of goods and services—and bads and nuisances—as before, substituting other resources (capital, design, management, care, etc.) for some of the energy we formerly used. When measures of this type use today's technologies, are advantageous today by conventional economic criteria, and have no significant effect on lifestyles, they are called "technical fixes."

In addition, or instead, we can make and use a smaller quantity or a different mix of the outputs themselves, thus to some degree changing (or reflecting ulterior changes in) our lifestyles. We might do this because of changes in personal values, rationing by price or otherwise, mandatory curtailments, or gentler inducements. Such "social changes" include car-pooling, smaller cars, mass transit, bicycles, walking, opening windows, dressing to suit the weather, and extensively recycling materials. Technical fixes, on the other hand, include thermal insulation, heat pumps (devices like air conditioners which move heat around—often in either direction—rather than making it from scratch), more efficient furnaces and car engines, less overlighting and overventilation in commercial buildings, and recuperators for waste heat in industrial processes. Hundreds of technical and semi-technical analyses of both kinds of conservation have been done; in the last two years especially, much analytic progress has been made.

Theoretical analysis suggests that in the long term, technical fixes *alone* in the United States could probably improve energy efficiency by a factor of at least three or four.[11] A recent review of specific practical measures cogently argues that with only those technical fixes that could be implemented by about the turn of the century, we could nearly double the efficiency with which we use energy.[12] If that is correct, we could have steadily increasing economic activity with approximately constant primary energy use for the next few decades, thus stretching our present energy supplies rather than having to add massively to them. One careful comparison shows that *after* correcting for differences of climate, hydroelectric capacity, etc., Americans would still use about a third less energy than they do now if they were as efficient as the Swedes (who see much room for improvement in their own efficiency).[13] US per capita energy intensity, too, is about twice that of West Germany in space heating, four times in transport.[14] Much of the difference is attributable to technical fixes.

Some technical fixes are already under way in the United States. Many factories have cut tens of percent off their fuel cost per unit output, often with practically no capital investment. New 1976 cars average 27 percent better mileage than 1974 models. And there is overwhelming evidence that technical fixes are generally much cheaper than increasing energy supply, quicker, safer, of more lasting benefit. They are also better for secure, broadly based employment using existing skills. Most energy conservation measures and the shifts of consumption which they occasion are relatively labor-intensive. Even making more energy-efficient home appliances is about twice as good for jobs as is building power stations: the latter is practically the least labor-intensive major investment in the whole economy.

The capital savings of conservation are particularly impressive. In the terms used above, the investments needed to save the equivalent of an extra barrel of oil per day are often zero to $3500, generally under $8000, and at most about $25,000—far less than the amounts needed to increase most kinds of energy supply. Indeed, to use energy efficiently in new buildings, especially commercial ones, the additional capital cost is often *negative*: savings on heating and cooling equipment more than pay for the other modifications.

To take one major area of potential saving, technical fixes in new buildings can save 50 percent or more in office buildings and 80 percent or more in some new houses.[15] A recent American Institute of Architects study concludes that, by 1990, improved design of new buildings and modification of old ones could save a third of our current *total* national energy use—and save money too. The payback time would be only half that of the alternative investment in increased energy supply, so the same capital could be used twice over.

A second major area lies in "cogeneration," or the generating of electricity as a by-product of the process steam normally produced in many industries. A Dow study chaired by Paul McCracken reports that by 1985 US industry could meet approximately half its own electricity needs (compared to about a seventh today) by this means. Such cogeneration would save $20–50 billion in investment, save fuel equivalent to 2–3 million barrels of oil per day, obviate the need for more than 50 large reactors, and (with flattened utility rates) yield at least

20 percent pretax return on marginal investment while reducing the price of electricity to consumers.[16] Another measure of the potential is that cogeneration provides about 4 percent of electricity today in the United States—but about 12 percent in West Germany. Cogeneration and more efficient use of electricity could together reduce our use of electricity by a third and our central-station generation by 60 percent.[17] Like district heating (distribution of waste heat as hot water via insulated pipes to heat buildings), US cogeneration is held back only by institutional barriers. Yet these are smaller than those that were overcome when the present utility industry was established.

So great is the scope for technical fixes now that we could spend several hundred billion dollars on them initially plus several hundred million dollars per day—and still save money compared with increasing the supply! And we would still have the fuel (without the environmental and geopolitical problems of getting and using it). The barriers to far more efficient use of energy are not technical, nor in any fundamental sense economic. So why do we stand here confronted, as Pogo said, by insurmountable opportunities?

The answer—apart from poor information and ideological antipathy and rigidity—is a wide array of institutional barriers, including more than 3000 conflicting and often obsolete building codes, an innovation-resistant building industry, lack of mechanisms to ease the transition from kinds of work that we no longer need to kinds we do need, opposition by strong unions to schemes that would transfer jobs from their members to larger numbers of less "skilled" workers, promotional utility rate structures, fee structures giving building engineers a fixed percentage of prices of heating and cooling equipment they install, inappropriate tax and mortgage policies, conflicting signals to consumers, misallocation of conservation's costs and benefits (builders vs. buyers, landlords vs. tenants, etc.), imperfect access to capital markets, and fragmentation of government responsibility.

Though economic answers are not always right answers, properly using the markets we have may be the greatest single step we could take toward a sustainable, humane energy future. The sound economic principles we need to apply include flat (even inverted) utility rate structures rather than discounts for large users, pricing energy according to what extra supplies will cost in the long run ("long-run marginal-cost pricing"), removing subsidies, assessing the total costs of energy-using purchases over their whole operating lifetimes ("life-cycle costing"), counting the costs of complete energy systems including all support and distribution systems, properly assessing and charging environmental costs, valuing assets by what it would cost to replace them, discounting appropriately, and encouraging competition through antitrust enforcement (including at least horizontal divestiture of giant energy corporations).

Such practicing of the market principles we preach could go very far to help us use energy efficiently and get it from sustainable sources. But just as clearly, there are things the market cannot do, like reforming building codes or utility practices. And whatever our means, there is room for differences of opinion about how far we can achieve the great theoretical potential for technical fixes. How far might we instead choose, or be driven to, some of the "social changes" mentioned earlier?

There is no definitive answer to this question—though it is arguable that if we are not clever enough to overcome the institutional barriers to implementing technical fixes, we shall certainly not be clever enough to overcome the more familiar but more formidable barriers to increasing energy supplies. My own view of the evidence is, first, that we are adaptable enough to use technical fixes *alone* to double, in the next few decades, the amount of social benefit we wring from each unit of end-use energy; and second, that value changes which could either replace or supplement those technical changes are also occurring rapidly. If either of these views is right, or if both are partly right, we should be able to double end-use efficiency by the turn of the century or shortly thereafter, with minor or no changes in lifestyles or values save increasing comfort for modestly increasing numbers. Then over the period 2010–2040, we should be able to shrink per capita primary energy use to perhaps a third or a quarter of today's.[18] (The former would put us at the per capita level of the wasteful, but hardly troglodytic, French.) Even in the case of fourfold shrinkage, the resulting society could be instantly recognizable to a visitor from the 1960s and need in no sense be a pastoralist's utopia—though that option would remain open to those who may desire it.

The long-term mix of technical fixes with structural and value changes in work, leisure, agriculture, and industry will require much trial and error. It will take many years to make up our diverse minds about. It will not be easy—merely easier than not doing it. Meanwhile it is easy only to see what not to do.

If one assumes that by resolute technical fixes and modest social innovation we can double our end-use efficiency by shortly after 2000, then we could be twice as affluent as now with today's level of energy use, or as affluent as now while using only half the end-use energy we use today. Or we might be somewhere in between—significantly more affluent (and equitable) than today but with less end-use energy.

Many analysts now regard modest, zero or negative growth in our rate of energy use as a realistic long-term goal. Present annual US primary energy demand is about 75 quadrillion BTU ("quads"), and most official projections for 2000 envisage growth to 130–170 quads. However, recent work at the Institute for Energy Analysis, Oak Ridge, under the direction of Dr Alvin Weinberg, suggests that standard projections of energy demand are far too high because they do not take account of changes in demographic and economic trends. In June 1976 the Institute considered that with a conservation program far more modest than that contemplated in this article, the likely range of US primary energy demand in the year 2000 would be about 101–126 quads, with the lower end of the range more probable and end-use energy being about 60–65 quads. And, at the further end of the spectrum, projections for 2000 being considered by the "Demand Panel" of a major US National Research Council study, as of mid-1976, ranged as low as about 54 quads of fuels (plus 16 of solar energy).

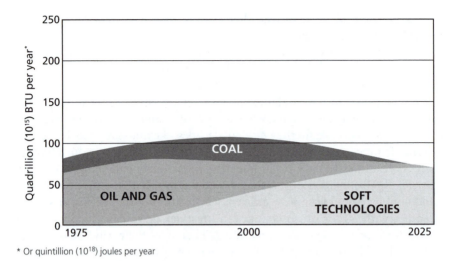

* Or quintillion (10^{18}) joules per year

Figure 6.2 *An alternative illustrative future for US gross primary energy use*

As the basis for a coherent alternative to the path shown in Figure 6.1 earlier, a primary energy demand of about 95 quads for 2000 is sketched in Figure 6.2. Total energy demand would gradually decline thereafter as inefficient buildings, machines, cars and energy systems are slowly modified or replaced. Let us now explore the other ingredients of such a path—starting with the "soft" supply technologies which, spurned in Figure 6.1 as insignificant, now assume great importance.

V

There exists today a body of energy technologies that have certain specific features in common and that offer great technical, economic and political attractions, yet for which there is no generic term. For lack of a more satisfactory term, I shall call them "soft" technologies: a textural description, intended to mean not vague, mushy, speculative or ephemeral, but rather flexible, resilient, sustainable and benign. Energy paths dependent on soft technologies, illustrated in Figure 6.2, will be called "soft" energy paths, as the "hard" technologies sketched in Section II constitute a "hard" path (in both senses). The distinction between hard and soft energy paths rests not on how much energy is used, but on the technical and sociopolitical *structure* of the energy system, thus focusing our attention on consequent and crucial political differences.

In Figure 6.2, then, the social structure is significantly shaped by the rapid deployment of soft technologies. These are defined by five characteristics:

1   They rely on renewable energy flows that are always there whether we use them or not, such as sun and wind and vegetation: on energy income, not on depletable energy capital.

2   They are diverse, so that energy supply is an aggregate of very many individually modest contributions, each designed for maximum effectiveness in particular circumstances.

3   They are flexible and relatively low-technology—which does not mean unsophisticated, but, rather, easy to understand and use without esoteric skills, accessible rather than arcane.

4   They are matched in *scale* and in geographic distribution to end-use needs, taking advantage of the free distribution of most natural energy flows.

5   They are matched in *energy quality* to end-use needs: a key feature that deserves immediate explanation.

People do not want electricity or oil, nor such economic abstractions as "residential services," but rather comfortable rooms, light, vehicular motion, food, tables, and other real things. Such end-use needs can be classified by the physical nature of the task to be done. In the United States today, about 58 percent of all energy at the point of end use is required as heat, split roughly equally between temperatures above and below the boiling point of water. (In Western Europe the low-temperature heat alone is often a half of all end-use energy.) Another 38 percent of all US end-use energy provides mechanical motion: 31 percent in vehicles, 3 percent in pipelines, 4 percent in industrial electric motors. The rest, a mere 4 percent of delivered energy, represents *all* lighting, electronics, telecommunications, electrometallurgy, electrochemistry, arc-welding, electric motors in home appliances and in railways, and similar end uses which now *require* electricity.

Some 8 percent of all our energy end use, then, requires electricity for purposes other than low-temperature heating and cooling. Yet, since we actually use electricity for many such low-grade purposes, it now meets 13 percent of our end-use needs—and its generation consumes 29 percent of our fossil fuels. A hard energy path would increase this 13 percent figure to 20–40 percent (depending on assumptions) by the year 2000, and far more thereafter. But this is wasteful because the laws of physics require, broadly speaking, that a power station change three units of fuel into two units of almost useless waste heat plus one unit of electricity. This electricity can do more difficult kinds of work than can the original fuel, but unless this extra quality and versatility are used to advantage, the costly process of upgrading the fuel—and losing two-thirds of it—is all for naught.

Plainly we are using premium fuels and electricity for many tasks for which their high energy quality is superfluous, wasteful, and expensive, and a hard path would make this inelegant practice even more common. Where we want only to create temperature differences of tens of degrees, we should meet the need with sources whose potential is tens or hundreds of degrees, not with a flame temperature of thousands or a nuclear temperature of millions—like cutting butter with a chainsaw.

For some applications, electricity is appropriate and indispensable: electronics, smelting, subways, most lighting, some kinds of mechanical work, and a few more. But these uses are already oversupplied, and for the other, dominant uses remaining in our energy economy this special form of energy cannot give us our money's worth (in many parts of the United States today it already costs $50–120 per barrel-equivalent). Indeed, in probably no industrial country today can additional supplies of electricity be used to thermodynamic advantage which would justify their high cost in money and fuels.

So limited are the US end uses that really require electricity that by applying careful technical fixes to them we could reduce their 8 percent total to about 5 percent (mainly by reducing commercial overlighting), whereupon we could probably cover all those needs with present US hydroelectric capacity plus the cogeneration capacity available in the mid-to-late 1980s.[19] Thus an affluent industrial economy could advantageously operate with no central power stations at all! In practice we would not necessarily want to go that far, at least not for a long time, but the possibility illustrates how far we are from supplying energy only in the quality needed for the task at hand.

A feature of soft technologies as essential as their fitting end-use needs (for a different reason) is their appropriate scale, which can achieve important types of economies not available to larger, more centralized systems. This is done in five ways, of which the first is reducing and sharing overheads. Roughly half your electricity bill is fixed distribution costs to pay the overheads of a sprawling energy system: transmission lines, transformers, cables, meters and people to read them, planners, headquarters, billing computers, interoffice memos, advertising agencies. For electrical and some fossil-fuel systems, distribution accounts for more than half of total capital cost, and administration for a significant fraction of total operating cost. Local or domestic energy systems can reduce or even eliminate these infrastructure costs. The resulting savings can far outweigh the extra costs of the dispersed maintenance infrastructure that the small systems require, particularly where that infrastructure already exists or can be shared (e.g. plumbers fixing solar heaters as well as sinks).

Small scale brings further savings by virtually eliminating distribution losses, which are cumulative and pervasive in centralized energy systems (particularly those using high-quality energy). Small systems also avoid direct diseconomies of scale, such as the frequent unreliability of large units and the related need to provide instant "spinning reserve" capacity on electrical grids to replace large stations that suddenly fail. Small systems with short lead times greatly reduce exposure to interest, escalation and mistimed demand forecasts—major indirect diseconomies of large scale.

The fifth type of economy available to small systems arises from mass production. Consider, as Henrik Harboe suggests, the 100-odd million cars in this country. In round numbers, each car probably has an average cost of less than $4000 and a shaft power over 100 kilowatts (134 horsepower). Presumably a good engineer could build a generator and upgrade an automobile engine to a reliable, 35-percent-efficient diesel at no greater total cost, yielding a mass-produced diesel generator unit costing less than $40 per kW. In contrast, the motive capacity in our central power stations—currently totaling about one-fortieth as much as in our cars—costs perhaps ten times more per kW, partly because it is not mass-produced. It is not surprising that at least one foreign car maker hopes to go into the wind-machine and heat-pump business. Such a market can be entered incrementally, without the billions of dollars' investment required for, say, liquefying natural gas or gasifying coal. It may require a production philosophy oriented toward technical simplicity, low replacement cost, slow obsolescence, high reliability, high volume and low markup; but these are familiar concepts in mass production. Industrial resistance would presumably melt when—as with pollution-abatement equipment—the scope for profit was perceived.

This is not to say that all energy systems need be at domestic scale. For example, the medium scale of urban neighborhoods and rural villages offers fine prospects for solar collectors—especially for adding collectors to existing buildings, of which some (perhaps with large flat roofs) can take excess collector area while others cannot take any. They could be joined via communal heat storage systems, saving on labor cost and on heat losses. The costly craftwork of remodeling existing systems—"backfitting" idiosyncratic houses with individual collectors—could thereby be greatly reduced. Despite these advantages, medium-scale solar technologies are currently receiving little attention apart from a condominium-village project in Vermont sponsored by the Department of Housing and Urban Development and the 100-dwelling-unit Mejannes-le-Clap project in France.

The schemes that dominate ERDA's solar research budget—such as making electricity from huge collectors in the desert, or from temperature differences in the oceans, or from Brooklyn Bridge-like satellites in outer space—do not satisfy our criteria, for they are ingenious high-technology ways to supply energy in a form and at a scale inappropriate to most end-use needs. Not all solar technologies are soft. Nor, for the same reason, is nuclear fusion a soft technology.[20] But many genuine soft technologies are now available and are now economic. What are some of them?

Solar heating and, imminently, cooling head the list. They are incrementally cheaper than electric heating, and far more inflation-proof, practically anywhere in the world.[21] In the United States (with fairly high average sunlight levels), they are cheaper than present electric heating virtually anywhere, cheaper than oil heat in many parts, and cheaper than gas and coal in some. Even in the least favorable parts of the continental United States, far more sunlight falls on a typical building than is required to heat and cool it without supplement; whether this is considered economic depends on how the accounts are done.[22] The difference in solar input between the most and least favorable parts of the lower 49 states is generally less than twofold, and in cold regions, the long heating season can improve solar economics.

Ingenious ways of backfitting existing urban and rural buildings (even large commercial ones) or their neighborhoods with efficient and exceedingly reliable solar collectors are being rapidly developed in both the private and public sectors. In some recent projects, the lead time from ordering to operation has been only a few months. Good solar hardware, often modular, is going into pilot or full-scale production over the next few years, and will increasingly be integrated into buildings as a multipurpose structural element, thereby sharing costs. Firms such as Philips, Honeywell, Revere, Pittsburgh Plate Glass, and Owens-Illinois, plus many dozens of smaller firms, are applying their talents, with rapid and accelerating effect, to reducing unit costs and improving performance. Some novel types of very simple collectors with far lower costs also show promise in current experiments. Indeed, solar hardware per se is necessary only for backfitting existing buildings. If we build new buildings properly in the first place, they can use "passive" solar collectors—large south windows or glass-covered black south walls—rather than special collectors. If we did this to all new houses in the next 12 years, we would save about as much energy as we expect to recover from the Alaskan North Slope.[23]

Secondly, exciting developments in the conversion of agricultural, forestry and urban wastes to methanol and other liquid and gaseous fuels now offer practical, economically interesting technologies sufficient to run an efficient US transport sector.[24] Some bacterial and enzymatic routes under study look even more promising, but presently proved processes already offer sizable contributions without the inevitable climatic constraints of fossil-fuel combustion. Organic conversion technologies must be sensitively integrated with agriculture and forestry so as not to deplete the soil; most current methods seem suitable in this respect, though they may change the farmer's priorities by making his whole yield of biomass (vegetable matter) saleable.

The required scale of organic conversion can be estimated. Each year the US beer and wine industry, for example, microbiologically produces 5 percent as many gallons (not all alcohol, of course) as the US oil industry produces gasoline. Gasoline has 1.5–2 times the fuel value of alcohol per gallon. Thus a conversion industry roughly 10 to 14 times the scale (in gallons of fluid output per year) of our cellars and breweries would produce roughly one-third of the present gasoline requirements of the United States; if one assumes a transport sector with three times today's average efficiency—a reasonable estimate for early in the next century—then the whole of the transport needs could be met by organic conversion. The scale of effort required does not seem unreasonable, since it would replace in function half our refinery capacity.

Additional soft technologies include wind-hydraulic systems (especially those with a vertical axis), which already seem likely in many design studies to compete with nuclear power in much of North America and Western Europe. But wind is not restricted to making electricity: it can heat, pump, heat-pump, or compress air. Solar process heat, too, is coming along rapidly as we learn to use the 5800°C potential of sunlight (much hotter than a boiler). Finally, high- and low-temperature solar collectors, organic converters, and wind machines can form symbiotic hybrid combinations more attractive than the separate components.

Energy storage is often said to be a major problem of energy-income technologies. But this "problem" is largely an artifact of trying to recentralize, upgrade, and redistribute inherently diffuse energy flows. Directly storing sunlight or wind—or, for that matter, electricity from any source—is indeed difficult on a large scale. But it is easy if done on a scale and in an energy quality matched to most end-use needs. Daily, even seasonal, storage of low- and medium-temperature heat at the point of use is straightforward with water tanks, rock beds, or perhaps fusible salts. Neighborhood heat storage is even cheaper. In industry, wind-generated compressed air can easily (and, with due care, safely) be stored to operate machinery: the technology is simple, cheap, reliable and highly developed. (Some cities even used to supply compressed air as a standard utility.) Installing pipes to distribute hot water (or compressed air) tends to be considerably cheaper than installing equivalent electric distribution capacity. Hydroelectricity is stored behind dams, and organic conversion yields readily stored liquid and gaseous fuels. On the whole, therefore, energy storage is much less of a problem in a soft energy economy than in a hard one.

Recent research suggests that a largely or wholly solar economy can be constructed in the United States with straightforward soft technologies that are now demonstrated and now economic or nearly economic.[25] Such a conceptual exercise does not require "exotic" methods such as sea-thermal, hot-dry-rock geothermal, cheap (perhaps organic) photovoltaic, or solar-thermal electric systems. If developed, as some probably will be, these technologies could be convenient, but they are in no way essential for an industrial society operating solely on energy income.

Figure 6.2 shows a plausible and realistic growth pattern, based on several detailed assessments, for soft technologies given aggressive support. The useful output from these technologies would overtake, starting in the 1990s, the output of nuclear electricity shown in even the most sanguine federal estimates. For illustration, Figure 6.2 shows soft technologies meeting virtually all energy needs in 2025, reflecting a judgment that a completely soft supply mix is practicable in the long run with or without the 2000–2025 energy shrinkage shown. Though most technologists who have thought seriously about the matter will concede it conceptually, however, some may be uneasy about the details. Obviously the sketched curve is not definitive, for although the general direction of the soft path must be shaped soon, the details of the energy economy in 2025 would not be committed in this century. To a large extent, therefore, it is enough to ask yourself whether Figure 6.1 or 6.2 seems preferable in the 1975–2000 period.

A simple comparison may help. Roughly half, perhaps more, of the gross primary energy being produced in the hard path in 2025 is lost in conversions. A further appreciable fraction is lost in distribution. Delivered end-use energy is thus not vastly greater than in the soft path, where conversion and distribution losses have been all but eliminated. (What is lost can often be used locally for heating, and is renewable, not depletable.) But the soft path makes each unit of end-use energy perform several times as much social function as it would have done in the hard path; so in a conventional sense, social welfare in the soft path in 2025 is substantially greater than in the hard path at the same date.

VI

To fuse into a coherent strategy the benefits of energy efficiency and of soft technologies, we need one further ingredient: transitional technologies that use fossil fuels briefly and sparingly to build a bridge to the energy-income economy of 2025, conserving those fuels—especially oil and gas—for petro-chemicals (ammonia, plastics, etc.), and leaving as much as possible in the ground for emergency use only.

Some transitional technologies have already been mentioned under the heading of conservation—specifically, cogenerating electricity from existing industrial steam and using existing waste heat for district heating. Given such measures, increased end-use efficiency, and the rapid development of biomass alcohol as a portable liquid fuel, the principal short- and medium-term problem becomes not a shortage of electricity or of portable liquid fuels, but a shortage of clean sources of heat. It is above all the sophisticated use of coal, chiefly at modest scale, that needs development. Technical measures to permit the highly efficient use of this widely available fuel would be the most valuable transitional technologies.

Neglected for so many years, coal technology is now experiencing a virtual revolution. We are developing supercritical gas extraction, flash hydrogena-tion, flash pyrolysis, panel-bed filters, and similar ways to use coal cleanly at essentially any scale and to cream off valuable liquids and gases as premium fuels before burning the rest. These methods largely avoid the costs, complex-ity, inflexibility, technical risks, long lead times, large scale, and tar formation of the traditional processes that now dominate our research.

Perhaps the most exciting current development is the so-called fluidized-bed system for burning coal (or virtually any other combustible material). Fluidized beds are simple, versatile devices that add the fuel a little at a time to a much larger mass of small, inert, red-hot particles—sand or ceramic pellets—kept suspended as an agitated fluid by a stream of air continuously blown up through it from below. The efficiency of combustion, of other chemical reactions (such as sulfur removal), and of heat transfer is remarkably high because of the turbulent mixing and large surface area of the particles. Fluidized beds have long been used as chemical reactors and for burning trash, but are now ready to be commercially applied to raising steam and operating turbines. In one system currently available from Stal-Laval Turbin AB of Sweden, eight off-the-shelf 70-megawatt gas turbines powered by fluidized-bed combusters, together with district-heating networks and heat pumps, would heat as many houses as a $1 billion-plus coal gasification plant, but would use only two-fifths as much coal, cost a half to two-thirds as much to build, and burn more cleanly than a normal power station with the best modern scrubbers.[26]

Fluidized-bed boilers and turbines can power giant industrial complexes, especially for cogeneration, and are relatively easy to backfit into old munici-pal power stations. Scaled down, a fluidized bed can be a tiny household device—clean, strikingly simple and flexible—that can replace an ordinary furnace or grate and can recover combustion heat with an efficiency over 80 percent.[27] At medium scale, such technologies offer versatile boiler backfits and improve heat recovery in flues. With only minor modifications they can burn practically any fuel. It is essential to commercialize all these systems now—not to waste a decade on highly instrumented but noncommercial pilot plants constrained to a narrow, even obsolete design philosophy.[28]

Transitional technologies can be built at appropriate scale so that soft technologies can be plugged into the system later. For example, if district heating uses hot water tanks on a neighborhood scale, those tanks can in the long run be heated by neighborhood solar collectors, wind-driven heat pumps, a factory, a pyrolyzer, a geothermal well, or whatever else becomes locally available—offering flexibility that is not possible at today's excessive scale.

Both transitional and soft technologies are worthwhile industrial investments that can recycle moribund capacity and underused skills, stimulate exports, and give engaging problems to innovative technologists. Though neither glamorous nor militarily useful, these technologies are socially effective—especially in poor countries that need such scale, versatility, and simplicity even more than we do.

Properly used, coal, conservation, and soft technologies together can squeeze the "oil and gas" wedge in Figure 6.2 from both sides—so far that most of the frontier extraction and medium-term imports of oil and gas become unnecessary and our conventional resources are greatly stretched. Coal can fill the real gaps in our fuel economy with only a temporary and modest (less than twofold at peak) expansion of mining, not requiring the enormous infrastructure and social impacts implied by the scale of coal use in Figure 6.1.

In sum, Figure 6.2 outlines a prompt redirection of effort at the margin that lets us use fossil fuels intelligently to buy the time we need to change over to living on our energy income. The innovations required, both technical and social, compete directly and immediately with the incremental actions that constitute a hard energy path: fluidized beds vs. large coal gasification plants and coal-electric stations, efficient cars vs. offshore oil, roof insulation vs. Arctic gas, cogeneration vs. nuclear power. These two directions of development are mutually exclusive: the pattern of commitments of resources and time required for the hard energy path and the pervasive infrastructure which it accretes gradually make the soft path less and less attainable. That is, our two sets of choices compete not only in what they accomplish, but also in what they allow us to contemplate later. Figure 6.1 obscures this constriction of options, for it peers myopically forward, one power station at a time, extrapolating trend into destiny by self-fulfilling prophecy with no end clearly in sight. Figure 6.2, in contrast, works backward from a strategic goal, asks what we must do when in order to get there, and thus reveals the potential for a radically different path that would be invisible to anyone working forward in time by incremental ad-hocracy.

VII

Both the soft and the hard paths bring us, each in its own way and at broadly similar rates, to the era beyond oil and gas. But the rates of internal adaptation meanwhile are different. As we have seen, the soft path relies on smaller, far simpler supply systems entailing vastly shorter development and construction time, and on smaller, less sophisticated management systems. Even converting the urban clusters of a whole country to district heating should take only 30–40 years. Furthermore, the soft path relies mainly on small, standard, easy-to-make components and on technical resources dispersed in many organizations of diverse sizes and habits; thus everyone can get into the act, unimpeded by centralized bureaucracies, and can compete for a market share through ingenuity and local adaptation. Besides having much lower and more stable operating costs than the hard path, the soft path appears to have lower initial cost because of its technical simplicity, small unit size, very low overheads, scope for mass production, virtual elimination of distribution losses and of interfuel conversion losses, low exposure to escalation and interest, and prompt incremental construction (so that new capacity is built only when and where it is needed).[29]

The actual costs of whole systems, however, are not the same as perceived costs: solar investments are borne by the householder, electric investments by a utility that can float low-interest bonds and amortize over 30 years. During the transitional era, we should therefore consider ways to broaden householders' access to capital markets. For example, the utility could finance the solar investment (leaving its execution to the householder's discretion), then be repaid in installments corresponding to the householder's saving. The householder would thus minimize his own—and society's—long-term costs. The utility would have to raise several times less capital than it would without such a scheme—for otherwise it would have to build new electric or synthetic-gas capacity at even higher cost—and would turn over its money at least twice as quickly, thus retaining an attractive rate of return on capital. The utility would also avoid social obsolescence and use its existing infrastructure. Such incentives have already led several US gas utilities to use such a capital-transfer scheme to finance roof insulation.

Next, the two paths differ even more in risks than in costs. The hard path entails serious environmental risks, many of which are poorly understood and some of which have probably not yet been thought of. Perhaps the most awkward risk is that late in this century, when it is too late to do much about it, we may well find climatic constraints on coal combustion about to become acute in a few more decades: for it now takes us only that long, not centuries or millennia, to approach such outer limits. The soft path, by minimizing all fossil-fuel combustion, hedges our bets. Its environmental impacts are relatively small, tractable and reversible.[30]

The hard path, further, relies on a very few high technologies whose success is by no means assured. The soft path distributes the technical risk among very many diverse low technologies, most of which are already known to work well. They do need sound engineering—a solar collector or heat pump can be worthless if badly designed—but the engineering is of an altogether different and more forgiving order than the hard path requires, and the cost of failure is much lower both in potential consequences and in number of people affected. The soft path also minimizes the economic risks to capital in case of error, accident, or sabotage; the hard path effectively maximizes those risks by relying on vulnerable high-technology devices each costing more than the endowment of Harvard University. Finally, the soft path appears generally more flexible—and thus robust. Its technical diversity, adaptability, and geographic dispersion make it resilient and offer a good prospect of stability under a wide range of conditions, foreseen or not. The hard path, however, is brittle; it must fail, with widespread and serious disruption, if any of its exacting technical and social conditions is not satisfied continuously and indefinitely.

VIII

The soft path has novel and important international implications. Just as improvements in end-use efficiency can be used at home (via innovative financing and neighborhood self-help schemes) to lessen first the disproportionate burden of energy waste on the poor, so can soft technologies and reduced pressure on oil markets especially benefit the poor abroad. Soft technologies are ideally suited for rural villagers and urban poor alike, directly helping the more than two billion people who have no electric outlet nor anything to plug into it but who need ways to heat, cook, light, and pump. Soft technologies do not carry with them inappropriate cultural patterns or values; they capitalize on poor countries' most abundant resources (including such protein-poor plants as cassava, eminently suited to making fuel alcohols), helping to redress the severe energy imbalance between temperate and tropical regions; they can often be made locally from local materials and do not require a technical elite to maintain them; they resist technological dependence and commercial monopoly; they conform to modern concepts of agriculturally based eco-development from the bottom up, particularly in the rural villages.

Even more crucial, unilateral adoption of a soft energy path by the United States can go a long way to control nuclear proliferation—perhaps to eliminate it entirely. Many nuclear advocates have missed this point: believing that there is no alternative to nuclear power, they say that if the United States does not export nuclear technology, others will, so we might as well get the business and try to use it as a lever to slow the inevitable spread of nuclear weapons to nations and subnational groups in other regions. Yet the genie is not wholly out of the bottle yet—thousands of reactors are planned for a few decades hence, tens of thousands thereafter—and the cork sits unnoticed in our hands.

Perhaps the most important opportunity available to us stems from the fact that for at least the next five or ten years, while nuclear dependence and commitments are still reversible, all countries will continue to rely on the United States for the technical, the economic, and especially the *political* support they need to justify their own nuclear programs. Technical and economic dependence is intricate and pervasive; political dependence is far more important but has been almost ignored, so we do not yet realize the power of the American example in an essentially imitative world where public and private divisions over nuclear policy are already deep and grow deeper daily.

The fact is that in almost all countries the domestic political base to support nuclear power is not solid but shaky. However great their nuclear ambitions, other countries must still borrow that political support from the United States. Few are succeeding. Nuclear expansion is all but halted by grass-roots opposition in Japan and The Netherlands; has been severely impeded in West Germany, France, Switzerland, Italy, and Austria; has been slowed and may soon be stopped in Sweden; has been rejected in Norway and (so far) Australia and New Zealand, as well as in two Canadian Provinces; faces an uncertain prospect in Denmark and many American states; has been widely questioned in Britain, Canada, and the USSR;[31] and has been opposed in Spain, Brazil, India, Thailand, and elsewhere.

Consider the impact of three prompt, clear US statements:

1 The United States will phase out its nuclear power program[32] and its support of others' nuclear power programs.
2 The United States will redirect those resources into the tasks of a soft energy path and will freely help any other interested countries to do the same, seeking to adapt the same broad principles to others' needs and to learn from shared experience.
3 The United States will start to treat nonproliferation, control of civilian fission technology, and strategic arms reduction as interrelated parts of the same problem with intertwined solutions.

I believe that such a universal, nondiscriminatory package of policies would be politically irresistible to North and South, East and West alike. It would offer perhaps our best chance of transcending the hypocrisy that has stalled arms control: by no longer artificially divorcing civilian from military nuclear technology, we would recognize officially the real driving forces behind proliferation; and we would no longer exhort others not to acquire bombs while claiming that we ourselves feel more secure with bombs than without them.

Nobody can be certain that such a package of policies, going far beyond a mere moratorium, would work. The question has received far too little thought, and political judgments differ. My own, based on the past nine years' residence in the midst of the European nuclear debate, is that nuclear power could not flourish there if the United States did not want it to.[33] In giving up the export market that our own reactor designs have dominated, we would be demonstrating a desire for peace, not profit, thus allaying legitimate European commercial suspicions. Those who believe such a move would be seized upon gleefully by, say, French exporters are seriously misjudging French nuclear politics. Skeptics, too, have yet to present a more promising alternative—a credible set of technical and political measures for meticulously restricting to peaceful purposes extremely large amounts of bomb materials which, once generated, will persist for the foreseeable lifetime of our species.

I am confident that the United States can still turn off the technology that it originated and deployed. By rebottling that genie we could move to energy and foreign policies that our grandchildren can live with. No more important step could be taken toward revitalizing the American dream.

IX

Perhaps the most profound difference between the soft and hard paths is their domestic sociopolitical impact. Both paths, like any 50-year energy path, entail significant social change. But the kinds of social change needed for a hard path are apt to be much less pleasant, less plausible, less compatible with social diversity and personal freedom of choice, and less consistent with traditional values than are the social changes that could make a soft path work. It is often said that, on the contrary, a soft path must be repressive; and coercive paths to energy conservation and soft technologies can indeed be imagined. But coercion is not necessary and its use would signal a major failure of imagination, given the many policy instruments available to achieve a given technical end. Why use penal legislation to encourage roof insulation when tax incentives and education (leading to the sophisticated public understanding now being achieved in Canada and parts of Europe) will do? Policy tools need not harm lifestyles or liberties if chosen with reasonable sensitivity.

In contrast to the soft path's dependence on pluralistic consumer choice in deploying a myriad of small devices and refinements, the hard path depends on difficult, large-scale projects requiring a major social commitment under centralized management. We have noted in Section III the extraordinary capital intensity of centralized, electrified high technologies. Their similarly heavy demands on other scarce resources—skills, labor, materials, special sites—likewise cannot be met by market allocation, but require compulsory diversion from whatever priorities are backed by the weakest constituencies. Quasi-war-powers legislation to this end has already been seriously proposed. The hard path, sometimes portrayed as the bastion of free enterprise and free markets, would instead be a world of subsidies, $100-billion bailouts, oligopolies, regulations, nationalization, eminent domain, corporate statism.

Such dirigiste autarchy is the first of many distortions of the political fabric. While soft technologies can match any settlement pattern, their diversity reflecting our own pluralism, centralized energy sources encourage industrial clustering and urbanization. While soft technologies give everyone the costs and benefits of the energy system he chooses, centralized systems allocate benefits to surburbanites and social costs to politically weaker rural agrarians. Siting big energy systems pits central authority against local auton-omy in an increasingly divisive and wasteful form of centrifugal politics that is already proving one of the most potent constraints on expansion.

In an electrical world, your lifeline comes not from an understandable neighborhood technology run by people you know who are at your own social level, but rather from an alien, remote, and perhaps humiliatingly uncontrol-lable technology run by a faraway, bureaucratized, technical elite who have probably never heard of you. Decisions about who shall have how much energy at what price also become centralized—a politically dangerous trend because it divides those who use energy from those who supply and regulate it.

The scale and complexity of centralized grids not only make them politi-cally inaccessible to the poor and weak, but also increase the likelihood and size of malfunctions, mistakes, and deliberate disruptions. A small fault or a few discontented people become able to turn off a country. Even a single rifle-man can probably black out a typical city instantaneously. Societies may therefore be tempted to discourage disruption through stringent controls akin to a garrison state. In times of social stress, when grids become a likely target for dissidents, the sector may be paramilitarized and further isolated from grass-roots politics.

If the technology used, like nuclear power, is subject to technical surprises and unique psychological handicaps, prudence or public clamor may require generic shutdowns in case of an unexpected type of malfunction: one may have to choose between turning off a country and persisting in potentially unsafe operation. Indeed, though many in the $100-billion quasi-civilian nuclear industry agree that it could be politically destroyed if a major accident occurred soon, few have considered the economic or political implications of putting at risk such a large fraction of societal capital. How far would govern-ments go to protect against a threat—even a purely political threat—a basket full of such delicate, costly and essential eggs? Already in individual nuclear plants, the cost of a shutdown—often many dollars a second—weighs heavily, perhaps too heavily, in operating and safety decisions.

Any demanding high technology tends to develop influential and dedicated constituencies of those who link its commercial success with both the public welfare and their own. Such sincerely held beliefs, peer pressures, and the harsh demands that the work itself places on time and energy all tend to discourage such people from acquiring a similarly thorough knowledge of alternative policies and the need to discuss them. Moreover, the money and talent invested in an electrical program tend to give it disproportionate influence in the councils of government, often directly through staff-swapping between policy- and mission-oriented agencies. This incestuous position, now well developed in most industrial countries, distorts both social and energy priorities in a lasting way that resists political remedy.

For all these reasons, if nuclear power were clean, safe, economic, assured of ample fuel, and socially benign per se, it would still be unattractive because of the political implications of the kind of energy economy it would lock us into. But fission technology also has unique sociopolitical side-effects arising from the impact of human fallibility and malice on the persistently toxic and explosive materials in the fuel cycle. For example, discouraging nuclear violence and coercion requires some abrogation of civil liberties;[34] guarding long-lived wastes against geological or social contingencies implies some form of hierarchical social rigidity or homogeneity to insulate the technological priesthood from social turbulence; and making political decisions about nuclear hazards which are compulsory, remote from social experience, disputed, unknown, or unknowable may tempt governments to bypass democratic decision in favor of elitist technocracy.[35]

Even now, the inability of our political institutions to cope with nuclear hazard is straining both their competence and their perceived legitimacy. There is no scientific basis for calculating the likelihood or the maximum long-term effects of nuclear mishaps, or for guaranteeing that those effects will not exceed a particular level; we know only that all precautions are, for fundamental reasons, inherently imperfect in essentially unknown degree. Reducing that imperfection would require much social engineering whose success would be speculative. Technical success in reducing the hazards would not reduce, and might enhance, the need for such social engineering. The most attractive political feature of soft technologies and conservation—the alternatives that will let us avoid these decisions and their high political costs—may be that, like motherhood, everyone is in favor of them.

X

Civilization in this country, according to some, would be inconceivable if we used only, say, half as much electricity as now. But that is what we did use in 1963, when we were at least half as civilized as now. What would life be like at the per capita levels of primary energy that we had in 1910 (about the present British level), but with doubled efficiency of energy use and with the important but not very energy-intensive amenities we lacked in 1910, such as telecommunications and modern medicine? Could it not be at least as agreeable as life today? Since the energy needed today to produce a unit of GNP varies more than 100-fold depending on what good or service is being produced, and since GNP in turn hardly measures social welfare, why must energy and welfare march forever in lockstep? Such questions today can be neither answered nor ignored.

Underlying energy choices are real but tacit choices of personal values. Those that make a high-energy society work are all too apparent. Those that could sustain lifestyles of elegant frugality are not new; they are in the attic and could be dusted off and recycled. Such values as thrift, simplicity, diversity, neighborliness, humility, and craftsmanship—perhaps most closely preserved in politically conservative communities—are already, as we see from the ballot box and the census, embodied in a substantial social movement, camouflaged by its very pervasiveness. Offered the choice freely and equitably, many people would choose, as Herman Daly puts it, "growth in things that really count rather than in things that are merely countable": choose not to transform, in Duane Elgin's phrase, "a rational concern for material well-being into an obsessive concern for unconscionable levels of material consumption."

Indeed, we are learning that many of the things we had taken to be the benefits of affluence are really remedial costs, incurred in the pursuit of benefits that might be obtainable in other ways without those costs. Thus much of our prized personal mobility is really involuntary traffic made necessary by the settlement patterns which cars create. Is that traffic a cost or a benefit?

Pricked by such doubts, our inflated craving for consumer ephemerals is giving way to a search for both personal and public purpose, to re-examination of the legitimacy of the industrial ethic. In the new age of scarcity, our ingenious strivings to substitute abstract (therefore limitless) wants for concrete (therefore reasonably bounded) needs no longer seem so virtuous. But where we used to accept unquestioningly the facile (and often self-serving) argument that traditional economic growth and distributional equity are inseparable, new moral and humane stirrings now are nudging us. We can now ask whether we are not already so wealthy that further growth, far from being essential to addressing our equity problems, is instead an excuse not to mobilize the compassion and commitment that could solve the same problems with or without the growth.

Finally, as national purpose and trust in institutions diminish, governments, striving to halt the drift, seek ever more outward control. We are becoming more uneasily aware of the nascent risk of what a Stanford Research Institute group has called "'friendly fascism'—a managed society which rules by a faceless and widely dispersed complex of warfare-welfare-industrial-communications-police bureaucracies with a technocratic ideology." In the sphere of politics, as of personal values, could many strands of observable social change be converging on a profound cultural transformation whose implications we can only vaguely sense: one in which energy policy, as an integrating principle, could be catalytic?[36]

It is not my purpose here to resolve such questions—only to stress their relevance. Though fuzzy and unscientific, they are the beginning and end of any energy policy. Making values explicit is essential to preserving a society in which diversity of values can flourish.

Some people suppose that a soft energy path entails mainly social problems, a hard path mainly technical problems, so that since in the past we have been better at solving the technical problems, that is the kind we should prefer to incur now. But the hard path, too, involves difficult social problems. We can no longer escape them; we must choose which kinds of social problems we want. The most important, difficult, and neglected questions of energy strategy are not mainly technical or economic but rather social and ethical. They will pose a supreme challenge to the adaptability of democratic institutions and to the vitality of our spiritual life.

XI

These choices may seem abstract, but they are sharp, imminent, and practical. We stand at a crossroads: without decisive action our options will slip away. Delay in energy conservation lets wasteful use run on so far that the logistical problems of catching up become insuperable. Delay in widely deploying diverse soft technologies pushes them so far into the future that there is no longer a credible fossil-fuel bridge to them: they must be well under way before the worst part of the oil and gas decline. Delay in building the fossil-fuel bridge makes it too tenuous: what the sophisticated coal technologies can give us, in particular, will no longer mesh with our pattern of transitional needs as oil and gas dwindle.

Yet these kinds of delay are exactly what we can expect if we continue to devote so much money, time, skill, fuel, and political will to the hard technologies that are so demanding of them. Enterprises like nuclear power are not only unnecessary but a positive encumbrance, for they prevent us, through logistical competition and cultural incompatibility, from pursuing the tasks of a soft path at a high enough priority to make them work together properly. A hard path can make the attainment of a soft path prohibitively difficult, both by starving its components into garbled and incoherent fragments and by changing social structures and values in a way that makes the innovations of a soft path more painful to envisage and to achieve. As a nation, therefore, we must choose one path before they diverge much further. Indeed, one of the infinite variations on a soft path seems inevitable, either smoothly by choice now or disruptively by necessity later; and I fear that if we do not soon make the choice, growing tensions between rich and poor countries may destroy the conditions that now make smooth attainment of a soft path possible.

These conditions will not be repeated. Some people think we can use oil and gas to bridge to a coal and fission economy, then use that later, if we wish, to bridge to similarly costly technologies in the hazy future. But what if the bridge we are now on is the last one? Our past major transitions in energy supply were smooth because we subsidized them with cheap fossil fuels. Now our new energy supplies are ten or a hundred times more capital-intensive and will stay that way. If our future capital is generated by economic activity fueled by synthetic gas at $25 a barrel-equivalent, nuclear electricity at $60–120 a barrel equivalent, and the like, and if the energy sector itself requires much of that capital just to maintain itself, will capital still be as cheap and plentiful as it is now, or will we have fallen into a "capital trap"? Wherever we make our present transition to, once we arrive we may be stuck there for a long time. Thus if neither the soft nor the hard path were preferable on cost or other grounds, we would still be wise to use our remaining cheap fossil fuels—sparingly—to finance a transition as nearly as possible straight to our ultimate energy-income sources. We shall not have another chance to get there.

Afterword

2011

Looking back from 2011, this article correctly anticipated many of today's key energy issues and outcomes, including climate change. Many commentators have misinterpreted its soft-path graph as a forecast, which it explicitly was not: it illustrated what could happen rather than trying to predict what would happen. As Prof. John Steinhart said, its intent was to fall "somewhere between a forecast and a fantasy, or between the unavoidable and the miraculous." Interestingly, though, a 2002 review of long-term energy forecasts[37] called it "remarkably accurate" in hitting "energy use at the end of the twentieth century almost exactly," unlike dozens of others. (My soft-path graph was 4.0 percent below actual 2000 US primary energy use, or 2.2 percent above it if adjusted to actual GDP growth. As far as I know, the only other long-term estimate to come close was by a Cornell econometrician who sketched price elasticities on the back of an envelope.) Here's how this 1976 article's soft-path graph compared with the actual unfolding of the US energy system during 1975–2010:

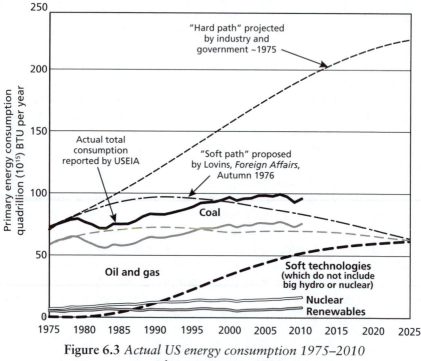

Figure 6.3 *Actual US energy consumption 1975–2010 with projections to 2025*

Demand evolved close to my 1976 sketch because energy use proved not to be inextricably locked to GDP, but decoupled as I'd anticipated. By 2010, the US was using 50 percent less primary energy, 60 percent less oil, 65 percent less natural gas, and 18 percent less electricity to produce a dollar of real GDP than in 1975. And by 2010, we knew that this achievement barely scratched the surface of how much efficiency was available and worth buying. By 2011, RMI's Reinventing Fire (www.reinventingfire.com) even showed how to run a 158 percent bigger 2050 US economy with no oil, coal, or nuclear power, one-third less natural gas, and $5 trillion less private cost (ignoring all "external" or hidden costs like carbon emissions)—led by business for profit.

Renewables were a different story. My soft-path graph presupposed "aggressive support" by public policy, but most of the next 35 years were the opposite. Even in 2010, as renewables continued to take over the global market for additional electricity, Congress severely disrupted the US windpower industry for the fourth time, halving its growth, by delaying renewal of its tax incentives (a convenient political pawn) while leaving untouched its fossil fuel and nuclear competitors' permanent and generally larger subsidies. By 2011, the US had become a relative laggard: Portugal, for example, had gone from 17 percent to 45 percent renewable electricity in the previous five years, while the US went from 9.2 percent to 10.5 percent.

These contrasting outcomes don't mean I was right about efficiency or wrong about renewables; neither was a forecast. Rather, it means that efficiency was better able than renewables to prevail despite often perverse policies, and that the two scenarios I outlined in 1976 remain a vital tension in today's energy system. The outcome remains contested, but market choices are tilting ever more strongly towards the soft path.

Notes

1   In this essay, the proportions assigned to the components of the two paths are only indicative and illustrative. More exact computations, now being done by several groups in the United States and abroad (notably the interim [autumn 1976] and forthcoming final [1976–1977] reports of the energy study of the Union of Concerned Scientists, Cambridge, Mass.), involve a level of technical detail which, though an essential next step, may deflect attention from fundamental concepts. This article will accordingly seek technical realism without rigorous precision or completeness. Its aim is to try to bring some modest synthesis to the enormous flux and ferment of current energy thinking around the world. Much of the credit (though none of the final responsibility) must go to the many energy strategists whose insight and excitement they have generously shared and whose ideas I have shamelessly recycled without explicit citation. Only the limitations of space keep me from acknowledging by name the 70-odd contributors, in many countries, who come especially to mind.

2   The foregoing data are from M. Carasso *et al.*, *The Energy Supply Planning Model*, PB-245 382 and PB-245 383, National Technical Information Service (Springfield, Va.), Bechtel Corp. report to the National Science Foundation (NSF), August 1975. The figures assume the production goals of the 1975 State

of the Union Message. Indirect labor requirements are calculated by C. W. Bullard and D. A. Pilati, CAC Document 178 (September 1975), Center for Advanced Computation, Univ. of Illinois at Urbana-Champaign.

3   I. C. Bupp and R. Treitel, "The Economics of Nuclear Power: De Omnibus Dubitandum," 1976 (available from Professor Bupp, Harvard Business School).

4   Computation concerning waste heat and projections to 2000 are based on data in the 1975 Energy Research and Development Administration Plan (ERDA–48).

5   B. Bolin, "Energy and Climate," Secretariat for Future Studies (Fack, S-103 10 - Stockholm); S. H. Schneider and R. D. Dennett, *Ambio 4*, 2:65–74 (1975); S. H. Schneider, *The Genesis Strategy*, New York: Plenum, 1976; W. W. Kellogg and S. H. Schneider, *Science 186*:1163–72 (1974).

6   Figure 6.1 shows only *nonagricutural* energy. Yet the sunlight participating in photosynthesis in our harvested crops is comparable to our total use of nonagricultural energy, while the sunlight falling on *all* US croplands and grazing lands is about 25 times the nonagricultural energy. By any measure, sunlight is the largest single energy input to the US economy today.

7   The capital costs for frontier fluids and for electrical systems can be readily calculated from the database of the Bechtel model (note 2 above). The electrical examples are worked out in my "Scale, Centralization and Electrification in Energy Systems," *Future Strategies of Energy Development* symposium, Oak Ridge Associated Universities, October 20–21, 1976.

8   The Bechtel model, using 1974 dollars and assuming ordering in early 1974, estimates direct construction costs totaling $559 billion, including work that is in progress but not yet commissioned in 1985. Interest, design, and administration—but not land, nor escalation beyond the GNP inflation rate—bring the total to $743 billion. Including the cost of land, and correcting to a 1976 ordering date and 1976 dollars, is estimated by M. Carasso to yield over $1 trillion.

9   M. Carasso *et al.*, op. cit.

10   E. Kahn *et al.*, "Investment Planning in the Energy Sector," LBL-4479, Lawrence Berkeley Laboratory, Berkeley, Calif., March 1, 1976.

11   American Institute of Physics Conference Proceedings No. 25, *Efficient Use of Energy*, New York: AIP, 1975; summarized in *Physics Today*, August 1975.

12   M. Ross and R. H. Williams, "Assessing the Potential for Fuel Conservation," forthcoming in *Technology Review*; see also L. Schipper, *Annual Review of Energy* I:455–518 (1976).

13   L. Schipper and A. J. Lichtenberg, "Efficient Energy Use and Well-Being: The Swedish Example," LBL-4430 and ERG-76–09; Lawrence Berkeley Laboratory, April 1976.

14   R. L. Goen and R. K. White, "Comparison of Energy Consumption Between West Germany and the United States," Stanford Research Institute, Menlo Park, Calif., June 1975.

15   A. D. Little, Inc., "An Impact Assessment of ASHRAE Standard 90–75," report to FEA, C-78309, December 1975; J. E. Snell *et. al.* (National Bureau of Standards), "Energy Conservation in Office Buildings: Some United States Examples," International CIB Symposium on Energy Conservation in the Built Environment (Building Research Establishment, Garston, Watford, England), April 1976; Owens-Corning-Fiberglas, "The Arkansas Story," 1975.

16   P. W. McCracken *et al.*, *Industrial Energy Center Study*, Dow Chemical Co. *et al.*, report to NSF, PB-243 824, National Technical Information Service

(Springfield, Va.), June 1975. Extensive cogeneration studies for FEA are in progress at Thermo-Electron Corp., Waltham, Mass. A pathfinding June 1976 study by R. H. Williams (Center for Environmental Studies, Princeton University) for the N.J. Cabinet Energy Committee argues that the Dow report substantially underestimates cogeneration potential.

17 Ross and Williams, *op. cit.*

18 A calculation for Canada supports this view: A. B. Lovins, *Conserver Society Notes* (Science Council of Canada, Ottawa), May/June 1976, pp. 3–16. Technical fixes already approved in principle by the Canadian Cabinet should hold approximately constant until 1990 the energy required for the transport, commercial, and house-heating sectors; sustaining similar measures to 2025 is estimated to shrink per capita primary energy to about half today's level. Plausible social changes are estimated to yield a further halving. The Canadian and US energy systems have rather similar structures.

19 The scale of potential conservation in this area is given in Ross and Williams, op. cit.; the scale of potential cogeneration capacity is from McCracken *et al.*, op. cit.

20 Assuming (which is still not certain) that controlled nuclear fusion works, it will almost certainly be more difficult, complex, and costly—though safer and perhaps more permanently fueled—than fast breeder reactors. See W. D. Metz, *Science* 192:1320–23 (1976), 193:38–40, 76 (1976), and 193:307–309 (1976). But for three reasons we ought not to pursue fusion. First, it generally produces copious fast neutrons that can and probably would be used to make bomb materials. Second, if it turns out to be rather "dirty," as most fusion experts expect, we shall probably use it anyway, whereas if it is clean, we shall so overuse it that the resulting heat release will alter global climate: we should prefer energy sources that give us enough for our needs while denying us the excesses of concentrated energy with which we might do mischief to the Earth or to each other. Third, fusion is a clever way to do something we don't really want to do, namely to find *yet another* complex, costly, large-scale, centralized, high-technology way to make electricity—all of which goes in the wrong direction.

21 Partly or wholly solar heating is attractive and is being demonstrated even in cloudy countries approaching the latitude of Anchorage, such as Denmark and The Netherlands (International CIB Symposium, op. cit.) and Britain (*Solar Energy: A U.K. Assessment*, International Solar Energy Society, London, May 1976).

22 Solar heating cost is traditionally computed microeconomically for a consumer whose alternative fuels are not priced at long-run marginal cost. Another method would be to compare the total cost (capital and life-cycle) of the solar system with the total cost of the other complete systems that would otherwise have to be used in the long run to heat the same space. On that basis, 100 percent solar heating, even with twice the capital cost of two-thirds or three-fourths solar heating, is almost always advantageous.

23 R. W. Bliss, *Bulletin of the Atomic Scientists*, March 1976, pp. 32–40.

24 A. D. Poole and R. H. Williams, *Bulletin of the Atomic Scientists*, May 1976, pp. 48–58.

25 For examples, see the Canadian computations in A. B. Lovins, *Conserver Society Notes,* op. cit.; Bent Sørensen's Danish estimates in *Science* 189:255–60 (1975); and the estimates by the Union of Concerned Scientists, note 1 above.

26 The system and its conceptual framework are described in several papers by H. Harboe, Managing Director, Stal-Laval (G.B.) Ltd., London: "District Heating

and Power Generation," November 14, 1975; "Advances in Coal Combustion and Its Applications," February 20, 1976; "Pressurized Fluidized Bed Combustion with Special Reference to Open Gas Turbines" (with C. W. Maude), May 1976. See also K. D. Kiang et. al., "Fluidized-Bed Combustion of Coals," GFERC/IC-75/2 (CONF-750586), ERDA, May 1975.

27 Small devices were pioneered by the late Professor Douglas Elliott. His associated firm, Fluidfire Development, Ltd. (Netherton, Dudley, W. Midlands, England), has sold many dozens of units for industrial heat treatment or heat recuperation. Field tests of domestic packaged fluidized-bed boilers are in progress in The Netherlands and planned in Montana.

28 Already Enköping, Sweden, is evaluating bids from several confident vendors for a 25-megawatt fluidized-bed boiler to add to its district heating system. New reviews at the Institute for Energy Analysis and elsewhere confirm fluidized beds' promise of rapid benefits without massive research programs.

29 Estimates of the total capital cost of "soft" systems are necessarily less well developed than those for the "hard" systems. For 100 percent solar space heating, one of the high-priority soft technologies, mid-1980s estimates are about $50,000–60,000 (1976 dollars) of investment per daily oil-barrel-equivalent in the United States, $100,000 in Scandinavia. All solar cost estimates, however, depend sensitively on collector and building design, both under rapid development. In most new buildings, passive-solar systems with negligible or negative marginal capital costs should suffice. For biomass conversion, the 1974 FEA Solar Task Force estimated capital costs of $10,000–30,000 per daily barrel equivalent—toward the lower end of this range for most agricultural projects. Currently available wind-electric systems require total system investment as high as about $200,000 per delivered daily barrel, with much improvement in store. As for transitional technologies, the Stal-Laval fluidized-bed gas-turbine system, complete with district-heating network and heat-pumps (coefficient of performance = 2), would cost about $30,000 per delivered daily barrel equivalent. See Lovins, op. cit., note 7.

30 See A. B. Lovins, "Long-Term Constraints on Human Activity," *Environmental Conservation 3*, 1:3–14 (1976) (Geneva); "Some Limits to Energy Conversion," Limits to Growth 1975 Conference (The Woodlands, Texas), October 20, 1975 (published in conference papers); J. W. Benson, "Energy and Reality: Three Perceptions," *Inst. Ecol. Policies* (Fairfax, Va.) 1977.

31 Recent private reports indicate the Soviet scientific community is deeply split over the wisdom of nuclear expansion. See also *Nucleonics Week*, May 3, 1976, pp. 12–13.

32 Current overcapacity, capacity under construction, and the potential for rapid conservation and cogeneration make this a relatively painless course, whether nuclear generation is merely frozen or phased out altogether. For an illustration (the case of California), see R. Doctor *et al.*, *Sierra Club Bulletin*, May 1976, pp. 4ff. I believe the same is true abroad. See Introduction to *Non-Nuclear Futures* by A. B. Lovins and J. H. Price, Cambridge, Mass.: FOE/Ballinger, 1975.

33 See *Nucleonics Week*, May 6, 1976, p. 7, and I. C. Bupp and J. C. Derian, "Nuclear Reactor Safety: The Twilight of Probability," December 1975. Bupp, after a detailed study of European nuclear politics, shares this assessment.

34   R. Ayres, *10 Harvard Civil Rights-Civil Liberties Law Review*, Spring 1975, pp. 369–443; J. H. Barton, "Intensified Nuclear Safeguards and Civil Liberties," report to USNRC, Stanford Law School, October 21, 1975.

35   H. P. Green, *43 George Washington Law Review*, March 1975, pp. 791–807.

36   W. W. Harman, *An Incomplete Guide to the Future*, Stanford Alumni Association, 1976.

37   Paul P. Craig, Ashok Gadgil and Jonathan G. Koomey, "What Can History Teach Us? A Retrospective Examination of Long-Term Energy Forecasts for the United States", *Annual Review of Energy and the Environment*, Vol. 27, November 2002, pp. 83–118.

Chapter 7

Cost–Risk–Benefit Assessments in Energy Policy

1977

Editor's note: *the mid-1970s was an era when public policy, especially policy aimed at rejuvenating worn-out urban areas, became a considerable force in the way society began to value the things it was supposedly "renewing." At the heart of this essay (which is reprinted with permission from* George Washington Law Review, *Vol. 45, No. 5, August 1977) is a discussion of the value of human life and, easily picked apart by Amory— and various other contemporary commentators mentioned in the piece— the way in which we judge the worth of a life becomes a chilling indictment of the potential injustice of cost–risk–benefit assessments. Not surprisingly, this piece reflects public policy discussions taking place in London (regarding a new London airport) as that was Amory's home at the time, though the article first appeared in a US-based journal.*

Energy policy is difficult and outwardly complex. More than other public policy choices, energy decisions combine intricate linkages with other concerns, pervasive uncertainties, very large and very small numbers, long time horizons, rancorous disagreements among experts, and issues of exquisite political sensitivity. Defining the energy problem, moreover, is itself an issue.[1] To some analysts, the problem is how to expand domestic supplies of energy to meet extrapolated demands; to others, the problem is how to meet heterogeneous end-use needs and accomplish social goals with an elegant frugality of energy and trouble. Energy paradigms diverge so fundamentally that the simplest substantive question, such as whether to build a big power station, can lead to an infinite regress of other questions: Why a power station? Why a big one? Why more electricity? Why electricity? Why more?[2] Nor is there consensus on the subject matter of energy policy.[3] To some, it is simply a set of severable technical and economic issues; to others, a web of sociopolitical and ethical questions that are inseparable from subordinate technical and economic issues. Statements about energy policy which seem self-evident to some analysts—"we need more energy," "one must either grow or decay," "nuclear power is cheap and safe," "nuclear power is expensive and dangerous," "life in a nuclear-armed crowd would be intolerable"—seem vacuous, nonsensical, tendentious, contingent, or false to others. Such basic disagreements cannot be resolved on scientific grounds. No wonder energy policy makes citizens protest, experts despair, politicians duck, and governments fall.

Spurred on by the requirements of the National Environmental Policy Act of 1964[4] and by the less formal but more insistent lure of jobs for unemployed systems analysts, students of energy policy have responded to the challenge with an abundance of calculations purporting to compare the costs, risks, and benefits of alternative energy policies. The principles of this analysis are hardly new. As Ida Hoos suggests, "we are witnessing a repeat performance of Molière's play *Le Bourgeois Gentilhomme*, in which an uneducated trades-man, bent on acquiring instant culture, hires a tutor and learns with delight that what he has been talking all his life is *prose*!"[5] Nevertheless, the impressive formalism of cost–risk–benefit assessments has given them remarkable credence among the credulous, especially those analysts and decisionmakers seeking an escape from genuinely perplexing problems.

I believe that this credence is misplaced, that cost–risk–benefit assessments are a part of the energy problem, rather than the solution, and that their aura of reliability, relevance, and cost-effectiveness is undeserved. As ritual, they are often useful, not for their results, but for the explicit data they elicit and the critical thought they may encourage. As science, however, such assessments tend to be subtly and irredeemably defective; as rhetorical art, sophistical; and as a decisionmaking tool, useless and dangerous. Like the street lamp under which the proverbial drunkard searched for his wallet, not because he lost it there but because that was the only place he could see, they only make the surrounding darkness darker.[6]

This article will outline the basis for this skepticism and examine the problems of evaluating benefits, risks, the monetary value of risks, and direct costs; the difficulty of comparing these quantities; and the use of these comparisons in decisionmaking. If my tone seems unduly critical, it is because the excesses now endemic in cost–risk–benefit assessments tend to degrade both energy decisions and democracy.

Benefits

Serious problems of both concept and practice arise in identifying benefits (and beneficiaries)[7] of an energy decision, assessing their quantum, and determining whether these benefits are accepted or rejected voluntarily. In some energy projects, the bearers of the costs are real while the beneficiaries are unidentifiable or imaginary. This difference leads to familiar political problems. The Roskill Commission's[8] recommendation on the siting of the third London airport is one illustration:

> *Making decisions that affect the lives of large numbers of people confirms one's own significance. But having one's life arbitrarily controlled by others is alienating. ... That some people were alienated ... is obvious; farmers, shop keepers, neighbours—people recognizable as individuals, real people whose pictures appeared in newspapers—were sufficiently incensed to burn an effigy of Justice Roskill. It was the job of the Roskill Commission to show that this opposition [to the airport], while certainly understandable, was narrow and selfish when placed in the context of the larger good that would be served by a new airport. But who were the people whose interests outweighed those of the airport's opponents? Where were the individuals whose desire to travel was so important? Nowhere. They were a statistical extrapolation; they were a graph of traffic rising [roughly 15-fold] to 300 million by the year 2006; they were an abstraction that could not be cross-examined.[9]*

The needs that energy or transport schemes are intended to meet, and the benefits they provide, are generally projections. Hence, their beneficiaries cannot participate in decisionmaking. These unreal people and their needs are often accorded a more tangible existence than real ones. One is commonly told, for example, that the growth of future energy needs to unprecedented levels is undeniable, while the alternative of using energy far more efficiently at lower consumption rates, an alternative which some have already chosen,[10] is the flimsiest conjecture.[11]

Even real beneficiaries of a decision may encounter awkward differences between gross and net benefits. Benefits can be illusory because they are balanced by remedial costs or transaction costs,[12] as when one must work to buy a car to get to work. In addition, benefits can increase costs perpetually by creating a self-fulfilling prophecy,[13] as when roads built to reduce local congestion, travel time, fuel waste, and accidents contribute to national growth in traffic, thus exacerbating those same problems.[14] Similarly, the substitution of energy-intensive and capital-intensive machines for human skill has led to exhortations to build more machines to fuel the economic growth deemed necessary to employ the people rendered jobless by that very process.

The problem of beneficiaries who do not desire benefits or do not even consider them to be benefits is more common and intractable. Negative perceptions may result from different values (is an artful new kind of synthetic dessert a benefit or a disgrace?) or from the view that the costs of realizing an uncontested benefit are prohibitive. For most citizens of affluent countries, the argument that gross benefits are really net costs can be made at the margin for personal mobility,[15] wealth, and technological marvels generally.[16] The argument also applies to cheap energy, the chief benefit claimed for most energy schemes. Paul Ehrlich, for example, notes that "one of the main purported 'benefits' of nuclear power, the availability of cheap and abundant energy, is, in fact, a cost [which] may run as high as the destruction of the ecological systems that are essential to the persistence of civilization."[17] The argument need not rest on the mischief that cheap and abundant energy enables us to do to the Earth or to each other. Because the mere presence of unrealistically cheap energy raises costs through side-effects and structural distortions elsewhere in the economy, it is actually very expensive:

> We use the apparently cheap energy wastefully, and thus continue to increase our dependence on imported oil, to the detriment of the Third World, Europe, Japan, and our own independence. We earn the foreign exchange to pay for the oil by running down domestic stocks of commodities, which is inflationary; by exporting weapons, which is inflationary, destabilizing, and immoral; and by exporting wheat and soybeans, which inverts Midwestern real estate markets, makes us mine groundwater unsustainably in Kansas, and raises our own food prices. Exported American wheat diverts Soviet investment from agriculture into defense, making us increase our own inflationary defense budget, which we have to raise anyhow to defend the sea lanes to bring in the oil and to defend the Israelis from the arms we sold to the Arabs. With crop exports crucial to our balance of payments, pressure increases for energy- and water-intensive agribusiness. This creates yet another spiral by impairing free natural life-support systems and so requiring costly, energy-intensive technical fixes (such as desalination and artificial fertilizers) that increase the stress on remaining natural systems while starving social investments. Excessive substitution of apparently cheap inanimate energy for people causes structural unemployment, which worsens poverty and inequity, which increase alienation and crime. Priorities in crime control and health care are stalled by the heavy capital demands of building and subsidizing the energy sector, which itself contributes to the unemployment and illness at which these social investments were aimed. The drift toward a garrison state at home (needed to protect the vulnerable energy system from strikes, sabotage, and dissent), failure to address rational development goals abroad, and the strengthening of oligopolies and of oil and uranium cartels all encourage international distrust and domestic dissent, both entailing further suspicion and repression. Energy-related climatic shifts could jeopardize marginal agriculture, even in the Midwestern breadbasket, endangering an increasingly fragile world peace. The nuclear weapons proliferated by the widespread use of nuclear power, the competitive export of arms, reactors, and inflation from rich to poor countries, and the

tensions of an ever more inequitable, hungry, and anarchic world
could prove a deadly combination.[18]

Centralized electrification provides another example of illusory benefits.[19] Assumed to be beneficial and essential, this process creates sociopolitical costs such as autarchy, centrism, vulnerability, technocracy, and inequity,[20] which only bureaucrats, dictators, saboteurs, technocrats, and millionaires might welcome. These costs are inseparable from the alleged benefits and should therefore be subtracted to obtain net benefits. Thus far no cost–benefit–risk assessment has mentioned these "costs of benefits" or considered the awkward problem that, if energy is an intermediate good which can be used as easily to make napalm as to light a concert hall, it is not beneficial per se, but only insofar as it furthers beneficial ends.

Whether we are threatened by too little energy too late or by too much too soon, whether it is dear or cheap energy that would worsen inflation and unemployment, whether it is the use or the abandonment of nuclear power that could send us back to caves and candles—such questions hinge on a whole universe of perceptions about the classification of means and ends.[21] If, instead of asking how energy can be supplied, one asks who will need how much of what kind of energy for what purposes for how long, more electrification becomes the lowest, rather than the highest, priority.[22]

Concentration on the ends rather than the means of production is, however, inconsistent with classical cost–benefit analysis, which considers rationality of means without regard to sanity of ends.[23] The purely instrumental character of the analysis is a grave weakness in this tool when its users lack consensus on ends.[24] Like any overly complex and inflexible tool, the method can mold the behavior of, and ultimately enslave, its users.

Moreover, any player of the cost–risk–benefit game must seek assiduously to maximize benefits, and in this pursuit may invent new and insubstantial kinds. The literature of energy cost–risk–benefit assessments illustrates the usual blunders: counting tautologous benefits (a project is good because it makes one feel proud because it is a good project),[25] double-counting benefits (as did the US Atomic Energy Commission (AEC) in assessing its fast breeder reactor program),[26] and counting as benefits not only future outputs but also past inputs (as did the AEC in ascribing benefits to "efficient use of the manpower and the facility resources committed to the breeder program").[27]

Once benefits have been identified, they must be quantified. The traditional method is to compute, for example, millidollars saved per unit of electricity generated by one means rather than another, and to multiply this saving by the number of units to be consumed. The logic of treating consumer savings as a benefit might seem unassailable to those who ignore countervailing structural costs.[28] Costs and prices, however, are not revealed truth.[29] They are determined by philosophical conventions, tacit value judgments, and the political forces behind rents, subsidies, taxes, and externalities. Such calculated fiscal benefits are not the result of the mechanical workings of a free market;

rather they are a negotiated result of the political process and are not objectively quantifiable.

Risks[30]

Risks, like benefits, must be identified, evaluated, and ascribed to particular people at particular times in particular states of knowledge and volition. Doing these tasks rigorously and defensibly is difficult or impossible.

Volition is the thorny question raised by those who consider what risks are or should be "acceptable." Risks are not necessarily "acceptable" simply because they are encountered. Many man-made risks are involuntary, or unknown to those who incur them, or wrongly believed to be negligible, or endured because of political impotence to abate them, or tolerated because of a lack of alternatives to incurring them.

Treatment of the time at which individuals will be exposed to a given risk is also troublesome. Noting that risks are converted eventually to costs, some commentators believe those costs should be discounted like any direct monetary costs.[31] But this approach is hazardous. Until recently risk discounting made it attractive to jerry-build British bridges and buildings that could fall on someone's head in 20 years, as a 20-year risk discounted at the 10 percent annual rate recommended by Her Majesty's Treasury was valued at 15 percent of an equivalent present risk. British authorities slowly realized, however, that safety and lives cannot be banked at interest as money can, and that discounting risks is neither morally nor theoretically sound—as Dutch dike designers appreciate.[32]

Quantifying risk may be easier, however, than identifying direct, energy-related risks.[33] The list of recently discovered but previously unforeseen risks is impressive: in coal combustion, synergistic effects of sulfates,[34] submicron particulates, and heavy metals; in coal conversion, powerfully carcinogenic by-products; in Arctic oil drilling, the melting of the sensitive Arctic sea-ice;[35] in liquefied natural gas (LNG) transport, several odd features of the gas itself[36] and the cryogenic structural failure of the tankers;[37] in nuclear power, releases of carbon-14,[38] climatic effects of krypton-85,[39] and some potential noncarcinogenic health effects of actinides.[40] Admittedly, many substances and technological devices have serious side-effects,[41] and the list will grow. As in the case of teratogenic and carcinogenic drugs such as thalidomide and diethyl stilbestrol, moreover, many energy-related illnesses, cancers, and genetic defects may be undiscovered for years or decades after their onset. Further, some risks are deliberately kept secret[42] for more or less persuasive reasons, and others are so alarming in their potential that scholars have advocated the suppression of research designed to determine their actual quantum.[43]

Finally, some technologies may present a large but subtle risk to the value structure of the society in which risk–benefit decisions are made. Technologies such as cloning and genetic engineering may have a dehumanizing effect on our understanding of what it means to be a person.[44] A theoretical and moral framework is necessary to evaluate these unprecedented risks.

Trans-Scientific Issues

Behind the identification and quantification of the most ordinary risks lurk trans-scientific issues, issues that "can be stated in the language of science, but appear for practical purposes to be unanswerable by science."[45] Apparently scientific disagreements about the nature and extent of risk in energy decision-making often reflect trans-scientific disagreements about perceptions and values. For example:

- Do problems of nuclear safety have engineering answers or are their solutions limited by human fallibility?[46]
- Are energy problems predominantly technical or social in character (if these categories are indeed distinct)?[47]
- Will present levels of nuclear safety improve with experience or deteriorate with haste, boredom, routine, corner-cutting, and the declining competence of personnel?
- Should assurances of safety rest on empirical demonstration, or only on engineering judgment?[48]
- Should the burden of proving an acceptable level of risk rest on the proponents of a given technological innovation, or should the burden of proving an unacceptable level of risk fall on the innovation's opponents?[49]
- Should an action be presumed safe until proven harmful, or potentially harmful until proven safe?[50]

Several of these trans-scientific issues are illustrated by the problem of assessing the risks of prolonged, low-level exposure to hazards such as radiation. As Harold P. Green notes:

> *In the real world, the lowest regions of the cause–effect curve are the most relevant, and reliable information can usually be derived only from statistical data. Statistical data cannot provide meaningful [information] unless and until the technology has been used long enough and widely enough to produce a body of data adequate for statistical analysis. Thus, if the technology is not adequately safe, this fact cannot be persuasively demonstrated until after many people have in fact sustained injury.*[51]

The dangers of low-level exposure to radiation can only be determined with certainty by encountering the danger that regulation and decision should prevent. Thus, there are risks in gathering data necessary for an accurate evaluation of those same risks. If a particular radiation-emitting technology is presumed to be dangerous, then a decision will be made on the basis of potential but unproven risk. If, on the other hand, the technology is presumed safe until proven dangerous, these potential risks will be encountered before they are evaluated. A choice between these presumptions, however, is an ethical rather than a scientific matter.

Thus, risks may be significant even though they cannot be identified with complete scientific accuracy. For example, the results of an experiment designed to determine the carcinogenic risk of a particular substance may be indistinguishable in principle from the effects of other, uncontrolled variables;[52] most cancers do not announce their cause. Controlled experiments which seek to isolate such causes are generally impracticable.[53] Yet, someone must decide whether undetectable risks are also insignificant. J. Martin Brown offers a useful analogy. He argues that although low-level radiation effects

> *are so small that they are masked by other environmental factors, and hence (one is led to conclude) of little significance, ... the same statement could be made about other causes of death (such as murder by handguns) which we do regard as of some significance. If we had no way of distinguishing death by murder from death by natural causes, the death rate from murder could increase manyfold before it became noticeable as an increase in the mortality from all causes. Such is the problem with cancers induced by radiation or by any other carcinogen in our environment.*[54]

A similar ethical dilemma arose in the fallout debate. The group seeking to minimize the potential risks cited the "small" percentage of deaths attributable to fallout while opponents focused on the "large" absolute number of deaths. In that instance, "the perception of a large number of eventual fatalities evidently won out, and led the politicians to sign a test ban treaty."[55] Nevertheless, the difference between the undetectable and the insignificant is a recurring theme in debates over low-level effects. One side urges that the problem be put in "perspective;" the other considers such advice as scant consolation for the victims, who should be protected "even if most of [them] would not know the original cause of their affliction."[56]

One way to estimate such risks is to assume, as all national and international radiation protection agencies do, that the relation between effect and dose is linear down to zero dose. Under some circumstances, however, "[t]here is even experimental and epidemiological evidence that the risks ... may ... actually increase with decreasing dose and dose rate."[57] Until we have a detailed understanding of the molecular mechanisms of carcinogenesis, we have no basis for saying whether the "linear hypothesis" assures an adequate margin of safety. Epidemiological data may not help in this inquiry: a recent review at Brookhaven National Laboratory found that though air pollution of certain types may be much more harmful than anyone thought, it is equally probable, on the same evidence, to be good for people![58]

The effects of inhaling microscopic doses of insoluble aerosols of plutonium and other actinides, elements created in large quantities by the nuclear fuel cycle, present a similar trans-scientific issue.[59] Notwithstanding many experiments with laboratory animals, our knowledge of these effects is unsatisfactory and will remain so at least until the long latency period of current experiments elapses.[60] What little we know is not reassuring. Inhalation by beagles of plutonium-239 dioxide dust, for example, caused lung cancer in virtually all animals tested, even at the lowest doses used.[61] If one assumes that risk depends on plutonium concentration (plutonium mass per gram of lung tissue) rather than on total plutonium amount, number of particles, or number of lung cells at risk, then that lowest beagle dose could be multiplied by the ratio of human to beagle lung mass to obtain the equivalent human dose. The result is 100 times the human lung burden of the same substance currently allowed in the workplace.[62] This 100-fold margin is not very large, as humans may be more radio-sensitive than beagles and lower doses may be sufficient to kill beagles. Nor do empirical data from accidents help with present decisions on the use of plutonium, as the figures are ambiguous[63] and do not resolve the lively theoretical controversy whether present exposure standards should be strengthened by two or three orders of magnitude.[64] Finally, the long-lived actinides are already considered so toxic that if present stringent standards were tightened by even a small factor, the maximum permitted air concentrations and lung burdens would become undetectable in principle, and thus operationally meaningless, so that neither compliance nor noncompliance with the standard could ever be proved.[65] The choice of assumptions under these conditions of uncertainty transcends the domain of science.

The carcinogenicity of plutonium and related elements is not an academic nicety, but the essence of major policy problems. Plutonium has been proposed as the main fuel of many industrialized countries. World trade in the element would then amount to thousands of metric tons per year. The amount of plutonium-239 virtually 100 percent carcinogenic in beagles is approximately three millionths of a gram,[66] and a gram in turn is a millionth of a metric ton. The ratio of these quantities—some 15 orders of magnitude—suggests a need for the most diligent containment throughout the geological periods[67] of plutonium's effective lifetime. Previous attempts to ensure such containment, however, have failed.[68] Moreover, our knowledge of the actinides' environmental pathways is sketchy for plutonium and virtually nil for the higher actinides.[69] A recent symposium on the subject reveals that much of our knowledge about actinide pathways is wrong.[70]

Indeed, the vexing problem of inadequate scientific knowledge arises frequently. Exploring the technical literature of health physics and radioecology over the past few years, I have been impressed by wide uncertainty and ignorance on the most basic matters. What appears to be a solid wall of meticulously verified empirical bricks proves on closer inspection to be a façade of holes strung together with bits of mortar.[71]

Many data needed for reliable cost–risk–benefit analyses are not just unknown but unknowable in principle.[72] For example, few social scientists would claim they can calculate the probability that malevolent persons will sabotage a reactor, build an atomic bomb, or blow up a liquefied natural gas

(LNG) tanker. Nevertheless, such probabilities are used routinely in cost–risk–benefit assessments. The arbitrary manner in which these unknowns are assessed is both disturbing and revealing.

Conceding that "data-oriented analytical techniques are not generally applicable" to risks such as terrorism, a Stanford Research Institute (SRI) study simply concluded that "we must rely on expert judgment, quantified using subjective probabilities."[73] The authors of the study accordingly asked an elite group of experts to pick numbers out of the air by using the same intuitive judgment the authors condemned as "irrational" when exercised by the public. The numbers were then "encoded" and embodied in an esoteric model from which no non-expert could disentangle them. The SRI authors did, however, suggest a remedy for this complexity:

> *Individuals who differ on their assessment of probability, consequence, or social values can suggest their own numbers, and the model is again used to see if their new information would cause society to select a different alternative. The same is true of model structure.*[74]

This is obviously an idle suggestion. The public does not have the expertise, time, patience, money, and computer access to alter the data or grasp the structure of these complex models. Further, the public may decline to play the "my number is better than your number" game by refusing to choose any numbers. The report's approach not only ignores the inherent uncertainties of certain risks but also smacks of elitism.[75] Decisions to let experts resolve such uncertainties, as well as choices about the criteria of certainty on which particular risk–benefit decisions will be made under conditions of uncertainty, are transscientific.

A final problem of quantifying risks is their sensitivity to varying technological assumptions. The changes in air pollution levels calculated by assuming that nuclear reactors are replaced by fossil-fuel power stations may be smaller than changes which could be obtained by using different combustion technologies, such as fluidized beds.[76]

Moreover, the variations associated with other approaches to meeting the same end-use needs,[77] such as cogeneration, district heating from combined-heat-and-power stations, solar technologies, and increased end-use efficiency, are even larger. Few cost–risk–benefit assessments make this point; instead, they predetermine their results by their choice of technical alternatives. Even those technologies which are assumed are frequently applied asymmetrically. Although uranium mill tailings,[78] for example, are assumed to be controlled at zero or negligible cost, the by-products of coal mining are not. Although nuclear releases are assumed to be restricted to the lowest possible levels, a corresponding assumption is not made for combustion emissions. The threat of war is considered a risk of oil imports but ignored as a potential risk of nuclear programs. Finally, "learning curves" are assumed to apply to the most complex, speculative, and intractable technologies but not to the simplest, surest, and most straightforward ones.[79]

Converting Risks to Costs

Once risks have been identified and quantified, their conversion to costs, i.e. their valuation on a scale commensurate with benefits, raises conceptual problems. At one time, these problems seemed less important than in this age of preoccupation with the quantification of costs and benefits. Recently a large technical literature on the valuation of human life has grown up, not to help the legal profession, for which valuation of life, liberty, and happiness is an ancient and familiar problem, but rather to help systems analysts who argue that explicit cash valuation of human life is an essential part of decisionmaking.[80] Other commentators note, however, that such valuation provides an escape from the genuine problems posed by the sacrifice of human life and "can only ... encourage the sort of mentality that can contemplate 'megadeaths' with equanimity."[81]

The valuation of human life raises insurmountable technical problems.[82] The method assumes that, although people may be unwilling to accept an infinite amount of money for certain death,[83] they will nevertheless accept cash compensation for encountering certain risks.[84]

These assumptions permit the decisionmaker to take actions that will kill a statistically known number of people as long as the victims' identities are not known in advance. The Pareto optimality criterion, one method frequently used to judge the legitimacy of such social choices, holds that a decision is an improvement on the status quo if it makes one or more persons better off without at the same time making other persons worse off. The criterion thus requires that the benefits of a given action, minus any hypothetical[85] compensation paid to those disadvantaged, must be a positive quantity. Because the beneficiaries and the bearers of risk value their own gains and losses,[86] difficulties in this approach arise when those at risk insist that their loss is not compensable by any amount of money.[87]

Under the Pareto criterion, one such person is able to veto a project. John Adams argues that this is proper:

> [A] difficulty ... arises when there is uncertainty about whether the net effect of a project, in terms of fatal accidents, is a benefit or a cost. If a project yields a benefit in terms of lives saved, then it is reasonable to ask how much people might be prepared to pay for it. Often the answer is little or nothing. This is because life is risky and death is certain. The number of "natural" risks to which we are exposed is very large relative to the protection we can afford to buy, and at the most we can only hope to buy a deferment of the inevitable. A common response to invitations to buy small extra bits of protection is a fatalistic shrug. But if a

project threatens to inflict death or serious injury, then the problem is different. In this case, if the cost–benefit game is being played honestly, the question that must be asked is how much money would people demand to be paid before they would willingly allow the threat to be imposed on them. The answer to this question is often a very large sum indeed, and if the risk is seen to be great, the answer is commonly a sum that exceeds all the money in the world. The most that can be said for [any particular value] is that it falls within the range of zero to infinity commonly uncovered by empirical research.[88]

Thus, one infinite selling price set by someone exposed to risk is sufficient to destroy the theoretical basis of the entire cost–benefit exercise. Unlike the plaintiff in a personal injury action, who is entitled to due process of law in fixing the amount his injury is "worth," the risk-bearer has no remedy except noncooperation with the exasperated analyst, who, defeated by one infinite selling price, quietly revises the test and asks risk-bearers what they would be prepared to *pay* to *avoid* their loss. Payment to avoid loss is a very different matter from payment to compensate for loss; it depends not only on willingness but also on *ability* to pay. The price someone is prepared to pay to protect something he or she values, however, is often a small fraction of the price considered fair compensation for its loss. Thus, the only legitimate measure of loss is compensation.[89]

Unfortunately for practitioners of cost–risk–benefit analysis, many people consider some person or thing to be priceless. The Roskill Commission, for example, used questionnaires to find out how much money people would accept as compensation for the loss of their homes; 8 percent in that survey, and 38 percent in a similar survey by the British Airports Authority, said that no amount of money could compensate them for such a loss.[90] The Commission's reaction was to impugn their honesty or rationality and to ask if they were merely trying to establish a favorable bargaining position for compensation, having mistaken a hypothetical question for a real one. The Commission responded by assuming that everyone has a price and by formulating an arbitrary one: three times the market value of the house. The Commission assumed that twice the market value was "generous" compensation for location, friendship, association, and the like.[91]

Such a Procrustean adaptation of people's real preferences to an intellectually bankrupt methodology produces deep hostility. Some people are concerned not only with the final result of the principle of distributive justice but also with the process that generates the result.[92] Any person may insist on some things as a matter of right, to be recognized and enjoyed because they are organically and historically a part of that person rather than assigned on contingent and managerial grounds of efficiency or utility.[93] To call such an assertion of rights a discontinuous preference function is to ignore the difficulty of reconciling human dignity with the Pareto criterion.

A favorite modern detour around respondents who have no neat prefer-ence function is to invent a risk-aversion function that expresses their willingness to pay to reduce the probability of catastrophic losses. This bit of "technical scrimshaw"[94] conceals the improper substitution of a buying bid for a selling bid[95] and is an unreal abstraction.[96]

The divergence between real people's risk perceptions and those calculated for them is a matter of psychological outlook. Someone having a purely techni-cal viewpoint who alleges the ability to store nuclear wastes in isolation from the biosphere despite geological[97] and social contingencies, for example, will be puzzled by public skepticism about his position. Conversely, someone who believes that past experience is not reassuring, that we know little about the pathways and effects of some important nuclides, and that we have no adequate measure of the hazard of the wastes[98] will wonder how any responsi-ble scientist can make such a claim and will consider it unjustified.[99]

Costs

Identifying costs means selecting some and excluding others. Judicious selec-tion of certain costs, to which values are carefully assigned from within a wide range of plausible values, can predetermine the outcome of a cost–risk–benefit analysis. Just as arbitrary selection of certain probabilities of nuclear accident or violence, or exclusion of the nuclear tailings risk,[100] can make any nuclear power station appear more attractive than a non-nuclear one, the arbitrary selection of a price elasticity of demand, or exclusion of some debatable or contingent category of monetary costs, can have the same effect.

The cost–benefit balance of any capital-intensive project whose alleged benefits are deferred but whose real costs are immediate is sensitive to the assumed discount rate. Accordingly, a favorite cost–benefit game is adjusting discount and inflation rates to achieve desired results. A large literature on such adjustments in the United States' fast breeder reactor program exists.[101] A change of a few percentage points in the assumed discount rate—the sort of change about which all the economists in the world, laid end to end, might never agree—can change tens of billions of dollars of benefit into a net cost, or vice versa.

The costs used to construct a balance between costs and benefits are deter-mined by policy, subsidy, externalities, tax treatment, depreciation methods, and philosophy. The value assigned to fossil fuels or any other resource in the ground is largely arbitrary.[102] Economic theory indicates only that this value lies somewhere between the technical cost of extraction, which is almost zero, and the cost of replacement, which is almost infinite. Although high discount rates[103] and approximation to the zero-cost boundary are traditionally used, long-run cost accounting methods and many economists favor the replacement cost convention. Thus, the choices between historical cost or replacement cost valuation, between private and social costs, and between solar energy and fossil fuels are philosophical.[104] The comparison between solar and fossil fuels suggests that the analyst's determination of costs should depend on comparing all long-term alternatives to dwindling oil and gas reserves—solar energy, conservation, nuclear power, coal-based synthetics—with each other, rather than comparing some options (nuclear, synthetics) with each other and others (solar, conservation) with the cheap fuels they are all meant to replace.[105] Yet this latter type of asymmetry is common in the literature.

Many costs and prices used in energy cost–benefit analyses are fuzzy. The price of OPEC oil, for example, is perhaps one-third of the real long-run cost of replacing it. The real costs of coal gas plants, oil shale plants, and many Arctic facilities have been doubling every few years. For a time, the capital cost projected for the proposed Clinch River, Tennessee prototype fast breeder reactor was doubling over a period which was itself halving every six months. Light-water reactors, hailed by some commentators[106] as a mature technology and already the object of worldwide investments, are consistently constructed in the United States at twice their expected real capital costs[107] and will probably operate at three to six times their expected fuel cycle cost.[108] Nuclear economics depend on some 50 main variables, of which few are known within a factor of two, all are disputed, and probably none are independent.[109] Thus, any desired cost of nuclear electricity within a factor of ten can be obtained from assumptions which some analyst will defend, however disingenuously.

The economics of other bulk electricity sources are similarly uncertain. For example, the costs and prices of mining and shipping coal, restoring mined land, obtaining and replacing water rights, operating stack gas scrubbers with high-sulfur and high-chlorine coal, and degrading air quality are all uncertain. The only bright spot in this pervasive economic obscurity is that proven soft and transitional technologies have a robust cost advantage over conventional energy systems at the margin. In addition, soft technology costs are likely to fall, while the costs of conventional systems will rise.[110]

The task of computing internal private costs is complicated by certain unsuspected terms. Economies of scale, for example, are often illusory because they are outweighed by larger diseconomies of scale which are omitted from traditional analyses.[111] Such effects are observable today in both power station reliability[112] and reserve margin (backup capacity) requirements.[113] Many costly operations now required in the nuclear fuel cycle were unexpected and thus not included in original cost calculations.[114] Increased knowledge of the effects of submicron particulates may have major effects on future operation of fossil-fueled power stations. Finally, greater knowledge of the nonlinear, perhaps irreversible climatic impacts of burning fossil fuel may transform our energy constraints within a few decades.[115]

Quantitative uncertainties in identified parameters—the cost of building a refinery, its efficiency, and so on—can be dealt with by post hoc sensitivity analysis.[116] The sheer number of such uncertainties requires that any useful sensitivity analysis be completely multivariate. Variables then can be manipulated in any combination through wide ranges.[117] This approach, however, may be impossible owing to nonorthogonalities and nonlinearities; it is certainly cumbersome and embarrassing, as it enables critics to probe the weaknesses of cost–benefit analyses constructed to support a particular position. For some combination of these reasons, such multivariate sensitivity analyses are never performed. The escape hatch of sensitivity analyses, moreover, is unavailable for costs that have not yet been identified.

Consequent Costs, Prediction, and Modeling

Consequential or secondary costs accompanying realization of risks are often neglected or inadequately measured by cost–risk–benefit analyses. Possible climatic change[118] induced or aggravated by combustion products is one such cost; it can dislocate world agriculture, food, health, and peace. Similarly, interruptions of oil supply, a risk posed by the lack of indigenous fuels and strategic reserves, could cause a wide range of global social, economic, and military disruptions. Efforts to prevent nuclear violence and coercion may infringe civil liberties,[119] while failure to prevent it may cause major psychological trauma[120] and political upheaval. Terrorism may encourage more terrorism. Accidental releases of radioactivity may lead to panic, evacuation and looting, international nuclear shutdowns, political instability, and mass demands for prophylactic thyroidectomies.[121]

Some important consequential costs have been recognized only recently. Sulfur oxides released by burning fossil fuels produce acid rain, which may increase soil acidity; but recent research has also shown that the increased acidity can increase the nitrous oxide yield of denitrifying soil bacteria, increasing the threat posed to the ozone layer by a forced nitrogen cycle (such as artificial nitrogen fixation in fertilizer).[122] In an interrelated world, consequential costs are beyond the wildest dreams of litigants who hope to prove that remote damages were proximately caused. The large aggregate of known and unknown consequences is not cumulatively negligible. Indeed, forgotten higher-order terms can dominate the analysis. Neglected as a cost in every published analysis, the intricate web of nuclear proliferation problems is nuclear power's most important cost.[123] Such consequences, however, are not assessed by traditional energy cost–risk–benefit assessments.

The difficulty of identifying, understanding, and quantifying consequential costs and risks and higher-order effects in a complex world leads to decisions in the realm that Wolf Häfele calls "hypotheticality."[124] One can be quickly transported to this ethereal realm by using computer modeling to simulate the supposed interactions of large numbers of estimated variables over long periods of time. Such models are inflexible; they are obliged to assume constants in lieu of slow variables, and their structure, boundary values, and objective functions do not evolve like those of a real society. Linear programming, widely used to simulate decisions under constraints, can veer violently between extremes in an unrealistic fashion. Indeed, it appears that there are few instances in which models are very helpful for policy:

Formal energy models can function only if stripped of surprises, but then they can say nothing about a world in which discontinuities and singularities matter more than the fragments of secular trend in between. [Moreover, one] makes a model, presumably, because some system is too complex for its behavior to be apprehended intuitively. The same, unfortunately, is then bound to be true of the model, and one will never know whether to believe it or not, nor how far it can be used for guidance, since one does not understand it and has no way to validate it.[125]

Thus, models complex enough to be interesting cannot be validated, and models complex enough to be valid cannot be written.

All too often, modelers will point out fatal flaws in their own models and then use the same models to draw policy conclusions. An econometric model often cited by the American utility industry as evidence that a nuclear moratorium would cost consumers hundreds of billions of discounted dollars is a good example.[126] The model generates perpetual rapid growth in nuclear capacity from a few simple assumptions: reactors are cheap, capital and political acceptance are unlimited, and electricity use will quickly increase despite significant price rises. The model predicts that the United States will install new thousand-megawatt reactors at a rate exceeding one per day in 2030. When I suggested to one of the model's authors that this implausible behavior suggested the unreliability of the model, his response was not to try to improve the model or to add caveats to his conclusions, but rather to say that because the long-term behavior of the model seemed to upset people, future editions of the report could simply omit graphs that went to the year 2030!

Assessments

The comparison of the uncertain quantities of costs, benefits, and risks raises additional problems. A practical problem of comparison is the choice between average costs, benefits, and risks and marginal forms of these quantities. Harold P. Green suggests that cost–risk–benefit assessments should balance "the incremental risks [and costs] attributable to marginal benefits,"[127] not total benefits and total risks and costs. Thus, the benefits and burdens of a single large refinery should be compared with those of several smaller refineries. On closer examination, however, the problem is more difficult, for the size of the refinery itself has profound implications for direct and indirect costs, risks, and benefits. Size affects reliability, vulnerability, technical flexibility, lead times (hence exposure to interest during construction, cost escalation, mistimed demand forecasts, and changes in political climate), infrastructure costs, and other issues that have not yet been properly studied.[128]

Because energy and especially nuclear choices are of the form "either/or" rather than "yes/no," they raise the question of forgone alternatives.[129] Indeed, forgone and foreclosed[130] alternatives are at stake in every cost–benefit assessment. Consider this analogy:

> *Suppose that you ask my advice about backing a horse to win the Derby, and I confidently recommend a particular name. Relying on my judgment, you place a bet of £1 on that horse to win, and sure enough the animal comes in last. The question then is: How much have you lost as a consequence of your decision to take my bad advice? The obvious answer may seem to be that you have lost £1; but this in fact may be the wrong answer. It will be correct if and only if, had you not taken my advice, you would not have placed your bet at all. But suppose that you would in any case have placed a single bet of £1 on some particular horse to win the race. Then whether and how much you lost by listening to me depends on which horse you would have chosen if you had not so listened, and on whether or not it won. If it is quite clear that you would have backed a different loser, then the loss from taking my advice is not £1 but zero. If on the other hand it is clear that but for me you would have backed the winner when the odds stood at ten to one, then the loss is £11.*[131]

As the putative purpose of energy investment is to supply projected energy demands, the problem of alternatives arises in virtually every case. Proper cost–risk–benefit assessments turn on choices covering the entire range of competing and exclusive energy alternatives. Often the range of choices is too narrow because of promotional zeal or lack of imagination or understanding. This narrowness, more than a poor choice among artificially constrained alternatives, is a central problem of cost–benefit analysis.

A third practical problem of energy cost–risk–benefit assessment is a concentration on micro-choices (which kind of power station to build at a particular site) at the expense of macro-choices (what kind of society people want, what its energy policy should be, what is the role of power stations in that policy, who decides). Micro-analysts tend to concentrate on micro-determinants and to omit the enormous range of issues that would be significant if the same micro-decision were replicated nationally. Archetypal micro-decisions, however, cannot be summed to make a sound macro-decision, because relevant factors change. Attempting to make micro-decisions in a macro-vacuum leads to macro-trouble, because a framework that cannot handle the sum of micro-decisions is not going to produce sound individual micro-decisions.[132] This "sum" is not a linear addition of independent elements, but rather an organic totality of complex effects whose essence lies in "the structural features that give them their total character."[133] Cost–risk–benefit assessment is not, on the other hand, holistic but reductionist, and the customary response to complaints that the heart of the problem has somehow been left out is to become more reductionist by including more detailed numbers. This response only exacerbates the problem.

In addition to these practical problems, the task of comparing costs, benefits, and risks raises several theoretical difficulties. One of the pioneers of the method has remarked:

> *It is my belief that benefit–cost analysis cannot answer the most important policy questions associated with the desirability of developing a large-scale, fission-based economy. To expect it to do so is to ask it to bear a burden it cannot sustain. This is so because these questions are of a deep ethical character. Benefit–cost analysis certainly cannot solve such questions and may well obscure them. ... Unfortunately, the advantages of fission are much more readily quantified in the format of a benefit–cost analysis than are the associated hazards. Therefore, there exists the danger that the benefits may seem more real.*[134]

The tension between cost–benefit analysis and ethical considerations arises because the method provides no way of choosing between different ends; the method can at best suggest the appropriate means to a given end. The mere fact that benefits exceed costs is not a sufficient reason to take a particular action; one must ask whether the same benefits could be obtained in another way at even lower costs. One must also ask whether greater and perhaps different benefits can be obtained in some other way. Cost–benefit analysis cannot answer such questions.

A second theoretical problem of cost–benefit analysis is the gap between the Peter who is robbed and the Paul who is paid: it is impractical for beneficiaries to compensate risk-bearers. The large literature in public welfare economics dealing with interpersonal utility comparisons provides no solution to this problem. Although the theoretical basis of cost–risk–benefit assessment requires that redistributional effects of a decision in time and space be insignificant, this condition is almost never satisfied, and distributive justice is ignored.[135]

Without using value judgments, moreover, cost–benefit analysts cannot bridge the gap between the incommensurable values which the method purports to compare. The Roskill Commission, for example, counting only the countable, compared and transformed into a money standard "decibels, minutes, houses, flights, schools, acres, private airfields, hospitals, miles, ... deaths," Norman churches,[136] beauty, and human relationships.[137] Harold P. Green suggests that the problems of quantifiability and commensurability are a mere artifact of overreliance on cost–benefit analysis:

> *Experts can perhaps agree on the numbers to be assigned to benefits, but the numbers will ... reflect nothing more than the experts' own value judgments. Even if the calculation of benefits*

rested on empirical data scientifically derived and analyzed, the calculation would reflect only the public values existing at a given moment in time and would not reflect the fact that values may shift mercurially.[138]

Similarly, Laurence H. Tribe concludes: "If analysis yields a result that still requires the decisionmaker in the end to make difficult intuitive choices between alternatives that differ in some incommensurable way,"[139] or whose quantification on one or both sides of the balance he does not trust, "he may conclude that the analysis was not worth the cost and that he might as well have relied on his intuition from the outset."[140]

John Adams suggests that "the intuition and prejudice that the Roskill Commission shunned"[141] have only been renamed rather than banished: "Computer models do not get rid of conjecture and unreasoning attachment to creed, rather they systematise it ... in a peculiarly bigoted way." They ignore any parameters that cannot be neatly quantified, and in so doing express a value judgment.[142] Thus, cost–benefit analysis does not eliminate subjective value judgments but rather moves them from the realm of political controversy into the expert's opaque model—a step with disquieting implications for the political process.[143] These judgments are made when the alternatives to and the criteria for the proposed action are determined.[144] Cost–benefit analysis founders on the "quaint fantasy"[145] that defining a problem, selecting alternative solutions, choosing relevant parameters, and estimating values for them can replace, rather than disguise, subjective judgment. Some commentators have suggested that this assumption conceals a desire to escape the frightening responsibilities and uncertainties of decisionmaking.[146]

Decisionmaking

Cost–risk–benefit assessment is not only intrinsically subjective from a theoretical viewpoint; in many instances, such assessments are commissioned not to aid impartial decisionmaking but rather to advocate a particular position.[147] Ida Hoos has identified this trend with the shift in role of many academic experts from asking questions to providing answers for specific clients. No actual collusion between assessors and clients is necessary, because both the experts and the audience know what answers the clients want. The academic quality of such assessments inevitably suffers:[148]

Obedient consultants in think-tanks and accommodating academics of all stripe constitute a reservoir of talent. ... They produce "data" to substantiate any position. They dutifully interpret "facts." As a result, several sets of experts can use the same body of information and reach totally different conclusions. What is more remarkable, the same *group of researchers, using the* very same data *that they themselves had gathered in a study for [the*

Federal Energy Administration], could come up with three different executive summaries.[149]

The results of energy cost–risk–benefit assessments often are perceived as absurd, not because they contradict themselves or because several studies disagree, but rather because they are discredited by subsequent events,[150] give politically unacceptable answers,[151] or are punctured by a trenchant criticism.[152] Energy cost–risk–benefit analyses have met all three of these fates with monotonous consistency. The result is public cynicism about studies by authors with a vested interest in their outcome. This attitude may rise to a general public suspicion in all political matters.[153] This destruction of political legitimacy is perhaps the greatest danger of the abuse of the method.

Three other dangers of abuse are noteworthy: that valid and useful methodologies will be discredited,[154] that a decisionmaker will believe and act on the results of advocacy cost–risk–benefit analyses, and that the method will tempt politicians to abrogate their political responsibilities and to shuffle them off onto a supposedly objective formalism whose limitations they imperfectly understand.[155] If the Roskill report had not been demolished,[156] many British officials would have been glad to have a "rational" defense for their folly, a safely insulated bin into which to toss the hot potato of the day. This provision of an escape hatch from accountability is a particularly pernicious distortion of the democratic process.

A typical exposition of the political theory of many proponents of cost–benefit analysis appears in a recent Stanford Research Institute report which obscures all questions of political process with fuzzy phrases about "decisionmakers" and how "society chooses."[157] The authors state their belief in their own role with unwonted crispness:

> [P]ublic decisions should be based on quantitative judgment of the experts who know the most about the technology. ... By assessing ... values in monetary terms, we are able to determine the monetary amount that should be spent on anti-pollution or safety systems, and at a higher level, to choose between entire energy systems. ... Once the alternatives, information, and preferences are established, society can make the decision using only the principles of logic.[158]

Cost–benefit ratios equal automatic decisions; no value judgments are needed; no politicians need apply. This approach, in which the political process becomes a mere appendix to expert analysis while computer printouts reign supreme, admirably reflects the view that political decisionmaking should be a kind of natural science,[159] even if its objectivity leads us "inexorably to ... decisions for which we are not yet ready."[160]

Many scientists and technologists bridle at the constraints of the political process because its characteristically "emotional debate" leads to results less "rational" than they would like.[161] Some scientists, for example, sincerely believe it is a fact that democratic institutions and even Western civilization will crumble if hundreds, and ultimately thousands, of nuclear reactors are not built. A democracy, however, need not accommodate its decisions to the authority of those scientific groups that currently enjoy credence and power,[162] nor to any other authority save that of the people and the Constitution. Democratic decisions should rest on common-sense judgments made by generalists who

> *have greater affinity with public values leading to greater [and more deserved] confidence in the conclusions reached. ... Any other approach smacks of government by the elite, a concept that is directly at variance with democratic principles.*[163]

Generalist politicians and regulators speak publicly "in the language of ordinary political discourse,"[164] thus preserving accountability through the ballot box and the courts.

Although "facts are often manipulated by the decisionmaker to produce the result he thinks correct under all of the circumstances, including political,"[165] the alternative—letting "ultimate value judgments ... be based to a substantial extent on the value judgments inherent in the risk–benefit calculus performed by the elite"[166]—would seriously inhibit political accountability and judicial review. If the Founding Fathers "had opted for decisions based on objective truth" rather than on the inevitability of error, "they would have handed down to us a government by elite intellectuals, mystics, medicine men, or divines who could present us with their own respective brands of orthodox fact."[167]

I believe that the process by which a decision is reached is more important than its factual correctness. Process is not a mechanical series of gestures; rather, it is a social end. To reduce process to purely incremental, instrumental terms distorts the concept "much as 'rights' are flattened"[168] by making them a contingent grant rather than an inherent part of a person.[169] The democratic values and processes that cost–risk–benefit analysis threatens are central to a free society. Although science does have a vital role in public decisionmaking, that role is defined by the democratic character of the American political system.

Formal cost–risk–benefit analysis, therefore, has no useful role in the formulation of energy policy. As a part of a legal and institutional process which elicits information that promoters of particular schemes are otherwise reluctant to give, the method's main utility is ancillary. It encourages specificity in stating the grounds on which a project is believed to be a good or bad idea and rigor in approaching the genuine limits of quantification—for numbers, kept in their place, are indeed important. These benefits can be obtained without deifying the results of analysis and making them into computer fodder to which policies and budgets are aligned, praises sung, and goats sacrificed daily.

Resources and effort could be saved by concentrating on disclosure and free exchange of information, and by nourishing and improving existing mechanisms of public discourse and adversarial review. If the money now devoted to cost–risk–benefit analysis were used to promote these more fundamental and pragmatic ends, policy decisions would be incomparably better. Indeed, we would have a much sounder energy system today if computers larger than hand calculators had never been invented, for then our clever analysts would have had to think about what they were doing. In our present state of ignorance about the basic matters of energy end use,[170] all the important questions of energy policy can best be addressed for a few thousand dollars each on the backs of large envelopes and decided by any citizen in words of one syllable. The intellectual challenge today lies in holistic simplicity, not in the simplistically reductionist complexity of cost–risk–benefit analysis.

It may be argued that abandonment of what some people (especially those skilled in milking the federal cow) regard as an essential tool of policy analysis would deprive society of many worthwhile and even necessary technical innovations. I do not believe it. The effect would be to blow away a lot of deliberately generated fog. If, on the other hand, a political process cast adrift from the massive moorings of cost–risk–benefit analysis tended to overestimate risks and to underestimate benefits, that would be

> *precisely the kind of inadequacy—or, indeed, error—that society is best able to tolerate, [for] the result may be a decision that bars or postpones a benefit; but this means only that society is deprived of an opportunity for improvement. On the other hand, the opposite approach may expose society to actual injury. It is not difficult, therefore, to find justification for an overweighting of risk relative to benefit.*[171]

The suspicion that subordinating cost–benefit analysis to the democratic process would inhibit alleged technical advances is itself a hint that we ought not to have such advances. The method is seen by its promoters as a powerful agent "to grease the tracks with rationality in order to avoid emotional impediments to technological advance" in a society that now makes "too many anti-technology social decisions."[172]

It is precisely because there is "no moral imperative that our government push for new benefits ... that it never had before," but rather "a moral imperative that society should not needlessly inflict injury through its governmental decisions," that we should err on the side of inhibiting technological change, for it "frequently—perhaps always—involves undesirable consequences."[173] If fear of the verdict of an undeceived people is the strongest argument that proponents of cost–risk–benefit analysis in energy decisionmaking can muster, it is long past time that they adjusted to the rigors of life in a democratic republic. And at a time when it is urged that the problems of cost–risk–benefit analysis can be overcome by its more zealous application, John Maynard Keynes' maxim is apposite: "While some people might admire the technical ingenuity and elegance of the contributions, others might reply that if something is not worth doing, it is not worth doing well."[174]

Science is too important to be left to the scientists,[175] and politics is too important to be left to anyone but the people. If we must choose between particular pieces of hardware and the right to make up our diverse minds about them without being smothered in pseudoscientific snow, the hardware should lose every time. It is fundamental to energy policy that our problems, far from being too complex and technical for ordinary people to understand, are on the contrary too simple and political for most experts to understand. In conveying to the technical experts that message, and in illuminating for them the elements of civics, the legal profession faces an important and urgent task.

Notes

1   See J. Reuyl, W. Harman, R. Carlson, M. Levine & J. Witter, "A Preliminary Social and Environmental Assessment of the EBDA Solar Energy Program 1975–2020", at VI-29 (July 1976) [hereinafter cited as J. Reuyl] (draft report by Stanford Research Institute), reprinted in *Alternative Long-Range Energy Strategies: Joint Hearings Before the Senate Select Comm. on Small Business and the Senate Interior and Insular Affairs Comm.*, 94th Cong., 2d Sess. 114–50, 613–26 (1976) [hereinafter cited as Joint Hearings].

2   See A. Lovins, *Soft Energy Paths: Toward a Durable Piece* (1977).

3   Id. See also A. Lovins, "Alternative Long-Range Energy Strategies", in Joint Hearings, *supra* note 1, Supplement (1977), Attach. 1 (response to Dr Lapp) 1.

4   42 USC §§ 4321–4347 (1970). See also Technology Assessment Act of 1972, 2 USC §§ 471–481 (Supp V. 1975).

5   I. Hoos, "The Assessment of Methodologies for Nuclear Waste Management" (1976) (typescript, University of California at Berkeley).

6   Id.

7   The problem of identifying those at risk and those who benefit from a particular course of action usually does not arise in private decisionmaking. In a private investment decision, for example, a corporation not only bears costs and financial risks but also receives the benefits of its decision. In public welfare economics, on the other hand, the political process decides what benefits and costs are appropriate for someone other than the decisionmaking entity.

8   HM Stationery Office, Commission on the Third London Airport, Report (1971) [hereinafter cited as Roskill Commission].

9   Adams, "Life in a Global Village", 4 *Environment & Plan.* 381, 389 (1972). Future costs and risks and their bearers also can be speculative. Debates concerning such problems as oil depletion, climate-altering carbon dioxide emissions, and nuclear wastes focus on future risks.

10   See, e.g., Schipper and Lichtenberg, "Efficient Energy Use and Well-Being: The Swedish Example", 194 *Science* 1001, 1004 (1976).

11   H. Daly, "On Thinking About Future Energy Requirements" (1976) (memorandum, Economics Department, Louisiana State University).

12   See I. Illich, *Tools for Conviviality* (1973); *Energy and Equity* (1974); A. Lovins, *supra* note 2, at 161–70.

13   J. Adams, "The Appraisal of Road Schemes; Half a Baby is Murder" (1977) (typescript, Geography Department, University College, London).

14   See comments of Buchanan, in 137 *The Geographical J.*, pt. 4, at 501 (1972), where Professor Sir Colin Buchanan remarks that the south of England was being raped to build airports that could fly British tourists to the coasts of Cyprus, Malta, and Bermuda so that those coasts could be urbanized and become exactly like the newly raped south of England.

15   Adams, *supra* note 9, at 382. See also I. Illich, *supra* note 12.

16   See *Ecclesiastes* 5: 9–12; J. Ellul, *The Technological Society* (1964); S. Freud, *Civilization and its Discontents* (1959); I. Illich, *Medical Nemesis: The Expropriation of Health* (1976); L. Mumford, *The Myth of the Machine: The Pentagon of Power* (1970); L. Winner, *Autonomous Technology* (1977). One of the most notable features of the hotly debated new technologies for recombinant DNA manipulations is the insubstantiality of the claimed benefits. I am not aware that any significant benefit has ever been proposed. One apparent benefit, namely the implantation of nitrogen-fixing genes into non-leguminous plants or bacteria symbiotic with them, is almost certainly not a benefit but a cost, owing to the irreversible climatic consequences of forcing the nitrogen cycle. See note 122 *infra*. See Lovins, "Nitrogen Fixation", 255 *Nature* 8 (1975).

17   Ehrlich, "An Ecologist's Perspective on Nuclear Power", 28 *Fed'n Am. Scientists Pub Interest Rep.* 1, 3–6. (May–June 1975). See also Budnitz & Holdren, "Social and Environmental Costs of Energy Systems", in 1 *Ann. Rev. Energy* 553, 560–62 (J. Hollander ed. 1976).

18   A. Lovins, *supra* note 2, at 10. These are only some of the second-order effects; the third- and higher-order effects are more complex. For a strikingly convergent analysis from the viewpoint of orthodox political economy, see *The Political Economy of Energy Policy: A Projection for Capitalist Society*, chs 1 & 6 (J Hammarlund & L Lindberg eds. 1976) (Institute for Environmental Studies, University of Wisconsin); *The Energy Syndrome: Comparing National Responses to the Energy Crisis* (L. Lindberg ed. 1977).

19   Lovins, "Energy Strategy: The Road Not Taken?", 55 *Foreign Aff.* 65 (1976). See also A. Lovins, *supra* note 2, at 54–57, 147–59.

20   Inequity arises because centralized energy systems automatically allocate their energy and their social costs to different people at opposite ends of the transmission line, rail line, or pipeline. This is advantageous to those who get the energy—generally politically powerful people in cities and suburbs—but less so to the poorer rural people on the other end.

21   See A. Lovins, *supra* note 2, at 3–24, 161–70; J. Reuyl, *supra* note 1, at VI-27.

22   A. Lovins, *supra* note 2, at 3–60, 73–103, 133–44.

23   H. Daly, *supra* note 11.

24   One end on which there is a degree of consensus among some powerful groups, however, is the pursuit of technological innovation for its own sake, an urge we have not yet seriously tried to sublimate. Many Faustians still adhere to Bacon's model: "The End of our Foundation is the knowledge of Causes, the secret motions of things; and the enlarging of the bounds of Human Empire, to the effecting of all things possible." F. Bacon, "New Atlantis", in *Selected Writings of Francis Bacon* 574, 579 (1955). See also note 21 *supra*.

25 Adams, *supra* note 9, at 382. Appeals to national prestige are often in this form, even though the argument can backfire: participation in and promotion of technological follies does "not increase national prestige, but diminishes it." Address by Henderson, "Two British Errors: Their Probable Size and Some Possible Lessons" (24 May 1976) (inaugural lecture. University College, London). Henderson's particular reference is to Concorde, which "has been a milestone on the road of technological advance, even though it may eventually prove to have been a milestone in a blind alley." Id. at 25. See also Mote, "The Concorde Calculus", 45 *Geo. Wash. L. Rev* 1037 (1977).

26 VII US Atomic Energy Comm'n, Wash-1535, Proposed Final Environmental Statement Liquid Metal Fast Breeder Reactor Program 53, at 51–52 (1974). The benefits double-counted include a saving of $67 billion that would otherwise be invested in the uranium fuel cycle (a cost already included in the uranium cost for the non-breeder case), providing "a premium market for plutonium produced" by light-water reactors (a contingent benefit already included in the real resource cost savings of the breeder—and also ascribed to the LWRs in *their* own cost–risk–benefit analysis as a plutonium credit!), and "assurance that available uranium reserves will be most efficiently used" and "utilization of the stockpile of depleted uranium."

27 Creation of jobs is often counted as a benefit of a program; this is correct only if "a reduction in expenditure on the program, leading to a fall in the numbers employed on it, would not be associated with an equivalent increase in expenditure and employment elsewhere." Address by Henderson, *supra* note 25, at 19. In fact, most energy projects are so capital-intensive that just the opposite is true. Every new thousand-megawatt power station, for example, costs the United States economy about 4000 net jobs. See B. Hannon, *Energy and Labor Demand in the Conserver Society* at Table 1 (1976). Henderson also notes that it is
conventional in economic analysis, and for good reasons, to treat the benefits ... from an expenditure ... as resulting from the output, while inputs form an item on the costs side, at any rate in so far as their use in the project implies that some other form of useful output has to be foregone. [I]t is both confusing and unnecessary to treat employment as possessing independent value; and it is ... inconsistent to argue ... that the money costs of the ... program should be set off against the gains of employment creation. Most of these costs consist (whether directly or indirectly) of ... wages and salaries [and, if jobs were the goal, would have to be counted as benefits]. (Address by Henderson, *supra* note 25, at 17)

28 See text accompanying notes 18–20 *supra*.

29 See text accompanying notes 102–104 *infra*.

30 In this paper, "risk" means the extent of the hazard of injury or destruction to persons or property. The term is usually expressed as the product of the probability and the magnitude of a set of consequences. Risk is not synonymous with uncertainty (which cannot be defined by a probabilistic statement, i.e. unfalsifiably), nor with the concept of cost, which refers to noncontingent, nonprobabilistic economic cost, whether direct and private or external and public. See W. Lowrance, *Of Acceptable Risk* (1976).

31 Comments of Brooks, in S. Barrager, *The Economic and Social Costs of Coal and Nuclear Electric Generation* 70 (1976) (Stanford Research Institute paper for National Science Foundation).

32   See generally Jonas, "Technology and Responsibility: Reflections on the New Tasks of Ethics," 40 *Soc. Research* 31 (1973). Some eminent economists, notably Tjalling Koopmans, have declared such discounting to be inappropriate in particular cases involving long periods (nuclear wastes) or irreversibility (extinction of whales), but it is unclear how other cases are to be distinguished. The foundations of the theory of interest (on which discounting also rests) are themselves shaky. There is no theoretical requirement that a society have a positive interest rate, i.e. a preference for money now over money later. In many cultures, interest rates are at zero because money is not considered a saleable commodity.

33   The uncertain effects of coal combustion provide an example. See S. Keeny, *Nuclear Power Issues and Choices*, 1 6–17 (1977). For a broader review, see Budnitz and Holdren, *supra* note 17.

34   See, e.g., "Research and Development Relating to Sulfates in the Atmosphere: Hearings Before the Subcomm. on the Environment and the Atmosphere of the House Comm. on Science and Technology", 94th Cong., 1st Sess. 39 (1975); "The Costs and Effects of Chrome Exposure to Low-Level Pollutants in the Environment: Hearings Before the Subcomm. on the Environment and the Atmosphere of the House Comm. on Science and Technology", 94th Cong., 1st Sess. 49 (1975).

35   See Campbell, & Martin, "Oil and Ice in the Arctic Ocean: Possible Large-Scale Interactions", 181 *Science* 56, 56–58 (1973). See also *Study of Man's Impact on Climate Inadvertent Climate Modification* § 6.72–74 (1971); note 115 *infra*.

36   The risks include the explosively violent boiling sometimes observed when substantial amounts spill onto water, and the "rollover" effect whereby a cargo of LNG can suddenly invert itself with a concomitant sudden rise in the over pressure of gas. Neither of these effects was discovered until LNG had been shipped commercially for some years, and the latter was unsuspected until many large tankers had been built.

37   See, e.g., R. Dobson, "Problems in the Design and Construction of Liquefied Gas Carriers" (LNG/LPG Conference, London, 21–22 March 1972); W. Thomas & A. Schwendtner, "LNG Carriers: The Current State of the Art" (LNG/LPG Conference, London, 21–22 March 1972), reprinted in 79 *Transactions Soc'y Naval Architecture & Marine Engineering* 40, 447 (1971). Some minor failures of ships' plating embrittled by the intense cold have occurred. The relevance of this problem is that the energy content of the LNG in a typical (125,000 cubic meter) LNG tanker is equivalent to that of approximately 55 Hiroshima bombs, though the energy would be released by a firestorm rather than instantaneously.

38   P Magno, "A Consideration of the Significance of Carbon-14 Discharges from the Nuclear Power Industry" (Aug 1974) (13th AEC Air Cleaning Conference, San Francisco, Cal.); B. Pohl, *Nuclear Energy: Health Impact of Carbon-14* (Physics Dept., Cornell University). The release of carbon-14 (whose half-life is 5730 years) from reactors and reprocessing plants has only been recognized in the past few years, though the corresponding hazard is one of the dominant terms of risk from routine releases. See also S. Keeney, *supra* note 33, at 176–77, 181.

39   William Boeck has shown that krypton-85 can measurably affect the electrical properties of the atmosphere at concentrations two orders of magnitude below those allowed by health criteria. Somewhat higher levels, such as could arise in several decades from a classically projected nuclear industry, could have substan-

tial effects of a currently unknown nature on world climate. Boeck "Meteorological Consequences of Atmospheric Krypton-85", 193 *Science* 195 (1976); note 115 *infra*.

40   See Martell, "Tobacco Radioactivity and Cancer in Smokers", 63 *Am. Sci.* 404, 409–10 (1975); "Basic Considerations in the Assessment of the Cancer Risks and Standards for Internal Alpha Emitters" (1975) (E.P.A. hearings on plutonium standards).

41   Id.; see. e.g., Brodeur, "Microwaves", *The New Yorker*, 13 December 1976, at 50; 20 December 1976, at 43.

42   Casper, "Technology Policy and Democracy", 194 *Science* 29, 29–35 (1976) (laser isotopic enrichment). I believe this subject should be classified, indeed abandoned, because of the risks of proliferating such knowledge; but at present, especially for laser fusion, classification is being used less to inhibit the release of sensitive information than to preclude public discussion of unclassified policy aspects of the technology and their safeguards implications. Cf. Arrow. "Social Responsibility and Economic Efficiency", 21 *Pub. Pol'y* 303 (1973) (importance of full information).

43   Recombinant DNA is perhaps the best current example. See, e.g., Wade, "Recombinant DNA: A Critic Questions the Right to Free Inquiry", 194 *Science* 303, 303–306 (1976).

44   Tribe, "Technology Assessment and the Fourth Discontinuity: The Limits of Instrumental Rationality", 46 *S. Cal. L. Rev.* 617, 640, 648–49 (1973); cf. note 24 *supra*.

45   Hohenemser, Kasperson, and Kates. "The Distrust of Nuclear Power", 196 *Science* 25, 32 (1977). The term is from Weinberg, "Social Institutions and Nuclear Energy", 177 *Science* 27 (1972).

46   A. Lovins & J. Price, *Non-Nuclear Futures: The Case for an Ethical Energy Strategy* 10–14 (1975); Edsall, "Hazards of Nuclear Fission Power and the Choice of Alternatives," 1 *Envt'l. Conservation* (1974); Hardin, "The Fallibility Factor", 23 *Anticipation* 20 (1976).

47   Ida Hoos notes engineers' difficulty in appreciating the social roots of problems. "Calling upon an engineer to cure them is much like asking an economist to treat a heart ailment because the patient became ill over money matters!" I. Hoos, *Systems Analysis in Public Policy: A Critique* 24 (1972).

48   For a nontechnical summary of one notable kind of "hypotheticality," fast breeder reactor safety, see Lovins, "Fast Reactors: The U.S. Debate Goes Critical", 61 *New Scientist* 693 (1974). See also note 71 *infra*. The standard of containment for which plutonium reprocessing and fabrication plants are designed is so stringent—typically releases are to be parts per billion—that it has not yet been empirically verified.

49   For a discussion of the burden of proof in regulatory decisionmaking concerning pesticides, see Note, "Pesticide Regulation: Risk Assessment and Burden of Proof", 45 *Geo. Wash. L. Rev.* 1066 (1977).

50   See Green, "The Risk-Benefit Calculus In Safety Determinations", 43 *Geo. Wash. L. Rev.* 791, 799 n.30 (1975).

51   Id. at 796 n.17.

52   This is the classic problem with detecting climatic change. We can only speculate at the nature of some of the uncontrolled variables and would find it extremely hard to measure many of them with the needed precision. See note 115 *infra*.

53   See Weinberg, *supra* note 45, at 29.

54   Brown, 195 *Science* 48 (1977) (letter to the editor). This question arises with the long-term population dose from radon gas released by uranium mill tailings. R. Pohl, *Nuclear Energy: Health Effects of Thorium-230* (May 1975) See Comey, "The Legacy of Uranium Tailings", *Bull. Atom. Scientists*, September 1975, at 42.

55   Hohenemser, *supra* note 45, at 28.

56   von Hippel, "Nuclear Reactor Accidents: Long Term Health Effects", 194 *Science* 479 (1976).

57   S. Keeney *supra* note 33, at 164 (referring chiefly to high-linear-energy-transfer (e.g., alpha) radiation). See also Little, "Lung Cancer Induced in Hamsters by Low Doses of Alpha Radiation from Polonium-210", 188 *Science* 737 (1975); Morgan, 195 *Science* 344 (1977) (letter to the editor). For concise surveys of some of the statistical complexities and hazards of low-level studies, see Diesendorf, "Low Level Ionizing Radiation and Man", 6 *Search* 328 (1975); Neyman, "Public Health Hazards from Electricity-Producing Plants", 195 *Science* 754 (1977).

58   Interview with Dr Sam Morris (1976).

59   Pigford and Ang, "The Plutonium Fuel Cycles", 29 *Health Physics* 451 (1975); Pigford, "Environmental Aspects of Nuclear Energy Production", 24 *Ann. R. Nucl. Sci.* 515 (1974).

60   Bair & Thompson, "Plutonium: Biomedical Research", 183 *Science* 715 (1974); Thompson, "Animal Data on Plutonium Toxicity", 29 *Health Physics* 511 (1975).

61   Park, "Progress in Beagle Dog Studies with Transuranium Elements at Battelle-Northwest", 22 *Health Physics* 803 (1972); Thompson, *supra* note 60.

62   See Hohenemser, *supra* note 45.

63   Most authors have not troubled to examine the original papers. Keeney, Weinberg, and Hohenemser, as well as the National Radiological Protection Board, Medical Research Council, and the Royal Commission on Environmental Pollution, appear to think that the 26 overexposed Los Alamos workers on whom many claims of safety are based all inhaled insoluble plutonium. In fact at most three did so, and what they inhaled is at least an order of magnitude outside the activity range of interest to most hypotheses. Inhalation of insoluble plutonium dioxide within that activity range by about 25 people fighting the 1965 Rocky Flats fire was indeed significant and, once the latency period of perhaps a further decade or two has elapsed, should be a valid (if uncertain) test of various hypotheses.

64   An order of magnitude is a factor of 10. Several independent and often inconsistent hypotheses have been put forward to suggest that some important kinds of actinide exposure standards should be tightened by one to three orders of magnitude (about 10 to 1000). These hypotheses, none of which has been either proved or disproved, are concerned with "hot particles" (see Lovins & Patterson, "Plutonium Particles: Some Like Them Hot", 254 *Nature* 278 (1975)); carcinogenic mechanisms (Martell, *supra* note 40); dosimetry (Morgan, "Suggested Reduction of Permissible Exposure to Plutonium and Other Transuranium Elements", 36 *Am. Indus Hygiene J.* 567 (1975)) and lung clearance models. See also Edsall, "Toxicity of Plutonium", *Bull. Atom. Scientists*, September 1976, at 27. A remarkable feature of the several official American reviews of one contro-

versial hypothesis—the "hot particle" theory advanced by Dean and Langham in 1969, Geesaman in 1968, and the Natural Resources Defense Council in 1974—is that the supposedly independent review groups have without exception been dominated by the scientists, mainly in the AEC/ERDA orbit, whose work the hypothesis brought into question and many of whom have publicly attacked it. Nonetheless, "there is now strong sentiment among the plutonium researchers that there should be a substantial downward revision in the allowed body burden for plutonium." Brown, "Health, Safety and Social Issues of Nuclear Power", in W. Reynold, *The California Nuclear Initiative* 149 (1976). See also Thompson, *supra* note 60.

65   See Lovins & Patterson, *supra* note 64. The calculations are straightforward. Since alpha radiation has a very short range, plutonium can be detected at a modest distance only by the X-rays it sometimes emits also. At low plutonium levels these weak X-rays cannot be distinguished from background in reasonable whole-body counting times (a few hours). For plutonium that has not been highly irradiated, detection is especially difficult. Air can be monitored for alpha activity—integrating over enormous volumes and long periods (many hours) that hide "spike" or "puff" releases, which may be localized in time and space—but only after the fact; it is impossible in principle to detect excessive plutonium in the air in a typical plutonium working area before the air has been inhaled or changed or both.

66   See note 61 *supra*.

67   The most plentiful fuel-cycle actinide, plutonium-239, decays roughly a million-fold in 20 half-lives, or about 490,000 years. Some other important isotopes such as the amercium-241–neptunium-237 chain, however, require isolation for tens of millions of years or more. A. Lovins and J. Price, *supra* note 46, at app. 1–2.

68   Every United States waste storage site, and probably every major American plutonium facility, has leaked significant amounts of activity. See, e.g., Meyer, "Preliminary Data on the Occurrence of Transuranium Nuclides in the Environment at the Radioactive Waste Burial Site, Maxey Flats, Kentucky", in *Transuranium Nuclides in the Environment* (1976) (EBDA & IAEA symposium).

69   Transplutonium nuclides such as americium and curium tend to be biochemically more active than plutonium, may be concentrated more in food chains, are often more radioactive, may produce plutonium as a decay daughter, and are produced in large amounts with long fuel exposure (as with plutonium recycle or fast breeder reactors).

70   Symposium, *supra* note 68. See Lovins, "Plutonium and Other Actinides", 265 *Nature* 390 (1977). Nevertheless, exceedingly elaborate computer modeling of hypothetical releases of radioactive wastes (especially of actinides) from geological or seabed disposal has been done in an effort to show that the consequences of such releases are small. See, e.g., Burkholder, Cloninger, Baker & Jansen, "Incentives for Partitioning High-Level Waste", 31 *Nuclear Tech.* 202 (1976).

71   An anecdote is illustrative. A few years ago I discussed plutonium hygiene with the responsible official at Electricité de France. He assured me that there was nothing to worry about: his workers had less than a tenth of the allowed lung burden. I asked him how he knew that, since such a small amount of low-burnup plutonium cannot be detected *in vivo* by any instrument. He agreed, and said he supposed that the lung burden had been computed from the air concentration. I showed him by a simple calculation that the air concentration was also unmeasurable at the

required sensitivity. See note 65 *supra*. He agreed, somewhat less complacently, and said the air concentration must have been inferred from the efficiency of the filters. I pointed out that he had not actually measured filter efficiency, but rather the efficiency in trapping a quite different material, and that the two could not be correlated because he lacked such data as the particle size distribution of the plutonium. He then retreated to a still more remote empiricism that also turned out to be unsatisfactory. In the end, his simple statement about the workers' lung burdens, which he and his colleagues had told each other so often that they thought they really knew it, emerged as a tissue of suppositions that might or might not be true. This typical state of affairs is unknown to cost–risk–benefit assessment practitioners who lack suspicious minds, the zeal to scrutinize original sources, and a strong background in experimental science. Yet it is precisely the lack of these qualities which predisposes them to use the method.

72 The 2400 page "Rasmussen Report" from the nearly four-year, four-million-dollar Nuclear Regulatory Commission Reactor Safety Study, which attempted to predict the absolute probabilities and consequences of all possible accidents in light-water reactors, is a prominent example. Though its complex methodology is widely imitated, it is unlikely that a more elaborate risk assessment will be done for any energy technology. As an illustration of the pitfalls of the genre of analysis, the limited independent reviews of the report to date have virtually demolished its findings. See, e.g., "Reactor Safety Study (Ramussen Report): Oversight Hearings on Nuclear Energy—Safeguards in the Domestic Nuclear Industry Before the Subcomm. on Energy and the Environment of the House Comm. on Interior and Insular Affairs", 94th Cong., 2d Sess. (1976); S. Keeney, *supra* note 33; Lewis, "Report to the American Physical Society by the Study Group on Light-Water Reactor Safety", 47 *Revs. Mod. Phy.* (supp. 1); von Hippel, "Looking Back on the Rasmussen Report", *Bull. Atom. Scientists* February 1977, at 42; Yellin, "The Nuclear Regulatory Commission's Reactor Safety Study", 7 *Bell J. Econ.* 317 (1976); D. Ford, "A History of Nuclear Safety Assessment: From WASH-1400 Through the Reactor Safety Study" (1977) (paper prepared for Union of Concerned Scientists, Cambridge, Mass.), summarized in Shapley, "Reactor Safety: Independence of Rasmussen Study Doubted", 197 *Science* 29 (1977); H. Kendall, "Preliminary Review of the AEC Reactor Safety Study" (23 December 1974) (a synopsis of these and other published critiques has been prepared by the author but was deleted from this article because of space limitations). The report's failure to justify the claims of accuracy, completeness, and objectivity made for it or to provide scientific support for its much-quoted conclusions shows the importance of waiting for diverse, detailed, and adversarial review before drawing policy conclusions from such studies. Moreover, the critiques show that for problems as complex as reactor accidents, where physical phenomenology is not well understood, reliable actuarial failure rates are unavailable for most components, and biophysical data are highly uncertain, any calculations are inevitably uncertain by orders of magnitude and hence useless as a basis for policy decisions. Finally, because uncertainties due to ignorance rather than to stochastic effects multiply through the chain of calculations, slight and easily justifiable modifications of the input data can profoundly alter the results. For example, changing each input to a 20-step chain by an unnoticeable factor of two can change the result a millionfold. Whether or not such "nickel and diming" abuse occurs, its very possibility makes

the results useless for policy, as calculations that lend themselves so ideally to fraudulent manipulation which defies detection and disproof can be neither verifiable nor publicly credible.

73   S. Barrage, *supra*, note 31.

74   Id.

75   Id. at 107. See notes 155–159 *infra* and accompanying text.

76   A. Lovins, *supra* note 2, at 46–49, 117–21.

77   Id. at 21–22, 133–44.

78   To obtain refined uranium, relatively large amounts of uranium ore must be mined, crushed, and chemically processed. The residue after chemical extraction contains little uranium but most of the radium-226 (and its parent, thorium-230), the bulk of the original ore's radiological hazard. This residue, the "mill tailings," must therefore be contained indefinitely as a dilute but potentially hazardous waste. See note 54 *supra*.

79   A. Lovins, *supra* note 2, at 15–18, 25–60, 85–103, 117–36.

80   Adams, "London's Third Airport", 137 *Geographic J.* 468, 477 (1972). A common method of valuation is to subtract lifetime consumption from lifetime earnings. For many poor, unemployed, and retired people the resulting "value of life" is nonsensically negative.

81   Id. For an example of the level of abstraction that can be attained in such studies, see J. Linnerhooth, *A Critique of Recent Modelling Efforts to Determine the Value of Human Life* (1975).

82   Adams, "… and How Much for Your Grandmother?", 6 *Environment & Plan A619*, A621 (1974).

83   See Mishan, "Evaluation of Life and Limb: A Theoretical Approach", 79 *J. Political Econ* 687 (1971).

84   See note 82 *supra*.

85   Actual compensation is impracticable, requires nonexistent institutions, and may boomerang by confirming people's suspicions that there is plenty of risk and cost for which there should be compensation.

86   This is often done through attitude surveys of people potentially affected. They are asked their buying price for benefits and their selling price for costs. Those who do not care to cooperate are often ignored.

87   J. Adams, *supra* note 13, at 9.

88   Id. at 2.

89   Id. at 9–10.

90   See note 19 *supra* and accompanying text.

91   Roskill Commission, *supra* note 8.

92   Tribe, *supra* note 44, at 629–30. See also Tribe, "Policy Science: Analysis or Ideology?", 2 *Philosophy & Pub. Aff.* 84 (1972).

93   Tribe, *supra* note 44, at 629–30.

94   I. Hoos, *supra* note 47, at 6.

95   See text accompanying notes 87–91 *supra*.

96   Brooks, *supra* note 31, at 71.

97   See A. Lovins & J. Price, *supra* note 46, at 33–34, 87–88; Carter, "Radioactive Wastes: Some Urgent Unfinished Business", 195 *Science* 661 (1977); Rochlin, "Nuclear Waste Disposal: Two Social Criteria", 195 *Science* 23 (1977). The basic problem is that geophysics cannot yet predict the behavior of any place on Earth over the periods required. Retrievable disposal (in case something goes wrong)

implies surveillance to see if anything goes wrong and to prevent unauthorized retrieval; nonretrievable disposal bets that nothing will go wrong, and relies mainly on the assumed (but disputed) insolubility and immobility of solidified wastes in case something does. The geological criteria for such disposal abound with surreal paradoxes. For example, if we assume that geological disposal is all right in a suitable site, the way we find out whether the site is suitable is to drill it full of holes, which make it unsuitable. We then try to plug up (grout) the holes again, but we can never be certain we succeeded, because we do not really know what is happening down there—which is why we drilled the holes in the first place.

98   Astonishingly, current ignorance of the pathways and effects of many isotopes (especially the higher actinides) is so great that there is no generally acceptable numerical index of hazard.

99   Lovins, *supra* note 19, at 93. A natural result is distrust and "technological fundamentalism." See E. Gray, *The Story of a Woman: Or, Whatever Happened to My Faith in Technology?* (1976).

100   See note 78 *supra*.

101   For a bibliography, see A. Lovins, *supra* note 2, at 213–14 nn.120–21. In a letter dated 7 July 1977 to Senator Dale Bumpers, W. D. Nordhaus of the Council of Economic Advisers notes that the gross benefits of the US fast breeder program are "approximately zero."

102   See note 11, *supra*.

103   Georgescu-Roegen, "Energy and Economic Myths", 41 *S. Econ. J.* 347 (1975) (future generations do not set to bid on the resources).

104   See H. Daly, *supra* note 11.

105   See Lovins, *supra* notes 2, 19.

106   For the origins of this misconception in a remarkable process of "mutual intoxication" among the AEC, vendors, and most utilities, see I. Bupp & J. Derian, *Light Water* (1977).

107   See I. Bupp, "Trends in Light Water Reactor Capital Costs in the United States: Causes and Consequences" (18 December 1974), summarized in *Tech. Rev.*, February 1975, at 15. See also I. Bupp and R. Treitel, "The Economics of Nuclear Power: De Omnibus Dubitandum" (1976) (typescript).

108   See, e.g., J. Harding, *The Deflation of Rancho Seco Two* (1975) (Friends of the Earth, Inc.).

109   For example, Harding (at the California Energy Resources Conservation & Development Commission, Sacramento) has shown that complex logistical, economic, and reactor physics problems can make attainable fuel burnup in light-water reactors dependent on capacity factor, contrary to the classical literature.

110   See A. Lovins, *supra* note 2, at 66–72, 117–44.

111   Id. at 85–103.

112   See, e.g., C. Komanoff, *Power Plant Performance* (1976); Olmsted, "Nineteenth Steam Station Cost Survey", *Electrical World*. 15 November 1975, at 43, 51.

113   For example, the Wisconsin Public Utility Commission finds that three 400-MW(e) coal-fired stations should be equivalent, as reliable sources of bulk power, to two 900-MW(e) nuclear stations. "Division of System Planning, Environmental Research & Consumer Analysis, Assessment of 1976 Advance Plans for Future Electric Power Facilities" (1977). Likewise, Edward Kahn finds that six 350-MW(e) coal-fired stations should be equivalent, on the same

assumption of equal reliability, to two 1200-MW(e) nuclear stations plus one 800-MW(e) coal station. Testimony of Edward Kahn before the Maryland Public Utility Commission and the New Jersey Public Service Commission (1977). These capacity ratios, respectively 0.68 and 0.70, arise because most of the larger number of smaller stations will probably be working at any one time, and the failure of one or two is not as embarrassing to the grid as the failure of one giant station would be. Less reserve margin is therefore needed.

114 For example, regulations requiring solidification of plutonium and of high-level waste at the proposed Barnwell Reprocessing Plant in South Carolina roughly doubled its cost to about $1 billion (some 14 times the original estimate) before the project was abandoned for lack of a federal bailout.

115 W. Kellogg, *Effects of Human Activities on Global Climate* (1976); S. Schneider, *The Genesis Strategy* (1976); Kellogg, "Global Influences of Mankind on the Climate", in J. Gribbin, *Climate Change* (1977). See A. Lovins, *supra* note 2, at 27–28 n.5.

116 The Rasmussen Report did not do this properly, See note 72 *supra*.

117 The real world is full of singularities, but few economic analysts are willing to admit that the previous one might not be the last and to plan accordingly. Thus, in the £100 million-plus Dinorwic pumped storage project in North Wales, nothing could shake the Central Electricity Generating Board's (CEGB) faith that a sensitivity analysis valid only for fuel cost variations of tens of percent to 2010 was adequate. To the surprise only of the CEGB, oil prices trebled a few months later. It seems fair to say now—as it did in 1973—that virtually every major economic assumption behind the project was grossly invalid before the ink was dry; most were not even included in the sketchy sensitivity analysis. Lovins, "Things that Go Pump in the Night", 58 *New Scientist* 564 (1973).

118 See note 115 *supra*.

119 M. Flood & R. Grove-White, *Nuclear Prospects: A Comment on the Individual, the State, and Nuclear Power* (1976); Ayres, "Policing Plutonium: The Civil Liberties Fallout", 10 *Harv. Cr.-Cl. L. Rev.* 369 (1975); J. Barton, *Intensified Nuclear Safeguards and Civil Liberties* (1975); P. Sieghart, *The Times* (London), 31 March 1977, at 15, col. 5 (letter to the editor).

120 See R. Lifton, *Death in Life* (1967).

121 People who ought to be unaffected by such a release may not believe this; people who are in fact affected may have trouble proving it; the two classes may be indistinguishable. The tort law doctrine of *res ipsa loquitur* would not apply for most cancers, so false claims could not be disproven. One possible approach is to establish a rebuttable presumption that observed cancers are caused by radiation.

122 Crutzen, "Upper Limits on Atmospheric Ozone Reductions Following Increased Application of Fixed Nitrogen to the Soil", 3 *Geophysical Research Letters* 169 (1976); McElroy, Elkins, Wofsy & Yung, "Sources and Sinks for Atmospheric N_2O", 14 *Revs. Geophysics Space & Physics* 143 (1976); McElroy, 277 *Philosophical Trans. Royal Soc. London* B159 (1977).

123 A. Lovins, *supra* note 2, at 171–218.

124 W. Häfele, *Hypotheticality and the New Challenges: The Pathfinder Role of Nuclear Energy* (1973).

125 A. Lovins, *supra* note 2, at 63–66. Of course, simulation modeling is still a valuable technical design tool.

126 P. Auer, A. Manne &, O. Yu, *Nuclear Power, Coal, and Energy Conservation* (1976).

127 Green, *supra* note 50, at 800 n.32.

128 See *Liggett Co. vs. Lee*, 288 U.S. 517, 567 (1933) (Brandeis, J., dissenting); A. Lovins, *supra* note 2 at 54–57, 85–103, 147–59.

129 W. Patterson, *The Fissile Society: Energy, Electricity, and the Nuclear Option* (1977) (Earth Resources Research Ltd., London).

130 See A. Lovins, *supra* note 2, at 59–60, which argues that soft and hard energy paths are logistically competitive, culturally incompatible, and institutionally antagonistic—so it does not much matter that they need not be technically incompatible.

131 Address by Henderson, *supra* note 25, at 27–28.

132 See comments of Lovins, in S. Barrager, *supra* note 31, at 108–09.

133 Tribe, *supra* note 44, at 625.

134 Kneese, "The Faustian Bargain", 44 *Resources* 1 (1973). See also Boulding, "The Ethics of Rational Decision", 12 *Management Sci.* 161 (1966); Kneese notes some further technical difficulties with using market costs and quasi-market benefits as surrogates for social costs and benefits. The former requires "a long string of assumptions, such as that all markets are in competitive equilibrium and there is no unemployment of resources in the system." The latter, even worse, "presents extremely difficult practical problems of measurement and requires some very 'heroic' assumptions about distributional impacts"—because outputs may be public goods which suffer from jointness of supply and have different marginal utility for different people. Thus allocative and distributional questions cannot be separated in the formalism, so distributional problems—essentially the noncongruence of economic efficiency and justice—invalidate the exercise. In the face of such theoretical barriers, bold (but tacit) assumptions reflecting the analyst's values are almost inevitable. A, Kneese, *Economics and Environment* 20–21, 180, 228–31 (1977).

135 See Kneese, *supra* note 134.

136 As there is not much of a market in Norman churches, the Commissioners used the value of the fire insurance.

137 See note 8 *supra*; Adams, *infra* note 141.

138 Green, *supra* note 50, at 798 n.26.

139 Tribe, *supra* note 44, at 627.

140 Id. Some people are refreshingly ready to do this. Henderson, who as a British Treasury official helped introduce cost–benefit analysis to Whitehall two decades ago, states that expert opinion "seems at present to regard the risk of damage [by Concorde to the ozone layer] as negligible. If, however, it seemed in the least likely that Concorde operations might affect the ozone layer, I would personally feel that the aircraft should be grounded forthwith, and without the formality of cost–benefit assessments using probability distributions of possible outcomes and with hypothetical values attributed to the incidence of skin cancer." Address by Henderson, *supra* note 25, at 22. Cf. Green, *supra* note 50, at 800 n.24.

141 Adams, "The National Health", 9 *Environment & Plan* 23, 27–29 (1977).

142 Adams, "You're Never Alone With Schizophrenia", 4 *Indus. Marketing Management* 441, 444 (1972).

143 See notes 155–161 *infra* and accompanying text.

144 I. Hoos, *supra* note 47, at 131; L. Winner, *supra* note 16, at 213.

145 Lipson, "Technical Issues and the Adversary Process", 194 *Science* 890 (1976). See also Green, note 160, *infra*.

146 Adams, *supra* note 142, at 444; Adams, *supra* note 141, at 32.

147 This is particularly true of the statutory assessments found in environmental impact statements. It is almost axiomatic, for example, that environmental impact statements are "written to justify decisions already made, rather than to provide a mechanism for critical review." Council on Environmental Quality, *Ann. Rep.* (1971).

148 I. Hoos, *supra* note 47, at 14. Cf. F. Baldwin, *The I Ching and Policy Analysis: A Proposal for Developing Oracular Capability in Public Administration* (1975); G. Brewer, *An Analyst's View of the Uses and Abuses of Modelling for Decisionmaking* (1975); B. Fischoff, *Cost-Benefit Analysis and the Art of Motorcycle Maintenance* (1976).

149 I. Hoos, *supra* note 47, at 132–33; see I. Hoos, *supra* note 5, at 8–9.

150 See S. Barrager, *supra* note 31, at 74 (discussing the Brown's Ferry fire); M. Grupp, *The NRC's Reactor Safety Study and the Case of the Brown's Ferry Fire* (1977).

151 F. Niehaus, *The Cost-Effectiveness of Remote Nuclear Reactor Siting* (1976). Considering only the biological effects of potential radiation exposure and the extra costs of transmission lines, this report concludes that "remote siting of nuclear power facilities would not seem to be a cost-effective way to control potential radiation exposures." Anyone who tried on this basis to site reactors in cities would be in for a shock.

152 See, e.g., Adams, "Westminster: The Fourth London Airport?", 1970 *Area* 1. Adams merely adds Westminster to the list of Roskill sites and straight-facedly turns the crank. Travel time saved to the year 2000 is worth, on the Roskill criteria, at least a discounted £9 billion. The costs—buying five square miles of central London, soundproofing and paying compensation for noise, moving Westminster Abbey, paying for accidents—would be several billion pounds less. Since publication of Adams' paper, not much has been heard of Roskill.

153 See, e.g., letter from Professor Keith Miller to Mr Stan Fabic, 7 May 1976, in *Oversight Hearings*, note 72 *supra*, at 202.

154 Many of the best American energy modelers have actually held a private symposium to discuss this concern and to try to protect their reputations from the fallout of incompetent or deliberately slanted studies.

155 See comments of Lovins, in S. Barrager, *supra* note 31, at 108.

156 See note 152 *supra*.

157 S. Barrager, *supra* note 31.

158 Quoted by Lovins, in S. Barrager, *supra* note 31, at 105; the first of the three quotations was deleted from the SRI final draft report.

159 Green, *supra* note 50, at 806–07.

160 Address by Green, "A Skeptical View of the Proposed Science Court," 8–9 (May 1976).

161 This is partly because many scientists think that the key issues are all or mainly technical and that technical issues are separable from social and value issues.

162 Perhaps the best example is the persistence even into 1977 of "monkey trials" over how schools should teach biological evolution.

163 Green, *supra* note 50, at 805.

164   Id.
165   Id. at 806.
166   Id.
167   Address by Green, *supra* note 160, at 8.
168   Tribe, *supra* note 44, at 633 n.54.
169   See text accompanying note 93, *supra*.
170   No country in the world appears to have a reliable database showing the types of energy required for end-use tasks and the unit scale and degree of geographical clustering of those tasks. See A. Lovins, *supra* note 2, at 73–85.
171   Green, *supra* note 50, at 807.
172   Address by Green, *supra* note 160, at 10.
173   Id.
174   Keynes, "Professor Tinbergen's Method", 49 *Econ. J.* 559 (1939).
175   B. Kennard, *We Are All Scientists Now: The New Case for Public Participation* (1976).

Chapter 8

Energy End-Use Efficiency

2005

Editor's note: *This piece was commissioned by Dr Steven Chu for the InterAcademy Council in Amsterdam, a consortium of about 90 National Academies of Science, as part of its 2005–2006 study "Transitions to Sustainable Energy Systems."*

1. Importance

Increasing energy end-use efficiency—technologically providing more desired service per unit of delivered energy consumed—is generally the largest, least expensive, most benign, most quickly deployable, least visible, least understood, and most neglected way to provide energy services. The 46 percent drop in US energy intensity (primary energy consumption per dollar of real GDP) during 1975–2005 represented by 2005 an effective energy "source" 2.1 times as big as US oil consumption, 3.4 times net oil imports, 6 times domestic oil output or net oil imports from OPEC countries, and 13 times net imports from Persian Gulf countries. US energy intensity has lately fallen by ~2.5 percent per year, apparently due much more to improved efficiency than to changes in behavior or in the mix of goods and services provided, and outpacing the growth of any fossil or nuclear source. Yet energy efficiency has gained little attention or respect. Indeed, since official statistics focus ~99 percent on physical energy supply, only the fifth of the 1996–2005 increase in US energy services that came from supply was visible to investors and policymakers; the four-fifths saved was not. The last time this incomplete picture led to strongly supply-boosting national policies, in the early 1980s, it caused a train wreck within a few years—glutted markets, crashed prices, bankrupt suppliers—because the market had meanwhile invisibly produced a gusher of efficiency. Savings were deployed faster than the big, slow, lumpy supply expansions, whose forecast revenues disappeared. Today we have two fast competitors: efficiency and micropower.

Physical scientists find that despite energy efficiency's leading role in providing new energy services today, it has barely begun to tap its profitable potential. In contrast, many engineers tend to be limited by adherence to past practice, and most economists by their assumption that any profitable savings must already have occurred. The potential of energy efficiency is also increasing faster through innovative designs, technologies, policies, and marketing methods than it is being used up through gradual implementation. The uncaptured "efficiency resource" is becoming bigger and cheaper even faster than have oil reserves lately through stunning advances in exploration and production. The expansion of this efficiency resource is also accelerating, as designers realize that whole-system design integration (Part 4) can often make very large (one or two orders of magnitude) energy savings cost less than small or no savings, and as energy-saving technologies evolve discontinuously rather than incrementally. Moreover, similarly rapid evolution and enormous potential apply also to marketing and delivering energy-saving technologies and designs; R&D can accelerate both.

A. Terminology

"Efficiency" means different things to the two professions most engaged in achieving it. To engineers, "efficiency" means a physical output/input ratio. To economists, "efficiency" means a monetary output/input ratio (though for practical purposes many use a monetary output/physical input ratio)—and also, confusingly, "efficiency" may refer to the economic optimality of a market transaction or process. This paper uses only physical output/input ratios, but the common use of monetary ratios confuses policymakers accustomed to economic jargon.

Wringing more work from energy via smarter technologies is often, and sometimes deliberately, confused with a pejorative usage of the ambiguous term "energy conservation." Energy *efficiency* means doing more (and often better) with less—the opposite of simply doing less or worse or without. This confusion unfortunately makes the honorable and traditional term "energy conservation" no longer useful in certain societies, notably the United States, and underlies much of their decades-long neglect or suppression of energy efficiency.

However, deliberately reducing the amount or quality of *energy services* remains a legitimate, though completely separate, option for those who prefer it or are forced by emergency to accept it. The 2000–2001 California electricity crisis ended abruptly when customers, exhorted to curtail their use of electricity, cut their peak load per dollar of weather-adjusted real GDP by 14 percent in the first half of 2001. Most of that dramatic reduction, undoing the previous 5–10 years' demand growth, was temporary and behavioral, but later became permanent and technological.

Even absent crises, some people do not consider an ever-growing volume of energy services to be a worthy end in itself, but seek to live more simply— with elegant frugality rather than involuntary penury—and to meet nonmaterial needs by nonmaterial means. (Trying to do otherwise is ultimately futile.) Such choices can save even more energy than technical improvement alone, though they are often considered beyond the scope of energy efficiency.

Several other terminological distinctions are also important:

- At least five different kinds of energy efficiency can be measured in at least five different stages of energy conversion chains (Part 1B), but this paper is only on *end-use* efficiency.
- Technical improvements in energy efficiency can be applied to new buildings and equipment, or installable in existing ones ("retrofitted"), or addable during minor or routine maintenance ("slipstreamed"), or conveniently added when making major renovations or expansions for other reasons ("piggybacked").
- *Efficiency* saves energy whenever an energy service is being delivered, whereas *load management* (sometimes called *demand response* to emphasize reliance on customer choice) only changes the *time when* that energy is used—either by shifting the timing of the service delivery or by, for example, storing heat or coolth so energy consumption and service delivery can occur at different times. In the context chiefly of electricity, *demand-side management*, a term coined by the [US] Electric Power Research Institute, comprises both these options, plus others that may even increase the use of electricity. Most efficiency options yield comparable or greater savings in peak loads; both kinds of savings are valuable, and both kinds of value should be counted. They also have important but seldom-recognized linkages: for example, because most US peak electric loads are met by extremely inefficient simple-cycle gas-fired combustion turbines, saving 1 percent of US electricity, including peak hours, reduces total natural-gas consumption by 2 percent and cuts its price by 3–4 percent (Lovins, Datta, *et al.*, 2004).
- Conflating three different things—*technological* improvements in energy efficiency (such as thermal insulation), *behavioral* changes (such as resetting thermostats), and the *price* or *policy tools* used to induce or reward those changes—causes endless confusion.
- The theoretical potential for efficiency gains (up to the maximum permitted by the laws of physics) exceeds the technical potential, which exceeds the economic potential based on social value, which exceeds the economic potential based on private internal value, which exceeds the actual uptake not blocked by market failures, which exceeds what happens spontaneously if no effort is made to accelerate efficiency gains deliberately; yet these six quantities are often not clearly distinguished.
- Energy statistics are traditionally organized by the economic sector of apparent consumption, not by the physical end uses provided or services sought. End uses were first seriously analyzed in 1976, rarely appear in official statistics even three decades later, and can be hard to estimate accurately. But end-use analysis can be valuable, because matching energy

supplies in quality and scale, as well as in quantity, to end-use needs can save much energy and money. Supplying energy of superfluous quality, not just quantity, for the task is wasteful and expensive. For example, the US now provides about twice as much electricity as the fraction of end-uses that economically justify this special, costly, high-quality form of energy— yet during 1975–2000, 45 percent of the total growth in primary energy consumption came from increased conversion and grid losses in the expanding, very costly, and heavily subsidized electricity system. Much of the electric growth, in turn, provided low-temperature heat—a physically and economically wasteful use of electricity, an extremely high-quality and costly carrier.

Many subtleties of defining and measuring energy efficiency merit but seldom get rigorous treatment, such as:

*   distribution losses downstream of end-use devices (an efficient furnace feeding leaky ducts or poorly distributing the heated air yields costlier delivered comfort);
*   undesired or useless services, such as leaving equipment on all the time (as many factories do) even when it serves no useful purpose;
*   misused services, such as space-conditioning rooms that are open to the outdoors;
*   conflicting services, such as heating and cooling the same space simultaneously (wasteful even if both services are provided efficiently);
*   parasitic loads, as when the inefficiencies of a central cooling system reappear as additional fed-back cooling loads that make the system less efficient than the sum of its parts;
*   misplaced efficiency, such as doing with energy-using equipment, however efficiently, a task that doesn't need the equipment—such as cooling with a mechanical chiller when groundwater or ambient conditions can more cheaply do the same thing; and
*   incorrect metrics, such as measuring lighting by raw quantity (lux) unadjusted for its visual effectiveness (Equivalent Sphere Illuminance), which may actually *decrease* if greater illuminance is improperly delivered, causing veiling reflections and uncomfortable glare.

To forestall a few other semantic quibbles:

*   Physicists (including the author) know that energy is not "consumed," as the economists' term "consumption" implies, nor "lost," as engineers refer to unwanted conversions into less useful forms. Energy is only converted from one form to another; yet the common metaphors are clear, common, and adopted here. Thus an 80 percent-efficient motor converts its electricity input into 80 percent torque and 20 percent heat, noise, vibration, and stray electromagnetic fields; the total equals 100 percent of the electricity

input, or roughly 30 percent of the fuel input at a classical thermal power station. (Note that this definition of efficiency combines engineering metrics with human preference. The motor's efficiency may change, with no change in the motor, if changing *intention* alters which of the outputs are desired and which are unwanted: the definition of "waste" is as much social or contextual as physical. An incandescent floodlamp used to keep plates of food warm in a restaurant may be effective for that purpose even though it is an inefficient source of visible light.)

- More productive use of energy is not, strictly speaking, a physical "source" of energy but is only a way to displace physical sources. This distinction is rhetorical, since the displacement or substitution is real and makes supply fully fungible with efficiency.

- Energy/GDP ratios are a very rough, aggregated, and sometimes misleading metric, because they combine changes in technical efficiency, human behavior, and the composition of GDP (a metric that problematically conflates goods and services with bads and nuisances, counts only monetized activities, and is an increasingly perverse measure of well-being). Yet the 46 percent drop in US energy intensity and the 52 percent drop in oil intensity during 1975–2004 reflect mainly better technical efficiency. Joseph Romm has also shown that an important compositional shift of US GDP—the information economy emerging in the late 1990s—has significantly decreased energy and probably electrical energy intensity, as bytes substituted for (or increased the capacity utilization of) travel, freight transport, lit and conditioned floorspace, paper, and other energy-intensive goods and services.

The aim here is not to get mired in word games, but to offer a clear overview of what kinds of energy efficiency are available, what they can do, and how best to consider and adopt them.

B. Efficiency along energy conversion chains

The technical efficiency of using energy is the product of efficiencies successively applied along the chain of energy conversions: the conversion efficiency of primary into secondary energy, times the distribution efficiency of delivering that secondary energy from the point of conversion to the point of end use, times the end-use efficiency of converting the delivered secondary energy into such desired energy services as hot showers and cold beer. Some analysts add another term at the upstream end—the extractive efficiency of converting fuel in the ground, wind or sun in the atmosphere, etc. into the primary energy fed into the initial conversion device—and another term at the downstream end—the hedonic efficiency of converting delivered energy services into human welfare (delivering junk mail with high technical efficiency is futile if the recipients didn't want it).

Counting all five efficiencies permits comparing ultimate means—primary energy tapped—with ultimate ends—happiness or economic welfare created. Focusing only on intermediate means and ends loses sight of what human purposes an energy system is to serve. Most societies pay attention to only three kinds of energy efficiency: extraction (because of its cost, not because the extracted fuels are assigned any intrinsic or depletion value), conversion, and perhaps distribution. End-use and hedonic efficiency are left to customers, are least exploited, and hence hold the biggest potential gains.

They also offer the greatest potential leverage. Since successive efficiencies along the conversion chain all multiply, they are often assumed to be equally important. Yet downstream savings—those nearest the customer—are the most important. Figure 8.1 shows schematically the successive energy conversions and losses that require about ten units of fuel to be fed into a conventional thermal power station in order to deliver one unit of flow in a pipe. But conversely, every unit of flow (or friction) saved in the pipe will save approximately *ten* units of fuel, cost, pollution, and "global weirding" at the power station. It will also make the pump's motor (for example) nearly two and a half units smaller, hence cheaper. To save the most primary energy and the most capital cost, therefore, efficiency efforts should start all the way downstream (Part 4B), by asking: How little flow can actually deliver the desired service? How small can the piping friction become? How small, well-matched to the flow regime, and efficient can the pump be made? Its coupling? Its motor? Its controls and electrical supplies?

Analyses of energy use should, but seldom do, start with the desired services or changes in well-being, then work back upstream to primary supplies. This maximizes the extra value of downstream efficiency gains and

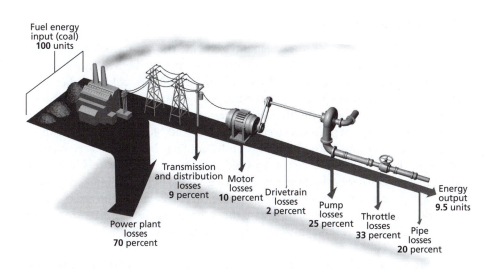

Figure 8.1 *Energy conversions in a conventional thermal power plant*

the capital-cost savings from smaller, simpler, cheaper upstream equipment. Unfortunately, most energy policy analysts analyze how much energy could be supplied before asking how much is optimally needed and at what quality and scale it could be optimally provided. This backwards direction (upstream to downstream) and supply orientation lie at the root of many if not most energy policy problems.

Even modest improvements in efficiency at each step of the conversion chain can multiply to large collective values. For example, suppose that during 2000–2050, world population and economic growth increased economic activity by 6–8 times, in line with conventional projections. But meanwhile, the carbon intensity of primary fuel, following a two-century trend, is likely to fall by at least 2–4 times as coal gives way to gas, renewables, and carbon offsets or sequestration. Conversion efficiency is likely to increase by at least 1.5 times with modernized, better-run, combined-cycle, and cogenerating power stations. Distribution efficiency should improve modestly. End-use efficiency could improve by 4–6 times if the intensity reductions sustained by many industrial countries when they were paying attention were sustained for 50 years (e.g., the US decreased its primary energy/GDP intensity at an average rate of 3.4 percent per year during 1979–1986 and 3.0 percent per year during 1996–2001). And the least-understood term, hedonic efficiency, might remain constant or might perhaps double as better business models and customer choice systematically improve the quality of services delivered and their match to what customers want. On these plausible assumptions, global carbon emissions from burning fossil fuel could decrease by 1.5–12 times despite the assumed 6–8 times grosser World Product. The most important assumption is sustained success with end-use efficiency, but the decarbonization and conversion-efficiency terms also appear able to take up some slack if needed.

C. Service redefinition

Some major opportunities to save energy redefine the service being provided. This is often a cultural variable. A Japanese person, asked why the house isn't heated in winter, might reply, "Why should I? Is the house cold?" In Japanese culture, the traditional goal is to keep the *person* comfortable, not to heat or cool empty space. Thus a modern Japanese room air conditioner may contain a sensor array and swiveling louvers that detect and blow air toward people's locations in the room, rather than wastefully cooling the entire space. Western office workers, too, can save energy (and can often see better, feel less tired, and improve aesthetics) by properly adjusting Venetian blinds, bouncing glare-free daylight up onto the ceiling, and turning off the lights. As Jørgen Nørgård remarks, "Energy-efficient lamps save the most energy when they are turned off;" yet many Westerners automatically turn on every light when entering a room.

This example also illustrates that energy efficiency may be hard to distinguish from energy supply that comes from natural energy flows. All houses are already ~98 percent solar-heated, because if there were no Sun (which provides

99.8 percent of the Earth's heat), the temperature of the Earth's surface would average approximately −272.6°C rather than +15°C. Thus, strictly speaking, engineered heating systems provide only the last 1–2 percent of the total heating required.

Service redefinition becomes complex in personal transport. Its efficiency is not just about vehicular fuel economy, people per car, or public transport alternatives. Rather, the underlying service should often be defined as access, not mobility. Typically the best way to gain access to a place is to be there already, so one needn't go somewhere else. This is the realm of spatial planning—no novelty in the US, where it's officially shunned yet practiced (zoning laws mandate *dispersion* of location and function, real-estate practices segregate housing by income, and other market distortions maximize unneeded and often unwanted travel). Obviously sprawl would decrease if not mandated and subsidized. Another way to gain access is virtually—moving just the electrons while leaving the heavy nuclei behind—via telecommunications, soon including realistic "virtual presence." Sometimes that's a realistic alternative to physically moving flesh. And if such movement is really necessary, it merits real competition, at honest prices, between all modes: personal or collective, motorized or human-powered, conventional or innovative. Creative policy tools can enhance that choice in ways that enhance property value, saved time, quality of life, and public amenity and security. Efficient cars can be an important part of efficient personal mobility and, importantly, are vehicles for emotions as well as for bodies, but also reducing the need to drive can save even more energy and yield greater total benefit.

D. Historic summaries of potential

People have been saving energy for centuries, even millennia; this is the essence of engineering. Most savings were initially in conversion and end use: pre-industrial households often used more primary energy than modern ones do, because fuelwood-to-charcoal conversion, inefficient open fires, and crude stoves burned much fuel to deliver sparse cooking and warmth. Lighting, materials processing, and transport end uses were also very inefficient. Billions of human beings still suffer such primitive conditions today. Developing countries' primary energy/GDP intensities *average* ~3 times those of industrialized countries. Corrected for purchasing power, China's energy intensity is ~3 times that of the US, ~5 times of the EU, and ~9 times of Japan. Fast-growing economies like China's have the greatest *need* and the greatest *opportunity* to leapfrog to efficiency. But even the most energy-efficient societies still have enormous and expanding room for further efficiency gains. Less than one-fourth of the energy delivered to a typical European cookstove ends up in food; less than 1 percent of the fuel delivered to a standard car actually moves the driver; US power plants discard waste heat equivalent to 1.2 times Japan's total energy use; and even Japan's economy doesn't approach one-tenth the efficiency that the laws of physics permit. Nor is energy efficiency the end of the story: for example, not only are Chinese shaft kilns an extremely energy-wasteful way to make cement, but the cement is of such poor and uncertain quality that many times more of it must be used to make each m^3 of concrete with a certain strength, so the energy leverage of a modern cement plant is these terms' *product*—and the carbon leverage is then multiplied by switching to any no- or low-carbon fuel.

Detailed and exhaustively documented engineering analyses of the scope for improving energy efficiency, especially in end-use devices, have been published for many industrial and some developing countries. By the early 1980s, those analyses had compellingly shown that most of the energy currently used was being wasted—i.e. that the same or better services could be provided using several times less primary energy by fully installing, wherever practical and profitable, the most efficient conversion and end-use technologies then available. Such impressive efficiency gains cost considerably less than the long-run, and often even the short-run, marginal private internal cost of supplying more energy. Most policymakers ignore both these analyses, well known to specialists, and the less-well-known findings showing even bigger and cheaper savings from whole-system design integration (Part 4). Despite much higher EU energy consciousness, policymakers still greatly underestimate efficiency's potential, while in the US, national policymakers in the past 20 years have forgotten more than they learned in the previous 20 years, leaving efficiency to be advanced instead by private-sector, local, and state choices.

Many published engineering analyses show a smaller saving potential because of major conservatisms, often deliberate (because the real figures seem too good to be true), or because they assume only partial adoption over a short period rather than examining the ultimate potential for complete practical adoption. For example, the American Council for an Energy-Efficient Economy estimates that just reasonable adoption of the top five conventional US opportunities—industrial improvements, 4.88 L/100 km light-vehicle standards, cogeneration, better building codes, and a 30 percent better central-air-conditioning standard—could save 530 million T/yr of oil-equivalent—respectively equivalent to the total 2000 primary energy use of Australia, Mexico, Spain, Austria, and Ireland. But the full long-term efficiency potential is far larger, much of it in many small terms. Saving energy is like eating an Atlantic lobster: there are big, obvious chunks of meat in the tail and the front claws, but there's also a similar total quantity of tasty morsels hidden in crevices and requiring some skill and persistence to extract.

The whole-lobster potential is best, though still not fully, seen in bottom–up technological analyses comparing the quantity of potential energy savings with their marginal cost. That cost is typically calculated using the Lawrence Berkeley National Laboratory methodology, which divides the marginal cost of buying, installing, and maintaining the more efficient device by its discounted stream of lifetime energy savings. The levelized cost in dollars of saving, say, 1 kWh then equals $Ci/S[1-(1+i)^{-n})]$, where C is installed capital cost (\$), i is annual real discount rate (assumed here to be 0.05), S is energy saved by the device (kWh/yr), and n is operating life (yr). Thus a \$10 device that saved 100 kWh/yr and lasted 20 yrs would have a levelized "cost of saved energy" (CSE) of 0.8¢/kWh. Against a 5¢/kWh electricity price, a 20-yr device with a 1-yr simple payback would have CSE = 0.4¢/kWh. It is then conventional for engineering-oriented analysts to represent efficiency "resources" as a supply curve, rather than as shifts along a demand curve (the convention among economists). CSE is methodologically equivalent to the cost of supplied energy (e.g. from a power station and grid): the price of the energy saved is *not* part of the calculation. Whether the saving is cost-effective depends on comparing the cost of achieving it with the avoided cost of the energy saved. (As Part 2 notes, this conventional engineering-economic approach materially understates the benefits of energy efficiency.)

On this basis, the author's analyses in the late 1980s found, from measured cost and performance data for more than 1000 electricity-saving end-use technologies, that their full practical retrofit could save about three-fourths of US electricity at an average CSE ~0.9¢/kWh (2004 $)—roughly consistent with a 1990 Electric Power Research Institute analysis, whose differences were mainly methodological rather than substantive. So many key technologies, now in Asian mass production, are now far cheaper yet more effective that today's potential is even larger and cheaper. (The analyses explicitly excluded the small financing and transaction costs. A huge literature accurately predicts and rigorously measures the empirical size, speed, and cost of efficiency improvements delivered by actual utility and government programs.)

Such findings are broadly consistent with equally or more detailed ones by European analysts: for example, that late-1980s technologies could save three-fourths of Danish buildings' electricity or half of all Swedish electricity at $0.024/kWh (2004 $), or four-fifths of German home electricity (including minor fuel-switching) with a ~40 percent per year after-tax return on investment. Such findings, with ever greater sophistication, have been published worldwide since 1979, but are generally rejected by non-technological economic theorists who argue that if such cost-effective opportunities existed, they'd already have been captured in the marketplace, even in planned economies with no marketplace or mixed economies with a distorted one. This mental model—"Don't bother to bend over and pick up that banknote lying on the ground, because if it were real, someone would have picked it up already"—often dominates government policy. It seems ever less defensible as more is learned about the reality of pervasive market failures (Part 5) and the astonishing size and cheapness of the energy savings empirically achieved by diverse enterprises (Part 3). But by now, the debate has become theological (Part 3)—about whether existing markets are essentially perfect, as most economic modelers assume for comfort and convenience, or whether market failures are at least as important as market function and lie at the heart of business and policy opportunity. This seems a testable empirical question.

It may soon be tested in the transport sector. The author's team's uncontested analysis in the Pentagon-cosponsored independent study *Winning the Oil Endgame*, published in September 2004, found that 52 percent of the officially forecast US 2025 oil use could be saved (half by then, half later as vehicle stocks turn over), at an average cost of just $12/bbl (2000 $). The remaining oil use could be displaced by saved natural gas and advanced biofuels at an average cost of $18/bbl. Thus, by the 2040s, the US could use *no* oil and revitalize its economy, led by business for profit, and encouraged by public policies not requiring mandates, fuel taxes, subsidies, or new national laws. Rather, the transition would be driven by competitive strategy in the car, truck, plane, and oil industries, plus military needs. These findings surprised many yet, a year later, remain unchallenged. Early sectoral progress with implementation has been encouraging.

E. Discontinuous technological progress

This engineering/economics divergence about the potential to save energy also reflects a tacit assumption that technological evolution is smooth and incremental, as mathematical modelers prefer. In fact, while much progress is as incremental as technology diffusion, discontinuous technological leaps, more like "punctuated equilibrium" in evolutionary biology, can propel innovation and speed its adoption, as perhaps with Hypercar® vehicles (Part 4A).

Technological discontinuities can even burst the traditional boundaries of possibility by redefining the design space:

- Generations of engineers learned that big supercritical-steam power plants were as efficient as possible—40-odd percent from fuel in to electricity out. But through sloppy learning or teaching, they'd overlooked the possibility of stacking two Carnot cycles atop each other. Such combined-cycle (gas-then-steam) turbines, based on mass-produced jet engines, can exceed 60 percent efficiency and are cheaper and faster to build, so in the 1990s, they quickly displaced the big steam plants. Fuel cells, the next innovation, avoid Carnot limits altogether by being an electrochemical device, not a heat engine. Combining both may soon achieve 80–90 percent fuel-to-electric efficiency. Even inefficient distributed generators can already exceed 90 percent system efficiency by artfully using recaptured heat.
- Pumps, fans, and other turbomachinery (the main uses of electricity) seemed a mature art until a novel biomimetic rotor, using laminar vortex flow instead of turbulent flow, proved substantially more efficient (www.paxscientific.com).
- The canonical theoretical efficiency limit for converting sunlight into electricity using single-layer photovoltaic (PV) cells is normally stated as 33–50+ percent using multicolor stacked layers, vs. lab values around 24 percent and 39 percent (lower in volume production). That's because semiconductor bandgaps were believed too big to capture any but the high-energy wavelengths of sunlight. But those standard data are wrong. A Russian-based team suspected in 2001, and Lawrence Berkeley National Laboratory proved in 2002, that indium nitride's bandgap is only 0.7 eV,

matching near-infrared (1.77 μm) light and hence able to harvest almost the whole solar spectrum. This may raise the theoretical limit to 50 percent for two-layer and to ~70 percent for many-layer thin-film PVs. Perhaps optical-dimension lithographed antenna arrays or quantum-dot PVs can do even better.

Caution is likewise vital when interpreting Second Law efficiency (the ratio of the least available work that could have done the job to the actual available work used to do the job). In the macroscopic world, the laws of thermodynamics are normally considered ineluctable—but the definition of the desired change of state can be finessed. Ernie Robertson notes that when turning limestone into a structural material, one is not confined to the conventional possibilities of cutting it into blocks or calcining it at ~1250°C into Portland cement. One can instead grind it up and feed it to chickens, whose ambient-temperature technology turns it into eggshell stronger than Portland cement. Were we as smart as chickens, we would have mastered this life-friendly technology. Extraordinary new opportunities to harness 3.8 billion years of biological design experience, as described by Janine Benyus in *Biomimicry*, can often make heat-beat-and-treat industrial processes unnecessary. So, in principle, can the emerging techniques of nanotechnology using molecular-scale self-assembly, as pioneered by Eric Drexler.

More conventional innovations can also bypass energy-intensive industrial processes. Making artifacts that last longer, use materials more frugally, and are designed and deployed to be repaired, reused, remanufactured, and recycled can save much or most of the energy traditionally needed to produce and assemble their materials (and can increase welfare while reducing GDP, which perversely swells when ephemeral goods are quickly discarded and replaced). Microfluidics can even shrink a big chemical plant to the size of a watermelon: millimeter-scale flow in channels etched into silicon wafers can control time, temperature, pressure, stoichiometry, and catalysis so exactly that a narrow product spectrum is produced, without the side-reactions that normally need most of the chemical plant to separate undesired from desired products.

Such "end-run" solutions—rather like the previous example of substituting sensible land use for better vehicles, or, better still, combining both—can greatly expand the range of possibilities beyond simply improving the narrowly defined efficiency of industrial equipment, processes, and controls. By combining many such options, it is now realistic to contemplate a long-run advanced industrial society that provides unprecedented levels of material prosperity with far less energy, cost, and impact than today's best practice. Part 4A, drawn from Paul Hawken *et al.*'s synthesis in *Natural Capitalism* and Ernst von Weizsäcker *et al.*'s earlier *Factor Four*, further illustrates recent breakthroughs in integrative design that can make very large energy savings cost less than small ones; and Part 6 summarizes similarly important discontinuities in policy innovation.

In light of all these possibilities, why does energy efficiency, in most countries and at most times, command so little attention and such lackadaisical pursuit? Several explanations come to mind. Saved energy is invisible. Energy-saving technologies may look and outwardly act just like inefficient ones, so they're invisible too. They're also highly dispersed—unlike central supply technologies that are among the most impressive human creations, inspiring pride and attracting ribbon-cutters and rent- and bribe-seekers. Many users believe energy efficiency is binary—you either have it or lack it—and that they already did it in the 1970s, so they can't do it again. Energy efficiency has relatively weak and scattered constituencies. And major energy efficiency opportunities are disdained or disbelieved by policymakers indoctrinated in a theoretical economic paradigm that claims big untapped opportunities simply cannot exist (Part 3).

2. Benefits

Energy efficiency avoids the direct economic costs and the direct environmental, security, and other costs of the energy supply and delivery that it displaces. Yet most literature neglects several key side-benefits (economists call them "joint products") of saving energy.

A. Indirect benefits from qualitatively superior services

Improved energy efficiency, especially end-use efficiency, often delivers better services. Efficient houses are more comfortable; efficient lighting systems can look better and help you see better; efficient motors can be more quiet, reliable, and controllable; efficient refrigerators can keep food fresher for longer; efficient cleanrooms can improve the yield, flexibility, throughput, and setup time of microchip fabrication plants; aerodynamically efficient chemical fume hoods can improve safety; airtight houses with constant controlled ventilation (typically through heat exchangers to recover warmth or coolth) have more healthful air than leaky houses that are ventilated only when wind or some other forcing function fortuitously blows air through cracks; efficient supermarkets can improve food safety and merchandising; retail sales pressure can rise 40 percent in well-daylit stores; students' test scores imply ~20–26 percent faster learning in well-daylit schools. Such side-benefits can be one or even two more orders of magnitude more valuable than the energy directly saved. For example, careful measurements show that in efficient buildings—where workers can see what they're doing, hear themselves think, breathe cleaner air, and feel more comfortable—labor productivity typically rises by about 6–16 percent. Since office workers in industrialized countries cost ~100 times more than office energy, a 1 percent increase in labor productivity has the same bottom-line effect as *eliminating* the energy bill—and the actual gain in labor productivity is ~6–16 times bigger than that. Practitioners can market these attributes without ever mentioning lower energy bills.

B. Leverage in global fuel markets

Much has been written about the increasing pricing power of major oil-exporting countries, especially Saudi Arabia, with its important swing production capacity. Yet the market power of the United States—the Saudi Arabia of energy waste—is even greater on the demand side. The US can save oil faster than OPEC can conveniently sell less oil. This was illustrated during 1977–1985, when US GDP rose 27 percent while total US oil use fell 17%, oil imports fell 50 percent, and imports from the Persian Gulf fell 87 percent. OPEC's exports fell 48 percent (one-fourth of this fall was due to US action), breaking its pricing power for a decade. The most important single cause of the US 5.2 percent per year gain in oil productivity was more efficient cars, each driving 1 percent fewer km on 20 percent less gasoline—a 7-mile/USgal gain in six years for new American-made cars—and 96 percent of those savings came from smarter design, only 4 percent from smaller size.

C. Buying time

Energy efficiency buys time. Time is the more precious asset in energy policy, because it permits the fuller and more graceful development and deployment of still better techniques for energy efficiency and supply. This pushes supply curves down towards the lower right (larger quantities at lower prices), postpones economic depletion, and buys even more time. The more time is available, the more information will emerge to support wiser and more robust choices, and the more fruitfully new technologies and policy options can meld and breed new ones. Conversely, hasty choices driven by supply exigencies almost always turn out badly, waste resources, and foreclose important options. Of course, once bought, time should be used wisely. Instead, the decade of respite bought by the US efficiency spurt of 1977–1985 was almost entirely wasted as attention waned, efficiency and alternative-supply efforts stalled, R&D teams were disbanded, and political problems festered. We all pay today the heavy price of that stagnation.

D. Integrating efficiency with supply

To first order, energy efficiency makes supply cheaper. But second-order effects reinforce this first-order benefit, most obviously when efficiency is combined with onsite renewable supplies, making them nonlinearly smaller, simpler, and cheaper. For example:

- A hot-water-saving house can achieve a very high solar-water-heat fraction (e.g. 99 percent in the author's home high in the Rocky Mountains) with only a small collector, so it needs little or no backup, partly because collector efficiency increases as stratified-tank storage temperature decreases.
- An electricity-saving house (the author's saves ~90 percent, using only ~110–120 average W for 372 m^2) needs only a few m^2 of PVs and a simple

balance-of-system (storage, inverter, etc.). This can cost less than connecting to the grid a few meters away.

- A passive-solar, daylit building needs little electricity, and can pay for even costly forms of onsite generation (such as PVs) by eliminating or downsizing mechanical systems.
- Such mutually reinforcing options can be bundled: e.g. 1.18 peak MW of photovoltaics retrofitted onto the Santa Rita Jail in Alameda County, California, was combined with efficiency and load management, so at peak periods when the power was most valuable, less was used by the jail and more sold back to the grid. This bundling yielded an internal rate of return over 10 percent including state subsidies, and a present-valued customer benefit/cost ratio of 1.7 without or 3.8 with those subsidies.

E. Gaps in engineering economics

Both engineers and economists conventionally calculate the cost of supplying or saving energy using a rough-and-ready toolkit called "engineering economics." Its methods are easy-to-use but flawed, ignoring such basic tenets of financial economics as risk-adjusted discount rates. Indeed, engineering economics omits 207 economic and engineering considerations that together increase the value of decentralized electrical resources by typically an order of magnitude. Many of these "distributed benefits," compiled in the author's team's *Small Is Profitable*, apply as much to end-use efficiency as to decentralized generation. Most of the literature on the cost of energy alternatives is based solely on accounting costs and engineering economics that greatly understate efficiency's value. Properly counting its benefits will yield far sounder investments.

End-use efficiency is also the most effective way to make energy supply systems more resilient against mishap or malice, because it increases the duration of buffer stocks, buying time to mend damage or arrange new supplies, and it increases the share of service that curtailed or improvised supplies can deliver. Efficiency's high "bounce per buck" makes it the cornerstone of any energy system design for secure service provision in a dangerous world.

3. Engineering vs. Economic Perspectives

Engineering practitioners and economic theorists view energy efficiency through profoundly different lenses, yet both disciplines are hard pressed to explain such phenomena as:

- During 1996–2001, US aggregate energy intensity fell at a near-record pace despite record-low and falling energy prices. (It fell faster only once in modern history, during the record-high and rising energy prices of 1979–1985.) Apparently something other than price was getting Americans' attention.
- During 1990–1996, when a kWh of electricity cost only half as much in Seattle as in Chicago, people in Seattle, on average, reduced their peak

electric load 12 times as fast, and their use of electricity ~3640 times as fast, as did people in Chicago—probably because the utility encouraged efficiency in Seattle but discouraged it in Chicago.

- In the 1990s, DuPont found that its European chemical plants were no more energy efficient than its corresponding US plants, despite long having paid twice the energy price—probably because all plants were designed by the same people in the same ways with the same equipment, and there's little room for behavior change in a chemical plant.

- In Dow Chemical Company's Louisiana Division during 1981–1993, nearly 1000 projects to save energy and reduce waste added $110 million per year to the bottom line and yielded returns on investment averaging over 200 percent per year—yet in the latter years, both the returns and the savings were trending upwards as the engineers discovered new tricks faster than they used up the old ones. (Economic theory would deny the possibility of so much "low-hanging fruit" that has fallen down and is mushing up around the ankles: such enormous returns, if real, would long ago have been captured. This belief was the main obstacle to engineers' seeking such savings, then after their discovery, persisting to save even more.)

- Only about 25–35 percent of apartment-dwellers, when told that their air conditioner and electricity are free, behave as economists would predict—turning on the air conditioner when they feel hot and setting the thermostat at a temperature at which they feel comfortable. The rest show no correlation between air-conditioning usage and comfort; instead, their cooling behavior is determined by at least six other variables: household schedules, folk theories about air conditioners (such as that the thermostat is a valve that makes the cold come out faster), general strategies for dealing with machines, complex belief systems about health and physiology, noise aversion, and wanting white noise to mask outside sounds that might wake the baby. Energy anthropology reveals that both the economic *and* the engineering models of air-conditioning behavior are not just incomplete but seriously misleading. People are complex—not mere cost–benefit calculating machines.

- By 1990, the United States had misallocated $1 trillion of investments to ~200 million refrigerative tons of air conditioners, and ~200 peak GW (two-fifths of total peak load) of power supply to run them, that would not have been bought if the buildings had been optimally designed to produce best comfort at least cost. This seems explicable by the perfectly perverse incentives seen by each of the 20-odd actors in the commercial-real-estate value chain, each systematically rewarded for inefficiency and penalized for efficiency. Each of these market failures is both a potential showstopper and a business opportunity.

- Not just individuals but also most firms, even large and sophisticated ones, routinely fail to make essentially riskless efficiency investments yielding many times their normal business returns: most require energy efficiency investments to yield ~6 times their marginal cost of capital, which typically applies to far riskier investments. This too is a huge business opportunity that smart firms are starting to exploit.

Many economists would posit some unknown error or omission in these descriptions, not in their theories. Indeed, energy engineers and energy economists seem not to agree about what is a hypothesis and what is a fact. Engineers take their facts from tools of physical observation. Three decades' conversations with noted energy economists suggest to the author that most think facts come only from observed flows of dollars, interpreted through indisputable theoretical constructs, and hence consider any contrary physical observations aberrant.

This divergence makes most energy economists suppose that buying energy efficiency faster than the "spontaneous" rate of observed intensity reduction (for 1997–2001, 2.7%/yr in the US, 1.4 percent per year EU, 1.3 percent per year world, and 5.3 percent per year China) would require considerably higher energy prices, because if greater savings were profitable at prevailing prices, they'd already have been bought; thus engineers' bottom–up analyses of potential energy savings must be unrealistically high. Economists' estimates of potential savings at current prices are "top–down" and very small, based on historic price elasticities that confine potential interventions to changing prices and savings to modest size and diminishing returns (otherwise the economists' simulation models would inconveniently explode). Engineers retort that high energy prices aren't necessary for very large energy savings (because they're so lucrative even at historically low prices, as at Dow Louisiana) but aren't sufficient either (because higher prices do little without enlarged ability to respond to them, as in the Seattle vs. Chicago example).

Of course, engineering-based practitioners agree that human behavior is influenced by price, as well as by convenience, familiarity, fashion, transparency, competing claims on attention, and many other marketing and social-science factors—missing from any purely technological perspective but central to day-to-day fieldwork. The main difference is that practitioners think these obstacles are "market failures" and dominate behavior in buying energy efficiency. Most economists deny this, and say efficiency's relatively slow adoption must be due to gross errors in the engineers' claims of how large, cheap, and available its potential really is.

This theological deadlock underlies the debate about climate protection. Robert Repetto and Duncan Austin showed in 1997 that all mainstream climate-economics models' outputs are hard-wired to their input assumptions, and that realistic inputs, conforming to the actual content of the Kyoto Protocol and its rules, show that climate protection *increases* GDP. Florentin Krause has shown that the main official US government analyses, taken together, concur. Yet the official White House position in 2005 is still that climate protection, even if desirable, cannot be mandated because it is too costly. By the time this view changes, private-sector and state- and local-government initiatives will have surpassed US Kyoto obligations, and may have already.

In fact, climate protection is not costly but profitable; its critics may have the amount about right, but they got the *sign* wrong. Before debating whether the cost of climate protection is a big or a small number, one must understand that it's a *negative* number. The clearest proof is in the behavior and achievements of the smart companies that are behaving *as if* the US had ratified the Kyoto Protocol, because energy efficiency costs less than the energy it saves. For example, such large firms as DuPont, IBM, and STMicroelectronics (the world's #7 chipmaker) have lately been raising their energy productivity by 6 percent per year with simple paybacks of a few years. DuPont expected by 2010 to cut its greenhouse gas emissions by 65 percent below the 1990 level; STM to zero (despite making 40 times more chips). DuPont has in fact saved 72 percent so far; by 2004 it had raised output 30 percent, cut energy use 7 percent, and saved over $2 billion net. Similarly, BP announced that its intended 10 percent carbon reduction by 2010 had been achieved eight years early at zero net cost; actually, the 10-year net-present-valued *saving* was $650 million.

Such examples abound (www.cool-companies.org, www.pewclimate.org). These famous companies, and more all the time, are hardly naïve or deluded. (In 2005, GE pledged to boost its energy efficiency 30 percent by 2012 to increase shareholder value.) Anyone ignoring this market reality is mistaking the econometric rear-view mirror for a windshield. Econometrics measures how human populations behaved under past conditions that no longer exist and that it is often a goal of energy policy to change. Where price is the only important explanatory variable, econometrics can be a useful tool for extrapolating history into how price may influence near-term, small, incremental changes in behavior. But limiting our horizons to this cramped view of technical possibilities and human complexity rules out innovations in policies, institutions, preferences, and technologies—treating the future like fate, not choice, and thus making it so.

4. Diminishing vs. Expanding Returns to Investments in Energy Efficiency

Among the most basic, most often skipped over, yet most simply resolved economic/engineering disagreements is whether investing in end-use efficiency yields expanding or diminishing returns. Economic theory says diminishing—the more efficiency we buy, the more steeply the marginal cost of the next increment of savings rises, until it becomes too expensive (Figure 8.2). But engineering practice often says expanding—big savings can cost *less* than small or no savings (Figure 8.3)—if the engineering is done unconventionally but properly.

A. Empirical examples

Consider, for example, how much thermal insulation should surround a house in a cold climate. Conventional design specifies just the amount of insulation

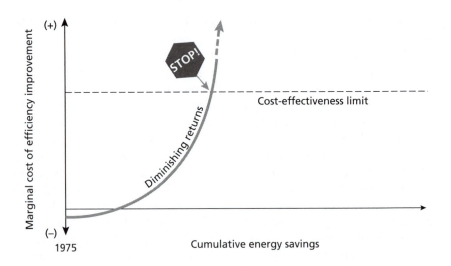

Figure 8.2 *Diminishing returns – greater energy savings incur greater marginal cost – can be true for some (not all) components, but need not be true for most systems*

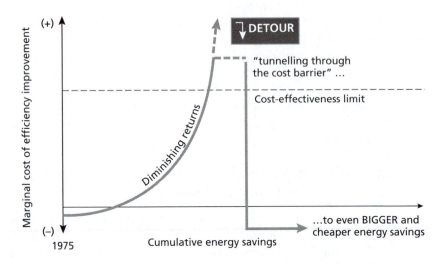

Figure 8.3 *Optimizing whole systems for multiple benefits, rather than isolated components for single benefits, can often "tunnel through the cost barrier" directly to the lower-right destination, making very large energy savings cost less than small or no savings; this has been empirically tested in a wide range of technical systems*

that will repay its marginal cost out of the present value of the saved marginal energy. But this is methodologically wrong, because the comparison omits the *capital cost of the heating system*—furnace, ducts, fans, pipes, pumps, wires, controls, and fuel source. The author's house illustrates that in outdoor temperatures down to −44°C, it is feasible to grow bananas (28 crops so far) at 2200 m elevation in the Rocky Mountains with no heating system, yet with *reduced* construction cost, because the superwindows, superinsulation, air-to-air heat exchangers, and other investments needed to eliminate the heating system cost less to install than the heating system would have cost to install. The resulting ~99 percent reduction in heating *energy* cost is an extra benefit, and it's better than free.

Similarly, Pacific Gas and Electric Company's Advanced Customer Technology Test for Maximum Energy Efficiency ("ACT^2") demonstrated in seven new and old buildings in the 1990s that the "supply curve" of energy efficiency generally bent downwards, as shown schematically in Figure 8.3. For example, an ordinary-looking new tract house was designed to save 82 percent of the energy allowed by the strictest US standard of the time (1992 California Title 24), yet PG&E estimated that if widely practiced, not a single experiment, it would cost ~$1800 less than normal to build and ~$1600 less in present value to maintain. It provided comfort with no cooling system in a climate that can reach 45°C; a similar house later did the same in a 46°C-peak climate. And the 350-m^2 Bangkok house of architect Professor Suntoorn Boonyatikarn provided normal comfort, with 16 percent of normal air-conditioning capacity, at no extra construction cost.

These examples illustrate how optimizing a house as a system rather than optimizing a component in isolation, and optimizing for lifecycle cost (capital plus operating cost, and preferably maintenance cost too), can make a super-efficient house cheaper to build, not just to run, by eliminating costly heating and cooling systems. Similarly, a retrofit design for a 19,000-m^2 curtainwall office building near Chicago found 75 percent energy-saving potential at no more cost than the normal 20-year renovation that saves nothing, because the $200,000 capital saving from making the cooling system 4× smaller (yet 4× more efficient), rather than renovating the big old system, would pay for the other improvements.

In a striking industrial example, a heat-transfer pumping loop originally designed to use 70.8 kW of pumping power was redesigned by Dutch engineer Jan Schilham (at Interface Nederland bv in Scherpenzeel) to use 5.3 kW—92 percent less[1]—with lower capital cost and better performance, using no new technologies but only two changes in the design mentality:

1 Use big pipes and small pumps rather than small pipes and big pumps. The friction in a pipe falls as nearly the fifth power (roughly −4.84) of its diameter. Engineers normally make the pipe just fat enough to repay its greater cost from the saved pumping energy. This calculation improperly omits the capital cost of the pumping equipment—the pump, motor,

inverter, and electricals that must overcome the pipe friction. Yet the size and roughly the cost of that equipment will fall as nearly the fifth power of pipe diameter, while the cost of the fatter pipe will rise as only about the second power of diameter. Thus conventionally optimizing the pipe as an isolated component actually pessimizes the system! Optimizing the whole system together will clearly yield fat pipes and tiny pumps, so total capital cost falls slightly and operating cost falls dramatically.

2 Lay out the pipes first, then the equipment. Normal practice is the opposite, so the connected equipment is typically far apart, obstructed by other objects, at the wrong height, and facing the wrong way. The resulting pipes have ~3–6× as much friction as they'd have with a straight shot—to the delight of the pipefitters, who are paid by the hour, mark up a profit on the extra pipes and fittings, and don't pay for the extra electricity or bigger pumping equipment. But the owner would do better with fat, short, straight pipes than thin, long, crooked pipes.

Together, these two design changes cut the measured pumping power by 12 times, with lower capital cost and better performance. They also saved 70 kW of heat loss with a two-month payback, because it's easier to insulate short, straight pipes. In hindsight, however, the design was still suboptimal, because it omitted seven additional benefits: less space, weight, and noise; better maintenance access; lower maintenance cost; higher reliability and uptime; and longer life (because the removed pipe elbows won't be eroded by fluid turning the corner). Properly counting these seven benefits too would have saved not 92 percent but nearer 98 percent of the energy and cost even less, so about a factor-four potential saving was left uncaptured.

Other recent design examples include a 97 percent reduction in air-conditioning energy for a California office retrofit, with attractive returns and better comfort; lighting retrofit savings upwards of 90 percent with better visibility and a one- to two-year payback; an energy cost reduction >40 percent with a three-year payback in retrofitting an already very efficient oil refinery; ~75 percent electrical savings in a new chemical plant, with ~10 percent lower construction time and cost; ~89 percent in a new data center at lower cost; ~70–90 percent in a new supermarket at probably lower cost; and 20 percent of energy and 33 percent of water in a new chip fab with 30 percent lower construction cost (the next fab, using two big options not tested quite in time for the first one, will save far more energy, yet cost even less to build).

The obvious lesson is that *optimizing whole systems for multiple benefits, not just components for single benefits, often boosts end-use efficiency by roughly an order of magnitude at negative marginal cost.* These enormous savings were not previously much noticed or captured, because of deficient engineering pedagogy and practice. (The author hopes to collect such examples into a *10×E—Factor Ten Engineering—*casebook to drive pedagogic reform.) Whole-system design integration isn't rocket science, but rediscovers the forgotten tradition of Victorian system engineering, before designers got specialized and forgot how components fit together.

It's not even generally true, as economic theory supposes, that greater end-use efficiency costs more at the level of components. For example, the most common type of industrial motor on the 1996 US market (1800-rpm TEFC, NEMA Design B) exhibited no empirical correlation whatever between efficiency and price up to at least 225 kW. (Premium-efficiency motors should cost more to build because they contain more and better copper and iron, but they're not priced that way.) The same is true for most industrial pumps and rooftop chillers, Swedish refrigerators, American televisions, and many other products. But even if it were not true, artfully combining components into systems can definitely yield expanding returns.

Perhaps the most consequential example is in light vehicles. A small private company completed in 2000 the manufacturable, production-costed virtual design (Lovins & Cramer, 2004) of a midsize-sport-utility-vehicle concept car that is uncompromised, cost-competitive, zero-emission, and quintupled-efficiency. It is so efficient (2.1 L/100 km [114 mile/USgal]) not just because its direct-hydrogen fuel cell is about twice as efficient as a gasoline-fueled Otto engine, but also because it is so lightweight (but crashworthy) and so low in aerodynamic drag and rolling resistance that it can cruise at highway speed on no more power to the wheels than a conventional SUV uses on a hot day just for air conditioning. This design suggests that cars, too, can "tunnel through the cost barrier," achieving astonishing fuel economy at no extra cost and with many other customer and manufacturing advantages. With aggressive licensing of existing intellectual property, such vehicles could start ramping up production as early as 2010. All major automakers have had parallel development programs under way, totaling ~$10 billion of related commitments through 2000, since the basic concept was put into the public domain in 1993 to maximize competition. Further analysis in *Winning the Oil Endgame* showed that ultralighting with carbon-thermoplastic composites (backstopped by ultra-light steels) would nearly double the efficiency of today's gasoline-hybrid cars, yielding 3.6 L/100 km [67 mile/USgal] for such a midsize SUV or ~2.6 L [90 mile/USgal] for a typical car (all with projected US 2025 size, acceleration, and safety). Yet their production cost would be the same as for hybrid cars today, because the composites' extra cost is offset by simpler automaking and smaller powertrain. The 3.6 L/100 km SUV would thus pay back in a year at today's EU or in two years at today's US gasoline prices. Its safety would improve, because its halved curb weight is more than offset by carbon thermoplastic structures' ~12 times greater crash energy absorption per kg than steel. The needed breakthrough in low-cost advanced-composite manufacturing is now being commercialized (www.fiberforge.com).

A full US fleet of such light vehicles of various shapes and sizes would save about as much oil as Saudi Arabia produces (~9–12 Mbbl/d); a global fleet, as much oil as OPEC sells. Moreover, if fuel-cell-powered, such vehicles can be designed to plug into the grid when parked (as the average car is ~96 percent of the time) as a power station on wheels, selling back enough electricity and ancillary services to repay most of its cost. A US fleet of light vehicles doing this would have ~6–12 times as much electric generating capacity as all power companies now own. This is part of a wider strategy that combines hydrogen-ready vehicles with integrated deployment of fuel cells in stationary and mobile applications to make the transition to a climate-safe hydrogen economy profitable at each step starting now (beginning with ultrareliable combined-heat-and-power in buildings). The resulting displacement of power plants, oil-fueled vehicles, and fossil-fueled boilers and furnaces could *decrease* net consumption of natural gas, could save about $1 trillion of global vehicle refueling investment over the next 40 years (compared with gasoline-system investment), and could profitably displace up to two-thirds of CO_2 emissions. It could also raise the value of hydrocarbon reserves, whose hydrogen is typically worth more without than with the carbon. It is favored by many leading energy and car companies today. And it is not too far off: more than two-thirds of the fossil-fuel atoms burned in the world today are already hydrogen, and global hydrogen production (~50 MT/yr), if diverted from its present uses, could fuel an entire fleet of superefficient US highway vehicles. (So could windpower on available land just in the Dakotas, or reforming much of the half of US natural gas that *Winning the Oil Endgame* found could be saved for <$1/GJ.) The business case for hydrogen is robust if and only if vehicles are superefficient. But if they are, then the hydrogen tanks for a normal driving range (530 km for the concept SUV above) become 3 times smaller—small enough to package well with no new storage technology—and the fuel cells, also 3 times smaller, become affordable far earlier: at an 80 percent experience curve, ~32 times less cumulative production is needed for competitive cost.

B. The right steps in the right order

Such breakthrough results need not just the right technologies but also their application in the right sequence. For example, most practitioners designing lighting retrofits start with more efficient luminaires—improving optics, lamps, and ballasts. But for optimal energy and capital savings, that should be step six, not step one. First come improving the quality of the visual task, optimizing the geometry and cavity reflectance of the space, optimizing lighting quality and quantity, and harvesting daylight. Then, after the luminaire improvements, come better controls, maintenance, management, and training.

Likewise, to deliver thermal comfort in a hot climate, most engineers retrofit a more efficient and perhaps variable-speed chiller, variable-speed supply fans, etc. But these should all be step five. The previous four steps are to expand the comfort range (by exploiting such variables as radiant temperature, turbulent air movement, and ventilative chairs); reduce unwanted heat gains within or into the space; exploit passive cooling (ventilative, radiative, ground-coupling); and, if needed, harness nonrefrigerative alternative cooling (evaporative, desiccant, absorption, and their hybrids like Pennington cycles). These preliminary steps can generally eliminate refrigerative cooling. If it's nonetheless still wanted, it can be made superefficient (COP 6.8 for supply fan through cooling tower in Singapore), then supplemented by better controls and coolth storage. Yet most designers pursue these seven steps in reverse order, worst buys first, so they save less energy, pay higher capital costs, yet achieve worse comfort and greater complexity.

Whole-system engineering optimizes for many benefits: there are, for example, 10 benefits of superwindows, and 18 of premium-efficiency motors or dimmable electronic lighting ballasts, not just the 1 normally counted. (The arch that holds up the middle of the author's house has 12 different functions but is paid for only once.) Superwindows cost more per window, but typically make the whole building cost less because they downsize or eliminate space-conditioning equipment. (They combine spectrally selective coatings with krypton fill to insulate as well as ~8–12+ sheets of glass, and for hot climates, they can almost perfectly admit light without heat.) Similarly, 35 retrofits to a typical industrial motor system, properly counting multiple benefits, can typically save about half its energy (not counting the larger and cheaper savings that should first be captured further downstream, e.g. in pumps and pipes) with a 16-month simple payback against a 5¢/kWh tariff (as of 1986; it costs less today). The saving is so cheap because buying first the correct 7 improvements yields 28 more as free by-products. Such motor-system savings alone, if fully implemented, could save ~30 percent of the world's electricity.

Such design requires a diverse background, deep curiosity, often a transdisciplinary design team, and meticulous attention to detail. Whole-system design is not what any engineering school appears to be teaching, nor what most customers currently expect, request, reward, or receive. But it represents a key part of the "overhang" of practical, profitable, unbought energy efficiency that so far remains missing from virtually all official studies.

5. Market Failures and Business Opportunities

In a typical US office, using one-size-fatter wire to power overhead lights would pay for itself within 20 weeks. Why wasn't that done? Because:

1 The wire size was specified by the low-bid electrician, who was told to "meet code," and the wire-size table in the [US] National Electrical Code is meant to prevent fires, not to save money. Saving money by optimizing resistive losses takes wire about twice as fat.
2 The office owner or occupant will buy the electricity, but the electrician bought the wire. An electrician altruistic enough to buy fatter wire isn't the low bidder and won't win the job. Correcting this market failure needs attention both to the split incentive *and* to the misinterpretation of a life-safety regulation as an economic optimum.

This micro-example illustrates the range and depth of market failures that any skilled practitioner of energy efficiency encounters daily. A 1997 compendium, "Climate: Making Sense *and* Making Money," organizes 60–80 such market failures into eight categories—and illustrates the business opportunity each can be turned into. Some arise in public policy, some at the level of the firm, some in individuals' heads. Most are glaringly perverse. For example, in all but two of the United States, and in almost every other country, regulated distribution utilities are rewarded for selling more energy and penalized for cutting customers' bills, so they're unenthusiastic about energy efficiency that hurts their shareholders. Nearly all architects and engineers, too, are paid for what they spend, not for what they save; "performance-based fees" have been shown to yield superior design, but are rarely used. Most firms set discounted-cash-flow targets for their financial returns, yet tell their operating engineers to use a simple-payback screen for energy-saving investments (typically less than two years), and the disconnect between these two metrics causes and conceals huge misallocations of capital. And when markets and bidding systems are established to augment or replace traditional regulation of energy supply industries, negawatts (saved watts) are rarely allowed to compete against megawatts. Indeed, dominant industries discourage competing investments by spreading disinformation. For example, from widely accepted but false claims that only nuclear power is big and fast enough to do much about climate change, one would hardly guess that its low- and no-carbon decentralized competitors already exceed its global output and capacity, and are adding manyfold more MW and TWh/yr every year—an order of magnitude more if electrical savings are included too (Lovins, 2005a).

In short, scores of market failures—well understood but widely ignored—cause available and profitable energy efficiency to get only a small fraction of the investment it merits. Thus most of the capital invested in the energy system is being misallocated. (The US Energy Policy Act of 2005, repeating the errors of the early 1980s, greatly increases the diversion of private capital from market winners to market losers via hugely distorting new subsidies to the losers.) The most effective remedy would be to put systematic "barrier-busting"—turning obstacles into opportunities, stumbling-blocks into stepping-stones—atop the policy agenda, so market mechanisms could work properly, as economic theory correctly prescribes.

Using energy in a way that saves money is not only a perquisite of the rich; it is also arguably the most powerful key to global economic development for the poor. Making quintupled-efficiency compact fluorescent lamps in Bombay or superwindows in Bangkok takes about a thousand times less capital than expanding the supply of electricity to produce the same light and comfort via inefficient lamps and office/retail air conditioners. The efficiency investment is also repaid about ten times faster. The resulting ~10,000 times decrease in capital requirements could turn the power sector—which now uses about one-fourth of global development capital—into a net *exporter* of capital to fund other development needs. (That's why it's vital to track and label, then tax, restrict, or at least stigmatize the global trade in very inefficient electricity-using devices—an invisible trade whose perverse capital leverage probably causes as much human misery as drugs.) Likewise at the micro-scale of a rural village, a package of photovoltaics and superefficient end-use devices (lamps, pumps, mobile phone, water purification, vaccine refrigerator, GPS mapper for land titling, etc.), with normal utility financing and no subsidy, often incurs debt service lower than the villagers were already paying for lighting kerosene, candles, and radio batteries, so they have a positive cash flow from day one.

Conversely, when Chinese authorities imported many assembly lines to make refrigerators more accessible, the saturation of refrigerators in Beijing households rose from 2 percent to 62 percent in six years, but the refrigerators' inefficient design created unintended shortages of power and of capital to generate it (an extra half-billion dollars' worth). A State Council member said this error must not be repeated: energy and resource efficiency must be the cornerstone of the development process. Otherwise resource waste will require supply-side investment of the capital meant to buy the devices that were supposed to use those resources. This realization contributed to China's emphasis on energy efficiency (halving primary energy/GDP elasticity in the 1980s, then nearly re-halving it again), laying the groundwork for the dramatic 1996 shift from coal-centric policy toward gas, renewables, and efficiency. This important contribution to reducing global carbon emissions was a by-product of two other domestic goals—de-bottlenecking China's development and improving public health. In June 2004, the State Council approved a visionary energy policy framework based on strong efficiency and leapfrog technologies. Least-cost implementation in this spirit would mean investing an order of magnitude more in efficiency than supply—not, as historically, the reverse—and would greatly accelerate Chinese development and set a strong example for OECD countries.

6. Old and New Ways to Accelerate Energy Efficiency

A. Old but good methods

In the 1980s and 1990s, practitioners and policymakers greatly expanded their toolkits for implementing energy efficiency. During 1973–1986, the US doubled its new-car efficiency, and during 1977–1985, cut national oil intensity 5.2 percent per year. In 1983–1985, ten million people served by Southern California Edison Company were cutting its decade-ahead forecast of peak load by 8.5 percent per year, at ~1 percent the long-run marginal cost of supply. In 1990, New England Electric System signed up 90 percent of a pilot market for small-business retrofits in two months. In the same year, Pacific Gas and Electric Company marketers captured a fourth of the new-commercial-construction market for design improvements in three months, so in 1991, PG&E raised the target—and got it all in the first nine days of January.

Such impressive achievements resulted from nearly two decades' refinement of program structures and marketing methods. At first, utilities and governments wanting to help customers save energy offered general, then targeted, information, and sometimes loans or grants. Demonstration programs proved feasibility and streamlined delivery. Standards knocked the worst equipment off the market. (Congress did this for household appliances without a single dissenting vote, because so many appliances are bought by landlords, developers, or public housing authorities—a manifest split incentive with the householders who'll later pay the energy bills.) Just refrigerator standards cut new US units' electricity usage fourfold during 1975–2001—5 percent per year—saving 40 GW of electric supply. In Canada, labeling initially did nearly as well. Utilities began to offer rebates—targeted, then generic—to customers, then to other value-chain participants—for adopting energy-saving equipment, or scrapping inefficient equipment, or both. Some rebate structures proved potent, such as paying a shop assistant a bonus for selling an energy-efficient refrigerator but nothing for selling an inefficient one. So did leasing (20¢ per compact fluorescent lamp per month, so you pay for it over time just like a power plant—but the lamp is far cheaper), paying for better design, and rewarding buyers for beating minimum standards. Energy-saving companies, independent or utility-owned, provided turnkey design and installation to reduce hassle. Sweden aggregated technology procurement to offer "golden carrot" rewards to manufacturers bringing innovations to market; once these new products were introduced, competition quickly eliminated their modest price premia.

These engineered-service-delivery models worked well, often spectacularly well. Steve Nadel's 1990 review of 237 programs at 38 US utilities found many costing <1¢/kWh (1988 $). During 1991–1994, the entire demand-side-management portfolio of California's three major investor-owned utilities saved electricity at an average program cost that fell from about 2.8 to 1.9 current ¢/kWh (1.2¢ for the cheapest), saving society over $2 billion more than the effort cost.

B. New and better methods

Since the late 1980s, another model has been emerging that promises even better results: not just marketing negawatts (saved watts)—maximizing how many customers save and how much—but also making markets *in* negawatts—thus maximizing competition in who saves and how, so as to drive quantity and quality up and cost down. Competitive bidding processes let saved and produced energy compete fairly. Savings can be made fungible in time and space, transferred between customers, utilities, and jurisdictions, and procured by "bounty-hunters." Spot, futures, and options markets can be expanded from just megawatts to embrace negawatts too, permitting arbitrage between them. Property owners can commit never to use more than x MW, then trade those commitments in a secondary market that values reduced demand and reduced uncertainty of demand. Efficiency can be cross-marketed between electric and gas distributors, each rewarded for saving *either* form of energy. Revenue-neutral "feebates" for connecting new buildings to public energy supplies—fees for inefficiency, rebates for efficiency—can reward continuous improvement. Standardized measurement and reporting of energy savings lets them be aggregated and securitized like home mortgages, sold promptly into liquid secondary markets, and hence financed easily and cheaply (www.ipmvp.org). Efficiency techniques can be conveniently bundled and translated to "vernacular" forms—easily chosen, purchased, and installed. Novel real-estate value propositions emerge from integrating efficiency with onsite renewable supply (part of the revolutionary shift now under way to distributed resources) so as to eliminate all wires and pipes, the trenches carrying them, and the remote infrastructure they connect to. Performance-based design fees, targeted mass retrofits, greater purchasing aggregation, and systematic barrier-busting all show immense promise. And aggressively scrapping inefficient devices—paying bounties to destroy them instead of reselling them—could both solve many domestic problems (e.g. oil, air, and climate in the case of inefficient vehicles) and boost global development by reversing "negative technology transfer."

Winning the Oil Endgame offers a similarly novel policy menu for saving oil. Revenue- and size-neutral "feebates" for widening the price spread between more and less efficient light vehicles in each size class—thus arbitraging the discount-rate spread between car-buyers and society—are far more effective than fuel taxes or efficiency standards, and can yield both consumer and producer surpluses. Tripled-efficiency heavy trucks and planes can be elicited, respectively, by "demand pull" from big customers (once they're informed of what's possible, as began to occur in 2005) and by innovative financing for insolvent airlines (on condition of scrapping inefficient parked planes). The first 25 percent fuel saving for trucks and 20 percent for planes (in Boeing's 787 Dreamliner) is *free*; the rest of the tripling of efficiency has very high returns.

A key player may be the military, which needs superefficient platforms for agile deployment and to cut the huge cost and risk of fuel logistics. Speeding ultralight, ultrastrong materials fabrication processes to market could transform civilian vehicle industries as profoundly as military R&D did to create the internet, GPS, jet engines, and chips. Only this time, that transformation could lead countries like the US off oil, making oil no longer worth fighting over.

Altogether, the conventional agenda for promoting energy efficiency—prices and taxes, plus regulation or deregulation—ignores nearly all the most effective, attractive, trans-ideological, and quickly spreadable methods. And it ignores many of the new marketing "hooks" just starting to be exploited: security (national, community, and individual), economic development and balance of trade, avoiding price volatility and costly supply overshoot, profitable integration and bundling with renewables, and expressing individual values.

Consider, for example, a good compact fluorescent lamp. It emits the same light as an incandescent lamp but uses 4–5 times less electricity and lasts 8–13 times longer, saving tens of dollars more than it costs. It avoids putting a ton of carbon dioxide and other pollutants into the air. But it does far more. In suitable volume—about a billion are now made each year—it can cut by a fifth the evening peak load that causes blackouts in overloaded Mumbai, can boost poor American chicken farmers' profits by a fourth, or can raise destitute Haitian households' disposable cash income by up to a third. As mentioned above, making the lamp requires 99.97 percent less capital than does expanding the supply of electricity, thus freeing investment for other tasks. The lamp cuts power needs to levels that make solar-generated power affordable, so girls in rural huts can learn to read at night, advancing the role of women. One light bulb does all that. You can buy it at the supermarket and install it yourself. One light bulb at a time, we can make the world safer. "In short," concludes Jørgen Nørgård, by pursuing the entire efficiency potential systematically and comprehensively, "it is possible in the course of half a century to offer everybody on Earth a joyful and materially decent life with a per capita energy consumption of only a small fraction of today's consumption in the industrialized countries."

C. De-emphasizing traditionally narrow price-centric perspectives

These burgeoning opportunities suggest that price may well become less important to the uptake of energy efficiency. Price remains important and should be correct, but is only one of many ways to get attention and influence choice; ability to respond to price can be far more important. End-use efficiency may increasingly be marketed and bought mainly for its qualitatively improved services, just as distributed and renewable supply-side resources may be marketed and bought mainly for their distributed benefits. Outcomes would then become decreasingly predictable from economic experience or using economic tools.

Meanwhile, disruptive technologies and integrative design methods are clearly inducing dramatic shifts *of*, not just along, demand curves, and are even making them less relevant by driving customer choice through non-price variables. Ultralight-hybrid Hypercar® vehicles, for example, would do an end-run around two decades of trench warfare in the US Congress (raising efficiency standards vs. gasoline taxes). They would also defy the standard assumption that efficient cars must trade off other desirable attributes (size, performance, price, safety), requiring government intervention to induce customers to buy the compromised vehicles. If, as now seems incontrovertible, light vehicles can achieve 3–5 times fuel savings as a by-product of breakthrough design integration, yet remain uncompromised and competitively priced, then the energy-price-driven "tradeoff" paradigm becomes irrelevant. People will prefer such vehicles because they're *better*, not because they're clean and efficient, much as most people now buy digital media rather than vinyl phonograph records: they're a superior product that redefines market expectations. This implies a world where energy price and regulation become far less influential than today, displaced by imaginative, holistic, integrative engineering and marketing.

In the world of consumer electronics—ever better, faster, smaller, cheaper— that world is upon us. In the wider world of energy efficiency, the master key to so many of the world's most vexing problems, it is coming rapidly over the horizon. We need only understand it and do it. And as inventor Edwin Land said, "People who seem to have had a new idea have often only stopped having an old idea." To think truly "outside the barrel," in the apt phrase of Rijkman Groenink (Chairman of ABN Amro's supervisory board), we all have much to unlearn.

Acknowledgments

The author is grateful to many colleagues, especially Drs. David B. Brooks, Jørgen S. Nørgård, and Joel N. Swisher, PE for their insightful comments, and to The William and Flora Hewlett Foundation for supporting the preparation of an *Encyclopedia of Energy* article (Lovins, 2004) from which this paper was adapted and updated with IAC's consent.

Note

1   As reported by Schilham to the author in 1997. Later evidence suggested 84% savings. Schilham has retired and isn't available to verify his initial report, so RMI is seeking by other means to check which figure is correct. The difference is immaterial to the lessons and value of the design story.

References

American Institute of Physics (1975). *Efficient Use of Energy*, AIP Conf. Procs. #25. AIP, New York.

Benyus, J.M. (1997). *Biomimicry: Innovation Inspired by Nature*. William Morrow, New York.

Brohard, G.J., Brown, M.W., Cavanagh, R., Elberling, L.E., Hernandez, G.R., Lovins, A., and Rosenfeld, A.H. (1998). "Advanced Customer Technology Test for Maximum Energy Efficiency (ACT²) Project: The Final Report." *Procs. Summer Study on Energy-Efficient Buildings*. American Council for an Energy-Efficient Economy, Washington, DC.

Daly, H.E. (1996) *Beyond Growth—The Economics of Sustainable Development*, Beacon Press, Boston.

Drexler, K.E. (1992). *Nanosystems: Molecular Machinery, Manufacturing, and Computation*. John Wiley and Sons, New York.

E SOURCE (2002). *Technology Atlas* series (6 vols.) and *Electronic Encyclopedia* CD-ROM. E SOURCE, Boulder, Colorado, www.esource.com.

Fiberforge, Inc. (Glenwood Springs, Colorado), advanced-composites manufacturing process information at www.fiberforge.com.

Fickett, A.P., Gellings, C.W., and Lovins, A.B. (1990). "Efficient Use of Electricity." *Sci. Amer.* **263**(3):64–74 (September).

Hawken, P.G., Lovins, A.B. and L.H. (1999). *Natural Capitalism: Creating the Next Industrial Revolution*. Little Brown, New York, www.natcap.org.

IPSEP (1989–99). *Energy Policy in the Greenhouse*. Report to Dutch Ministry of Environment, International Project for Sustainable Energy Paths, El Cerrito CA 94530, www.ipsep.org.

Johansson, T.B., Bodlund, B., and Williams, R.H., (Eds.) (1989): *Electricity*. Lund University Press, Lund, Sweden, particularly Bodlund *et al.*'s chapter "The Challenge of Choices."

Koplow, D., "Energy Subsidy Links Pages," Earthtrack (Washington, DC), 2005, http://earthtrack.net/earthtrack/index.asp?page_id=177&catid=66>http://earthtrack.net/earthtrack/index.asp?page_id=177&catid=66.

Krause, F., Baer, P., and DeCanio, S. (2001). "Cutting Carbon Emissions at a Profit: Opportunities for the U.S." IPSEP, El Cerrito, California, www.ipsep.org/latestpubs.html.

Lovins, A.B. (1992). *Energy-Efficient Buildings: Institutional Barriers and Opportunities*. Strategic Issues Paper II. E SOURCE, Boulder, Colorado, www.esource.com.

Lovins, A.B. (1994). "Apples, Oranges, and Horned Toads." *Electricity J.* 7(4):29–49.

Lovins, A.B. (1995). "The Super-Efficient Passive Building Frontier." *ASHRAE J.* 37(6):79–81, June; RMI Publication #E95-28, Rocky Mountain Institute, Snowmass, Colorado, www.rmi.org.

Lovins, A.B. (2003). "Twenty Hydrogen Myths," www.rmi.org/rmi/Library/E03-05_TwentyHydrogenMyths.

Lovins, A.B. (2004). "Energy efficiency, taxonomic overview," *Encyc. of Energy* 2:382–401, Elsevier, www.rmi.org/rmi/Library/E04-02_EnergyEfficiencyTaxonomicOverview.

Lovins, A.B. (2005). "More Profit With Less Carbon," *Sci. Amer.* 293:74–82, Sept., www.rmi.org/rmi/Library/C05-05_MoreProfitLessCarbon.

Lovins, A.B. (2005a). "Nuclear economics and climate-protection potential," Rocky Mountain Institute, 11 Sept. 2005, www.rmi.org/rmi/Library/E06-14_NuclearPowerEconomicsClimateProtection.

Lovins, A.B. and Cramer, D.R. (2004). Hypercars®, Hydrogen, and the Automotive Transition. *Intl. J. Veh. Design*, 35(1/2):50–85, www.rmi.org/rmi/Library/T04-01_HypercarsHydrogenAutomotiveTransition.

Lovins, A.B., Datta, E.K., Feiler, T., Rábago, K.R., Swisher, J.N., Lehmann, A., and Wicker, K. (2002). *Small Is Profitable: The Hidden Economic Benefits of Making Electrical Resources the Right Size*. Rocky Mountain Institute, Snowmass, Colorado, www.smallisprofitable.org.

Lovins, A.B., Datta, E.K., Bustnes, O.-E., Koomey, J.G., and Glasgow, N. (2004). *Winning the Oil Endgame*, Rocky Mountain Institute, Snowmass, Colorado, www.oilendgame.com.

Lovins, A.B., and Gadgil, A. (1991). "The Negawatt Revolution: Electric Efficiency and Asian Development." RMI Publication #E91-23. Rocky Mountain Institute, Snowmass, Colorado, www.rmi.org/rmi/Library/E91-23_NegawattRevolutionElectricEfficiencyAsianDevelopment.

Lovins, A.B. and L.H. (1991). "Least-Cost Climatic Stabilization." *Annual Review of Energy and the Environment* 16:433–531.

Lovins, A.B. and L.H. (1996). "Negawatts: Twelve Transitions, Eight Improvements, and One Distraction." *Energy Policy* 24(4):331–344, www.rmi.org/rmi/Library/U96-11_NegawattsTwelveTransitions.

Lovins, A.B. and L.H. (1997). "Climate: Making Sense *and* Making Money." Rocky Mountain Institute, Snowmass, Colorado, www.rmi.org/rmi/Library/C97-13_ClimateSenseMoney.

Lovins, A.B. and L.H. (2001). "Fool's Gold in Alaska." *Foreign Affairs* 80(4):72–85, July/August; annotated at www.rmi.org/rmi/Library/E01-04_

FoolsGoldAlaskaAnnotated.

Lovins, A.B. and L.H., Krause, F., and Bach, W. (1982). *Least-Cost Energy: Solving the CO₂ Problem*. Brick House, Andover, Massachusetts. Reprinted 1989 by Rocky Mountain Institute, Snowmass, Colorado.

Lovins, A.B., and Williams, B.D. (1999). "A Strategy for the Hydrogen Transition." *Procs. 10th Ann. Hydrogen Mtg.*, 8 April, National Hydrogen Assn., Washington, DC, www.rmi.org/rmi/Library/T99-07_StrategyHydrogenTransition.

Nadel. S. (1990). *Lessons Learned*. Report #90-08. New York State Energy R&D Authority (Albany), with NY State Energy Office and Niagara Mohawk Power Corp. American Council for an Energy-Efficient Economy, Washington, DC.

Nørgård, Jørgen S. (2002). "Energy Efficiency and the Switch to Renewable Energy Sources. Natural Resource System Challenge II: Climate Change. Human Systems and Policy" (ed. A. Yotova), *UNESCO Encyclopedia of Life Support Systems*, EOLSS Publisher Co., Oxford, UK.

Reddy, A.K.N., Williams, R.H., and Johansson, T.B. (1997). *Energy After Rio: Prospects and Challenges*. United Nations Development Program, New York.

Repetto, R., and Austin, D. (1997). "The Costs of Climate Protection: A Guide to the Perplexed." World Resources Institute, Washington, DC, www.wri.org/wri/climate.

Romm, J.J., Rosenfeld, A.H., and Herrmann, S. (1999). "The Internet Economy and Global Warming: A Scenario of the Impact of E-Commerce on Energy and the Environment." Center for Energy and Climate Solutions, Washington, DC, www.cool-companies.org.

Romm, J.J., and Browning, W.D. (1994). "Greening the Building and the Bottom Line: Increasing Productivity Through Energy-Efficient Design." Rocky Mountain Institute, Snowmass, Colorado, www.rmi.org/rmi/Library/D94-27_GreeningBuildingBottomLine.

Swisher, J.N. (2002). "The New Business Climate." Rocky Mountain Institute, Snowmass, Colorado.

Swisher, J.N., Jannuzzi, G., and Redlinger, R. (1998). *Tools and Methods for Integrated Resource Planning: Improving Energy Efficiency and Protecting the Environment*, UNEP Collaborating Centre on Energy and Environment, Denmark, www.uccee.org/IRPManual/index.htm.

von Weizsäcker, E.U., Lovins, A.B. and L.H. (1995/1997). *Factor Four: Doubling Wealth, Halving Resource Use*. Earthscan, London.

Wilson, A., Uncapher, J., McManigal, L., Lovins, L.H., Cureton, M., and Browning, W. (1998). *Green Development: Integrating Ecology and Real Estate*. John Wiley & Sons, New York; Green Developments CD-ROM (2002), Rocky Mountain Institute, Snowmass, Colorado.

Section 3

Nuclear Power: Fission and Confusion

As noted earlier, around 1969, Amory met the famed environmentalist and founder of Friends of the Earth, David R. Brower (1912–2000), whom many consider the greatest conservationist of the 20th century. Brower had been thrown out of the Sierra Club in 1969 for urging the environmental group take a stand against the Diablo Canyon nuclear power plant in southern California (two of those opposing Brower were legendary photographer Ansel Adams and Richard Leonard, one of the fathers of modern American climbing). Brower then established Friends of the Earth in several countries, including England, where he met Amory and where he continued to fight nuclear power. Amory, a "student" of nuclear power since the 1960s—when in high school he won national awards from, among others, Westinghouse, General Electric, and the Chairman of the US Atomic Energy Commission (AEC), Nobel laureate Dr Glenn Seaborg—had been developing his own opinions. The thing that struck Amory was that nuclear energy was not only a bad investment that posed safety risks like its coal counterparts: nuclear energy had an especially malignant side-effect, the spread of ingredients for nuclear weapons. One of the most significant early Brower–Lovins collaborations centered on the Clinch River [fast] breeder reactor, for which Congress made initial appropriations in 1972. At the time, Brower was in Washington attending hearings on nuclear reactor safety. As Brower recalled, Amory

> had been leading a boys' [camp] group in the mountains of Maine, which he does every summer; he was on his way back to London to work for Friends of the Earth ... and stopped by Washington to help us out. In the course of stopping by, he went around to see some of the physicists he knew in town, settled down to the typewriter about eight or nine at night, wrote through the night. By eight o'clock the next morning, he had produced a ten-

thousand-word analysis of the liquid metal fast breeder reactor proposal, which was to be the Friends of the Earth position on it. It was most ably presented. I found, I think, two typographical errors. In the ten thousand words he wrote, I saw two or three places where he might have changed the phrasing a little. This just came out of him, intact in first draft. It was so good that the Bulletin of the Atomic Scientists *ran it, only slightly shortened, in a subsequent issue.*[1]

Eventually, Amory summed up problems with a variety of reactor types and fuel cycles (as the letters herein illustrate), as well as his argument about nuclear proliferation, when he coauthored a 1980 *Foreign Affairs* piece "Nuclear Power and Nuclear Bombs" that explained: *"[E]very* form of every fissionable material in every nuclear fuel cycle can be used to make military bombs, either on its own or in combination with other ingredients made widely available by nuclear power." (That piece, at www.rmi.org/rmi/Library/S80-02_NuclearPower NuclearBombs, is not included here, as its main points are nicely summarized in "On Proliferation, Climate, and Oil: Solving for Pattern," which appeared in *Foreign Policy* magazine in 2010 and is at p. 308.) However, while he has continued to warn of the downside of nuclear energy, the marketplace has made his fundamental economic point for him, as he shares in "Mighty Mice," which originally appeared in the trade journal *Nuclear Engineering International* in 2005. By 2010, as the unabridged version of that *Foreign Policy* paper (www.rmi.org/rmi/Library/2010-02_ProliferationOilClimatePattern) shows, nuclear power's rout in the market was nearly complete—though its advocates, as he pointedly showed in a 2010 article in the conservative magazine *The Weekly Standard*, keep on seeking ever larger, even unlimited, public subsidies.

Note

1   From the Regional Oral History Office of The Bancroft Library, University of California. The interview was conducted by Susan Schrepfer during 1974–1978. The *Bulletin* article, available on Google Books, is at **29**:29–35 (March 1973).

Chapter 9

Clean Energy
or a Choice of Poisons?

1972

Editor's note: *This article was first published in the* New York Times, *19 January 1972.*

To the Editor:

Lelan Sillin's "30-Year Plan for Pollution-Free Energy Sources" (Op-Ed 13 December) is misleading. Mr Sillin may blow trumpets in all directions, hailing the Second Coming of Prometheus, but it is not true: the "abundant pollution-free energy sources" he seeks cannot exist.

Thermodynamics guarantees that all the energy we generate and use, however we do so, will end as heat in the biosphere. This form of pollution matters very much, since too much extra heat will cause drastic instabilities in the Earth's delicately balanced climate. Nobody knows exactly when or how this might happen, but a good guess is that an annual growth rate in world energy use anywhere near the present 5.7 percent will land us in serious global trouble in 50 years, give or take 20. Local and regional climate is already disturbed—e.g. in Manhattan, an average square foot of which receives about 60 watts from man and 8.7 (net) from the sun.

Mr Sillin's generous enthusiasm for energy suggests that if he and his colleagues could get it from a "clean" source, they would prescribe so much of it that society would risk discovering climatic heat-limits empirically. But we actually have no "clean" source of energy—only a choice of poisons. Nuclear fission, which Mr Sillin and the AEC favor, may be the dirtiest possible choice.

Fission reactors emit no soot or sulfur dioxide, but they do create huge amounts of genetically dangerous and physiologically active radionuclides that someone will have to look after for the next millennium or so. Despite much effort, nobody has yet devised a safe method for the perpetual storage of these high-level wastes. This problem may have no technical solution.

Fast breeder reactors, Mr Sillin's and the AEC's "most important current ... goal," would stretch our dwindling uranium supplies and ensure a continuing heavy commitment to normal fission reactors. Since even those relatively simple and well-worked-out devices seem to have intractable safety problems (with emergency core cooling), it is hard to view with equanimity the prospect of breeder reactors cooled by molten sodium (several thousand gallons of it a second) and containing tons of plutonium—arguably the most dangerous substance on Earth.

Nuclear fusion, research on which is being starved in the US and UK, would be cleaner than any major energy source we now have. Yet fusion is not "virtually pollution-free" even in a nonthermal sense: it produces radioactive hydrogen (costly to isolate) and some high-level wastes.

Most of Mr Sillin's curiously all-electric research program is aimed at the wrong problem—at increasing supply, not controlling demand. This cosmetic approach is no longer good enough: we must start learning how to live within our means. As I suggested in a letter on 4 November, there are more important targets for energy reform than "frivolous" appliances (cf. Irwin Stelzer's letter, 10 December).

Neither faith in unspecified technological advances nor a vague hope that ingenuity will triumph over physical law can substitute for insight into the implications of the round-Earth theory. We need this insight now. It is one of nature's rules that those who won't play by the rules won't play at all.

Chapter 10

Why Nuclear Safety Is Unattainable

1973

Editor's note: *This article was first published in the* New York Times, *8 June 1973.*

To the Editor:

Too many of your correspondents are still missing the point about nuclear safety. By stressing the great care taken in fission technology, they evade the central question: Are the safety problems of fission too difficult to solve? If they are, then (as Professor Alfvén points out) one cannot claim that they are solved by pointing to all the efforts made to solve them.

It is impossible to *prove*, except by experiment, whether or not the safety problems of widely proliferated fission technology are too difficult to solve. In assessing the risks of a complex technology in which "no acts of God can be permitted" (Alfvén), we can only rely on analogies with other highly engineered systems that have far smaller risks. Such analogies suggest that nuclear safety is limited not by our care or ingenuity but by our inescapable fallibility: limited by the gap between intention and performance. Thus nuclear safety is not a mere engineering problem that can be solved by sufficient care but rather a wholly new type of problem that can be solved only by infallible people. Infallible people are not now observable in the nuclear or any other industry.

Fallible but committed people are now making some basic moral decisions for us: e.g. the decision not to stop at 92 (chemical elements, that is) but to commit the Earth forever to a plutonium economy, for which there seems to be no economic need whatever.

Laymen are often told that they cannot understand these technical matters; but perhaps a wise democracy should not be making decisions too complex for ordinary people to understand. Laymen will understand all too well that whereas there are one to three tons of plutonium-239 in a fast breeder reactor, a piece the size of a grapefruit (about enough for a crude but convincing homemade A-bomb) is sufficiently toxic to contain a lethal dose for everyone on Earth: clearly a candidate for infallible and perpetual containment.

The rate and magnitude problems of energy supply are truly formidable. Many people who should but don't appreciate these problems are still heralding Prometheus laden with plutonium. There are no such easy answers on a planet subject to biological as well as purely physical constraints.

Utopian visions encourage us to think that energy is unlimited and free; they do not teach us thrift or moral responsibility. Yet it is these qualities that must persuade us not to sign the Faustian bargain by proxy—after we have had a chance to read the fine print.

Chapter 11

Out of the Frying-Pan into the PWR

1978

Editor's note: *This opinion appeared in* Nature, *Vol. 271, 5 January 1978, and is reprinted with permission. This essay does not reflect current Friends of the Earth campaigns.*

The juggernaut advance of light-water reactors (LWRs), and especially pressurized-water reactors (PWRs), through world nuclear markets has seemed to their promoters a sort of manifest destiny. When Britain rejected PWRs for the second time, in 1974, the president of Westinghouse, which first developed them, replied magisterially that this was only "an interim solution, as we are not convinced that it addresses Britain's long-range energy needs." Since then, British electricity use has risen by less than 2 percent, while excess output capacity above a reasonable 20 percent reserve margin has risen to 20 percent of peak output and may soon be 30 to 45 percent. Closer analysis of energy use has also revealed not only that far more efficient use and even proven renewable sources are practicable and advantageous, but also that Britain already has twice as much electricity as is needed for the premium uses that can give value for money from this very costly, high-quality form of energy. Yet Westinghouse and its allies—including European vendors, the National Nuclear Corporation (NNC), and the Central Electricity Generating Board (CEGB)—are now back again. They argue that the British preference for advanced gas-cooled reactors—the last major indigenous program of thermal reactors outside Canada—is a sentimental anomaly that now, bereft of further excuses, must give way to technical and economic rationality.

In an open letter to the Prime Minister of the United Kingdom, James Callaghan, on 21 December, Friends of the Earth document a counter-argument. They suggest that the worldwide dominance of PWRs has arisen not from merit but from aggressive salesmanship and "a process of 'mutual intoxication' whereby US promotional institutions persuaded each other, then their counterparts abroad, that LWRs' supposed merits were real and had been demonstrated when in fact they had not and still have not been realized; the distinction between promotional prospectus and critical evaluation was completely obscured." Thus, according to a forthcoming study,[1] the gas-graphite system displaced in France by PWRs was in retrospect superior, but policymakers were caught up in a skillfully propagated and wholly unfounded PWR euphoria. Systematically mistaking hope for fact, they thought they *knew* how much PWRs would cost to build and to run, how reliable they would be, and what fuel burnup they would attain. US experience has fallen far short of these expectations. Real capital costs have averaged more than twice as high as promised; real fuel-cycle costs will be about six times as high. Capacity factor has averaged 20 to 30 percent lower and burnup 20 to 40 percent lower than forecast. The picture is not getting brighter. Two recent studies, for example, show PWR capital costs rising by 20 percent per year or $188 per kW(e) per year in *constant* 1976 dollars—about three times as fast as coal-fired stations. Much European and Japanese experience is similarly disappointing. Yet NNC's latest report on reactor choice is not an analysis of independently established facts so much as a seller's advertisement of alleged virtues. It is remarkably like the promotional paper that US AEC staffers drafted for the European Community's "Three Wise Men" when they recommended 20 years ago that Europe switch to LWRs.

In the US, the euphoria has worn off abruptly. The doubtful economics of real (as opposed to paper) PWRs, uncertainties about demand, reliability, and safety, and the macroeconomic problems of a capital intensity 10 to 30 times that of new North Sea oil capacity have together led to what FOE call "the most dramatic collapse of a major industrial enterprise in history." US yearly domestic orders in 1972–1976 (net of cancellations ascribed to the year of original ordering) were respectively 28, 38, 17, 5, and 3 reactors. In 1975–1976, 7 reactors were ordered, 8 cancelled, and 13 deferred (6 indefinitely). In 1977, 2 were ordered, 4 cancelled, and 26 deferred. This quickening disintegration has led Dr Schlesinger's deputy to state that "the nuclear option has essentially disappeared" in the US.

Official forecasts of US nuclear capacity in the year 2000 have been falling so quickly that, extrapolating linearly, the 1978 forecast for 2000 should be zero (in fact the asymptote might be as high as 2.5 percent of present US delivered energy use, or about the level of firewood). Nuclear expectations are likewise plunging in France, Germany, the UK, Japan, and elsewhere at such a rate that the 1979 forecast of 1985 OECD nuclear capacity should be zero. In Canada, which has had none of the US regulatory problems, forecasts have dived just as steeply as in the US, suggesting that the cause is not some political artifact but fundamental market forces—which President Carter's energy policy does nothing to discourage and much to reinforce. In short, PWRs are proving all but unsaleable throughout the industrial world. Staunching the rapid hemorrhage of money and staff from reactor vendors is requiring proliferative exports to developing countries, lavishly subsidized by exporting governments. This is not much of a vote of confidence in PWRs or in Britain's prospects for profitably exporting them. Indeed, reports in the financial press suggest that no LWR vendor has sold reactors at a profit: vendors' cumulative losses are said to exceed £1000 million in the US (over half of it for Westinghouse alone), £250 million in Japan, over £200 million in West Germany, and large but undisclosed sums in France and elsewhere. This hardly seems a promising line for Britain to follow—salvaging and refloating other countries' lame ducks.

Further, while PWR advocates consider the vexed question of pressure-vessel rupture adequately resolved, FOE (like most US observers) have long put this issue rather far down a very long list of serious safety problems. Their letter documents ten compendia showing over 200 unresolved major safety problems of PWRs, most of them officially acknowledged. British advocates' optimism about PWR safety may rest not on the detailed knowledge which they claim, but on the lack of it which they have in the past displayed (notably in the 1973–1974 controversy) and which the uncharitable can infer from the 1977 NNC report. It is possible that the NNC, CEGB, DOE, and Nuclear Inspectorate staff have in fact all done their homework. There is, however, no published reason to believe this is the case. They may still not have consulted the main original sources, preferring to rely, as in 1973–1974, on summaries prepared by US parties that could hardly be considered disinterested. FOE have therefore proposed that specified papers well known in the US debate be the subject of public colloquy in Britain between highly qualified US critics and their British adversaries.

Only four years ago, the NNC and CEGB were pressing for an urgent program of 32 PWRs—which, if adopted, might well by now have pushed the British electronuclear industry over the brink of ruin. The advocates of haste at that time were wrong; the advocates of caution were right. So it may be again.

Note

1   *Light Water: How the Nuclear Dream Dissolved* (New York: Basic Books. March 1975), by Professor I.C. Bupp (Harvard Business School) and Dr J.-C. Derian (an official of the French regional development ministry DATAR).

Chapter 12

Nuclear Follies

1985

Editor's note: *This unpublished letter to* Forbes *magazine (from 8 February 1985) was in response to James Cook's cover-story article of the same name.*

James Cook ("Nuclear Follies") vividly describes a tragedy many of us predicted and tried to avert for over a decade. He mistakes only its causes and alternatives.

Mismanagement and premature commitments to immature technologies were indeed horrific. But if they were the root causes of US nuclear power's plight, foreign programs (where they exist—many countries have none) would all be flourishing. They're not. In every market economy in the world, they're collapsing about as fast as here—even where utilities are unregulated, as in West Germany, or where nuclear regulation poses no obstacle, as in Canada. Nuclear commitments persist only in the centrally planned economies, notably France and the USSR; and even those programs are in deep economic trouble, with cost escalation now approaching US rates. French reactors did look relatively cheap until about 1983, but they've still cost enough to create a fifth of France's huge foreign debt and to drive Électricité de France into profound financial crisis.

The real problem, then, lies deeper: nuclear plants, like coal plants, are *fundamentally uneconomic*. Their stiffest competitors are new electricity-saving technologies—now able to save three-fourths of our electricity more cheaply than just *running* nuclear plants. Such technologies can save utilities billions of dollars' operating costs. As I showed before the Maine and New Hampshire PUCs, those savings can pay off the sunk costs of an abandoned Seabrook plant while *lowering* the rates.

Much small power production can also compete today. During 1981–1984, 12 GW (12 huge plants' worth) of coal power and zero nuclear plants were ordered in the US, while 77 GW of coal and nuclear plants were cancelled. But meanwhile, 20+ GW of cogeneration and 20+ GW of renewables, mainly small hydro and windpower, were ordered. By September 1984, 14.6 GW of privately financed small power production was on offer in California alone—all cheaper than central plants, over half renewable, and 4.7 GW added in a single quarter.

Only such cheap, fast, small options can tolerate the risks of uncertain demand. Utilities which buy or finance such options will prosper. Those which continue to project demand growth, build big plants to meet it, and raise rates to pay for those plants will disappear—priced right out of the energy-service market.

Nuclear power was made possible not by "the private enterprise system" but by its perversion via intense government pressure and $75+ billion (1983 $) in federal subsidies. To prevent a rerun, we should not sacrifice our open society but restore our market system, by stopping the subsidies (still $10+ billion per year) which seduced utilities into such folly. In a free market, nuclear investments—equal just in 1982 to twice the total investment in the car, truck, iron, and steel industries—would never have begun.

Chapter 13

Mighty Mice

2005

Editor's note: *This essay is reprinted with permission from the 21 December 2005 issue of* Nuclear Engineering International, *a UK-published trade journal of the global nuclear industry.*

Two men on a wild and barren plain suddenly spy a huge bear charging towards them. One man immediately starts putting on his running shoes. "How futile!", the other exclaims, "you'll never outrun that bear!" His companion drily replies: "I don't need to outrun the *bear*."

In any race, it's vital to understand whom you need to outrun and what it takes to win. Yet an incomplete picture of the competitive landscape may be the nuclear industry's greatest impediment to sound strategic planning, profitable investment, and credible public discourse.

This knowledge gap is understandable because the industry has been working so hard to achieve impressive progress in so many areas at once: operational consistency and reliability, simpler and cheaper designs, better inherent safety, streamlined siting and approvals, stronger government support, and other prerequisites for nuclear revival. But while these demanding tasks have taken so much attention, our bear has gained speed, approaching from behind.

Steve Kidd, the World Nuclear Association's head of strategy and research, asked in *NEI* (September 2005): "How can new nuclear power plants be financed?" He predicted this would "prove very challenging" in the private capital market, even though several studies found circumstances in which new nuclear build could compete with "building gas- or coal-powered generating capacity of similar magnitude."

Investors, he suggested, remain concerned about public opposition, siting and licensing, quick construction at predictable cost, safety, security, liability, nonproliferation, waste, decommissioning, and smooth operation. And he felt nuclear power's economic merits would emerge if we had "power markets where different technologies can compete on a level playing field and where long-term investment in capacity is incentivized."

These issues remain important and challenging, yet the market reality is even more complex. Resolving all perceived risks wouldn't ensure nuclear power's market success. Rather, new nuclear plants and central coal- or gas-fired power plants are all uncompetitive with three other options whose status, prospects, and value propositions are not well understood within the nuclear industry: certain decentralized renewables, combined-heat-and-power (CHP), and efficient end use of electricity. In a rapidly evolving energy marketplace full of disruptive technologies, nuclear power's biggest challenges are not political but economic.

Most nuclear advocates consider the various "micropower" and "negawatt" (electricity saving) alternatives necessary and desirable but relatively small, slow, immature, uncertain, and futuristic—complementing central thermal stations without threatening their primacy. In this view, nuclear power will predominate within a balanced low-carbon electricity mix, and generation will remain overwhelmingly centralized, because nothing smaller could scale up enough to power a growing global economy. As the World Nuclear Association (WNA) website states: "Only nuclear power offers clean, environmentally friendly energy on a massive scale." Yet this view is hard to reconcile with recently compiled industry data.

Decentralized Competitors

The World Alliance for Decentralised Energy's (WADE's) March 2005 compilation from industry equipment sales and project data estimated that decentralized resources in 2004 generated 52 percent of the electricity in Denmark, 39 percent in The Netherlands, 37 percent in Finland, 31 percent in Russia, 18 percent in Germany, 16 percent in Japan, 16 percent in Poland, 15 percent in China, 14 percent in Portugal, and 11 percent in Canada. WADE's definition includes CHP gas turbines up to 120 MWe, CHP engines up to 30 MWe, CHP steam turbines only in China, windpower and photovoltaics (PVs), but no hydropower, no other renewables, no generators below 1 MWe, and no end-use efficiency.

Figure 13.1 shows the annual output of low- and no-carbon micropower compared with nuclear power. No hydroelectric dams over 10 MWe are included. Average nuclear capacity factor (load factor) is assumed to rise linearly from 84.1 percent in 1982 to 88.5 percent in 2010. Up- and down-ratings, new units commissioned, and permanent retirements are shown consistently for all technologies.

These data show that micropower has already eclipsed nuclear power in the global marketplace. About 65 percent of micropower's capacity and 77 percent of its output in 2004 was fossil-fueled CHP, which was about two-thirds gas-fired, and emitted 30 percent to 80 percent less carbon (averaging at least 50 percent less) than the separate power plants and boilers or furnaces it replaced. The rest of the micropower was diverse renewables, whose operation, like nuclear power's (neglecting enrichment), releases no fossil-fuel carbon. Micropower's output lags its capacity by three years due to typically

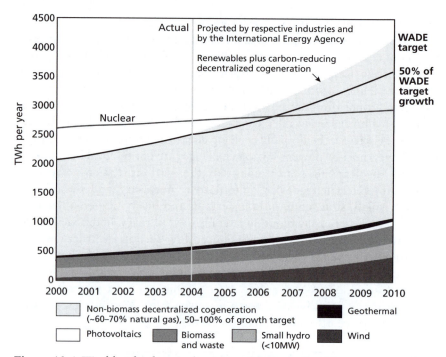

Figure 13.1 *Worldwide electrical output of decentralized low- or no-carbon generators (except large hydro)*

lower capacity factors for small hydro (~46 percent), windpower (~25–40 percent), and PVs (~17 percent) than for CHP (~83 percent), biofueled generation (~70 percent), and geothermal (~75 percent).

Worldwide, low- and no-carbon decentralized generators surpassed nuclear power's total installed capacity in 2002 and its annual output in 2005. In 2004 they added 5.9 times as much net capacity and 2.9 times as much annual output as nuclear power. The respective industries project that in 2010, micropower will add 136–184 times as much capacity as nuclear power will add, depending on CHP, wind and PV estimates (see Figure 13.2). Such projections are quite uncertain, but qualitatively clear. After 2010, whether the aging reactor fleet declines as projected by Schneider and Froggatt (see *NEI* June 2005, p. 36) or more slowly as predicted by the International Energy Agency (IEA), even with major new nuclear build in countries like China, micropower will continue to pull ahead.

Figure 13.2 shows net capacity added by each technology in each year since 1990. It also includes a leading indicator for nuclear power: construction starts through 2004. Their unknown size thereafter shouldn't materially affect 2010 completions. In 2004, windpower just in Germany and Spain added 2 GWe each, matching the average global net addition of nuclear capacity per year during 2000–2010. Worldwide nuclear construction starts will soon probably add fewer GWe per year than PV installations.

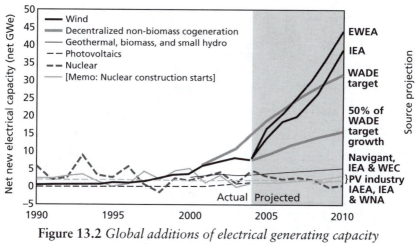

Figure 13.2 *Global additions of electrical generating capacity by year and technology*

Note: The total effect of supply-side competitors shown is the sum of their individual curves. In 2004, that sum was 28 GW, vs. nuclear power's 4.7 GW. In 2010, it's a forecast addition of 67–87 GW, vs. nuclear power's 0.48 GW.

These comparisons omit another key decentralized competitor—saved electricity—that is seldom properly tracked but clearly substantial. At constant capacity factor, the 2.0 percent and 2.3 percent decreases in US electricity consumed per dollar of GDP during 2003 and 2004 would respectively correspond to saving 14 and more than 16 peak GWe, plus 1 GWe per year of utility load management resources added and used. That's 6–8 times US utilities' declared 2.2 GWe of peak savings achieved in 2003 by demand-side management. Since the US uses only one-quarter of global electricity, and more efficient end-use is a global trend, worldwide electrical savings almost certainly exceed global additions of micropower (24 GWe in 2003, 28 GWe in 2004). Global additions of supply-side plus demand-side decentralized electrical resources are thus already an order of magnitude larger than global net additions of nuclear capacity (4.7 GWe in 2004).

Few investors and policymakers realize this, because most official statistics under-report decentralized and non-utility-owned resources, show only physical energy supply, and pay little attention to drops in energy intensity, whatever their cause (in most countries, chiefly more efficient end-use technologies). Per dollar of GDP, US primary energy consumption has lately been falling by about 2.5 percent per year, electricity by 2.0 percent per year Only 22 percent of the 1996–2005 increase in delivered US energy services was fueled by increased energy supply, 78 percent by reduced intensity—yet the latter four-fifths of market activity remains dangerously invisible.

That invisibility lately led US merchant firms to lose ~$100 billion by building ~200 GWe of combined-cycle gas plants for which there was no demand.

This calamity for investors could soon recur on a larger scale and not only in the power sector. The US Energy Policy Act of 2005 greatly increased subsi-

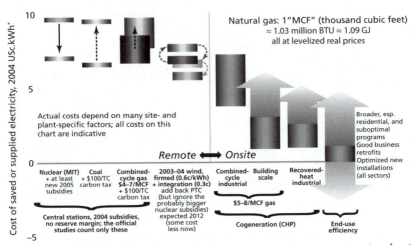

Figure 13.3 *Nuclear power's competitors on a consistent accounting basis; levelized cost of delivered electricity or end-use efficiency (at 2.7 ¢/kWh delivery cost for remote sources)*

Note: Savings: 12-year av. life, 4 percent per year real discount rate; Supply: merchant cash-flow model or market empirical; wind: 30 year life, 4 percent per year real; cogeneration: 25 year life, 4 pecent per year real

dies and regulatory aid for energy supply while largely ignoring demand-side resources. Yet "negawatts" expand as energy prices rise and as policies that have held per-capita electricity use flat for 30 years in California and are decreasing it in Vermont spread to other US states.

Like micropower, efficiency tends to be installed more quickly than supplies. If it continues to reach customers and grab revenues first, it will glut markets, crash prices, and bankrupt producers, just as it did under similar conditions in the mid-1980s. This would intensify investors' risk aversion.

Many factors tug energy outcomes in diverse directions. Windpower, for example, is heavily subsidized in the UK, where it has yet been slowed onshore by local opposition and offshore by two years of government debate on how to finance its links to the grid. Similarly, US windpower gets a production tax credit (PTC) but its erratic and brief renewals by Congress have repeatedly bankrupted leading wind turbine producers. Overall, the correlation between renewable installation rates and government subsidies is not clear-cut. Neither are per-kWh subsidies' relative sizes for renewables versus central plants, particularly nuclear power. Nor is it obvious whether relative subsidies are more or less important than the barriers that in most countries still block fair competition. This analytic fog makes it dangerous to assume that micropower's success is subsidy-driven, or that its obscure implementation obstacles are less important or tractable than nuclear's familiar ones.

A simpler explanation for micropower's market success might be superior basic economics. Figure 13.3 supports this hypothesis by comparing the cost of a kWh delivered to the retail meter from various marginal sources.

In concluding that nonhydro renewables are unsuitable "for large-scale

COMPARATIVE COST

The standard studies to which Steve Kidd referred (MIT, University of Chicago, IAEA, OECD, amongst others) all compare only the busbar costs of central stations—nuclear, coal, and combined-cycle gas. The assumptions and findings of MIT's 2003 analysis, *The Future of Nuclear Power*, are adopted here. However, to compare central stations (or remote windpower) fairly with onsite CHP and efficiency one must add to the former a delivery cost, conservatively assumed here to be $0.0275/kWh—the 1996 embedded average for US investor-owned utilities.

The MIT study found that a new 1 GWe advanced LWR with a 40-year life, 85 percent capacity factor and merchant financing has a busbar cost of $0.0702/kWh (in 2004 $), equivalent to $0.0977/kWh delivered. If its capital cost fell by 25 percent (from $2094/kWe to $1570/kWe overnight cost, compared to ~$2200/kWe for the new Finnish plant, an apparent loss-leader), its construction time fell from five to four years, the capital market attached zero nuclear risk premium, and fuel plus O&M cost dropped from $0.0157 to $0.0136/kWh (the lowest-quartile recent US value), the delivered cost could decrease to as little as $0.0715/kWh.

Imposing a high price on carbon emissions ($100/t CO_2) could raise the nominal cost of new delivered coal power from $0.072/kWh to $0.097/kWh (burning $1.33/GJ coal), and that of new combined-cycle gas power from $0.067–0.086/kWh to $0.078–0.098/kWh (at a levelized gas price of $3.6–7.6/GJ, equivalent to escalating those initial constant-$ gas prices at 5 percent per year). Figure 13.3 shows how these changes could shift the central plants' relative costs.

However, the standard studies ignore decentralized competitors, perhaps in the erroneous belief they're too small or slow to matter. Let's consider three kinds. (There are more, notably the diverse non-windpower renewables whose observed uptake bespeaks economic merit, but to avoid complex site-specific comparisons, and because windpower's siting and intermittence make it a difficult case, let's use it as a surrogate for all decentralized renewables).

Lawrence Berkeley National Laboratory reported in August 2005 that more than 2.7 GWe of US windpower projects installed during 1999–2005 had busbar costs, including PTC, ranging from $0.015 to $0.058/kWh (excluding one outlier), with a capacity-weighted average of $0.0337/kWh. Western US utilities' resource plans use levelized costs as low as $0.023/kWh, and the lowest 2003 non-firm wind energy contract price was $0.029/kWh, but we conservatively assume $0.030–0.035/kWh. The 2005 spike in wind turbine prices, 25–50 percent above 2003's, appears to reflect temporary imbalances: spot shortages that have filled all makers' books through 2006 are due largely to PTC-related postponement of US projects from 2004 to 2005–2006, while high steel prices will also boost central-station costs. On the contrary, industry and government expect windpower's costs to fall by ~$0.01/kWh during 2003–2012—more than the $0.0086/kWh levelized post-tax value of the PTC. For illustration, Figure 13.3 optionally adds back windpower's PTC but not the pre-2005 subsidies received by central stations, especially nuclear power. Those nuclear subsidies are complex, diverse and disputed, but the most authoritative independent US expert, Doug Koplow, estimates ~$0.0079–0.0422/kWh, increased by another ~$0.034–0.040/kWh in the Energy Policy Act of 2005 for at least the next 6 GWe ordered. For comparability with central stations, we assume that making windpower fully dispatchable costs $0.009/kWh—two-thirds for hydroelectric or other firming, one-third for grid integration. We conservatively adopt that extra cost, higher than most western US utilities pay or assume, partly in case some remote sites need extra transmission.

Conversely, central stations are assumed to incur no reserve-margin nor spinning-reserve costs, though their larger unit sizes make them tend to fail in larger chunks and for longer. Intermittence[2] does need attention and sound engineering, but it's not unique to renewables: every source of electricity is intermittent, differing only in why they fail, how often, how big, how long, and how predictably.

Grid operators' recent assessments confirm that windpower's intermittence even at high penetrations—about 14 percent for Germany, 20–25 percent for several US grids, and 30 percent for west Denmark—would be manageable at modest cost, typically a few $/MWh, if renewables are properly diversified, dispersed, forecasted, and integrated with the existing grid and with demand response.

The WNA's latest (February 2005) renewables webpage disagrees: it ignores technological and siting diversity and demand response. The WNA therefore concludes that intermittent renewables "cannot directly be applied as economic substitutes for coal or nuclear power" and will require "reliable duplicate sources of electricity, or some [unavailable] means of electricity storage on a large scale"—"almost 100 percent" backup—raising windpower's cost to twice the "generation cost" of nuclear or coal.

Highly intermittent supplies were long assumed to be limited to 5–10 percent of grid capacity, then 20 percent; the WNA claims 10–20 percent. Yet with better forecasting, grid integration, distribution automation, and smart power electronics, such supposed limits continue to recede. Windpower penetrations today are 20 percent in Denmark and up to 30 percent in three German states. On windy, light-load days in certain regions of Denmark, Germany, and Spain, windpower can exceed 100 percent of load, foreseeably and manageably. Yet windpower's grid integration costs are proving negligible or very modest. The corresponding costs of integrating other resources, all with nonzero forced outage rates, are of course already borne unnoticed. Nor are "reliable duplicate sources" proposed for nuclear plants, which in 2003 suffered prolonged large-scale curtailments in Europe's heatwave, slow restart after the USA/Canada blackout, and Tokyo Electric's safety shutdown.

CHP is a far more conventional and reliable resource already common in many countries. Figure 13.3 shows US costs for three arrangements, the first two based on actual projects by a leading US developer, Primary Energy, with 0.9 GWe of operating projects. Conventional gas-fired combined-cycle industrial CHP—with levelized gas prices of $5.4–8.7/GJ, a 10 percent per year return over 25 years, and unit sizes of 28-64 MWe—delivers new electricity for $0.038–0.073/kWh. Recovered industrial heat previously wasted can be worth more than CHP's other operating and capital costs, making its net cost of delivered electricity negative (−$0.021 to −$0.047/kWh) in the three 60–160 MWe projects evaluated. We graph instead their positive all-in electricity price ($0.011–$0.026/kWh), with the possibility of costs up to ~$0.04/kWh in less favorable cases. Well-integrated into a commercial building and with demand-side management, gas-fired "trigeneration" of power, heat, cooling, and perhaps other services can deliver electricity at a net cost around $0.01–0.03/kWh, or up to about $0.07/kWh with suboptimized designs.

The final major competitor shown in Figure 13.3 is efficient end use of electricity. Carefully evaluated programs of many US utilities have yielded reliable, durable, and accurately predicted savings at societal costs ~$0.01/kWh or less in commercial and industrial retrofits. Less optimized programs or those emphasizing homes can incur average costs up to ~$0.03–0.05/kWh. Alternatively, integrative design techniques well demonstrated in many buildings and industrial sectors often achieve very large savings at reduced capital cost, hence at a negative "cost of saved energy" (investment divided by the discounted stream of lifetime electricity savings).

See www.rmi.org/rmi/pid113 for documention.

power generation where continuous, reliable supply is needed," the WNA commits two common fallacies: supposing that making large amounts of electricity requires large generating units, and forgetting that *ceteris paribus* many small units near customers are more reliable than fewer, bigger units far away. Central thermal stations are no longer the cheapest or most reliable source of delivered electricity, because generators now cost less than the grid and have become so reliable that 98–99 percent of US power failures originate in the grid. Thus the cheapest, most reliable power is typically produced at or near customers. Three-quarters of US residential and commercial customers use electricity at an average rate not exceeding 1.5 and 12 kWe, respectively— severely mismatched to central plants' GWe scale. The WNA acknowledges a debate about scale, but ignores its profound implications and assumes central plants will remain dominant. Prudent investors favor micropower.

Comparative Potential

Of course, if decentralized resources had little potential to meet the world's rising needs for energy services, they'd be of minor competitive concern: one should worry about a bear, but hardly about a mouse. Yet a mighty swarm of mice is another matter. The modern literature suggests that decentralized resources' collective practical potential has been understated, as if the stunning technological and economic advances in conventional energy supply didn't apply to its rivals. To the contrary, such progress tends to be faster in decentralized resources. For example:

- At less than the delivered cost of just *operating* a zero-capital-cost nuclear plant (~$0.04/kWh), potential US electricity savings range from two to four times nuclear power's 20 percent share of the US electricity market, according to bottom–up assessments summarized by the Electric Power Research Institute (EPRI) and Rocky Mountain Institute's joint *Scientific American* article (September 1990). EPRI's Clark Gellings confirmed in 2005 that the US electric end-use efficiency resource is probably now even bigger and cheaper, because better mass-produced technologies more than offset savings already captured. Utility-specific data confirm a broad downward trend in the unit cost of "negawatts."
- CHP potential in industry and buildings is very large if regulators allow it. Waste-energy CHP alone is preliminarily estimated by Lawrence Berkeley National Laboratory to have a technical potential nearly as large as today's US nuclear capacity, though cost and feasibility are very site-specific.
- Modern windpower's US potential on readily available rural land is at least twice national electrical usage.
- Other renewable sources of electricity are also collectively important— small hydro, biomass power (especially CHP), geothermal, ocean waves, currents, solar-thermal, and PVs. These sources and windpower also tend to be statistically complementary, working well under different weather conditions. All renewables together (excluding big hydro), plus solar

technologies that indirectly displace electric loads (daylighting, solar water heating, passive heating and cooling), have a practical economic potential many times total US electricity consumption—at least an order of magnitude greater than nuclear power provides today.

- Even at such a scale, a diversified renewable portfolio needn't raise land-use concerns. For example, a rather inefficient PV array covering half of a sunny area 160 × 160 km could meet all annual US electricity needs. In practice, since sunlight is distributed free, PVs would be integrated into building surfaces, and installed on roofs, over car parks, and along roads, both to save land and to make the power near loads. Specious claims persist comparing (say) the footprint of a nuclear reactor with the (generally miscalculated) land area of which a fraction—a few percent for wind turbines—is physically occupied by energy systems and infrastructure. In fact, total fuel-cycle land use is roughly comparable for solar, coal, and nuclear.

Thus renewables clearly have a very large global potential. The IEA's *World Energy Outlook 2004* foresees a 2030 renewable potential of ~30,000 TWh per year (less than a quarter of it from hydropower). Such massive production would become far easier with CHP and efficient end use. It still wouldn't be easy, but neither would central stations of similar output—especially for serving the two billion people not now on any grid.

Comparative Speed

But might decentralized supply- and demand-side resources be too slow to deploy, requiring central stations to provide enough reliable power, quickly enough, to meet burgeoning demand? This widely held view seems inconsistent with observed market behavior. As shown above, micropower and efficient end use, despite many obstacles, are already adding an order of magnitude more GWe per year than nuclear power worldwide. Their brisk deployment reflects short lead times, modularity and economies of mass production (they're more like cars than cathedrals); usually mild siting issues (except in some unusual windpower cases); and the inherently greater speed of technologies deployable by many diverse market actors without complex regulatory processes, ponderous enterprises, or unique institutions.

Of course every energy option faces specific obstacles, barriers, and hence risk of slow or no implementation at scale. Efficiency, for example, faces some 60–80 market failures, many arcane, that have left most of it unbought. Yet US electric intensity has declined at an unprecedented average rate of 1.5 percent per year since 1996 even though electricity is the form of energy most heavily subsidized, most prone to split incentives, least priced on the margin, and sold by distributors widely rewarded for selling more kWh. Such firms as DuPont and IBM routinely cut their energy intensity by 6 percent per year with attractive profits and no apparent constraints.

Letting all decentralized resources really compete risks not a dry hole but a gusher. Just during 1982–1985, when California's three investor-owned utilities offered a relatively level playing field, fair competition elicited 23 GWe of efficiency plus 21 GWe of generation (13 GWe of it actually bought) rising by 9 GWe per year. The resulting glut, 144 percent of the 1984 peak load of 37 GWe, forced bidding suspension in 1985, lest every fossil and nuclear plant be displaced (which in hindsight could have been valuable).

Investors appreciate that diversification is wise but must be intelligent. The strategic virtue of a diversified portfolio doesn't justify buying every technology or financial asset on offer. The sweeping claim that "we need every energy technology"—as if we had infinite money and no need to choose—is often made but cannot withstand analysis. The WNA's website doesn't mention demand-side resources, and denies the existence of a large and compelling literature of nuclear-free, least-cost, long-term scenarios published over decades (in 1989, for example, Vattenfall published a roadmap for rapid economic growth, full nuclear phaseout, one-third power-sector CO_2 reduction, and $1 billion per year cheaper energy services). But investors with similarly limited vision are in for a shock. As all options compete and as increasingly competitive power markets clear, any supply investment costlier than end-use efficiency or alternative supplies risks being stranded by retreating demand.

Oil, Climate, and Strategy

A major argument often made for new nuclear build is oil displacement; yet this has already been largely completed. Only 3 percent of US electricity is made from oil and less than 2 percent of US oil makes electricity. Worldwide, these figures are around 7 percent and falling. Most of that oil, too, is residual, not distillate, and is burnt on relatively small grids by smaller plants with low capacity factors, unsuited to nuclear displacement. Both oil and fungible natural gas can be far more cheaply displaced by other means, mainly by doubled end-use efficiency.

A more compelling need is displacing coal-fired electricity to protect the Earth's climate. Yet nuclear power's dubious competitive economics could make it counterproductive, for four reasons:

1 Most of the carbon displacement should come from end-use efficiency, because it's profitable—cheaper than the energy it saves—and quick to deploy.
2 End-use efficiency should save not just coal but also oil, particularly in transport. Comprehensive energy efficiency addresses 2.5 times as much CO_2 emission as any electricity-only initiative.
3 Supply-side carbon displacements should come from a diverse portfolio of short-lead-time, mass-producible, widely applicable and accessible, benign, readily sited, rapidly deployable resources.
4 The total portfolio of carbon displacements should be both fast and effective.

This last point highlights a troublesome implication of Figure 13.3's cost comparison. Buying a costlier option, like nuclear power, instead of a cheaper

CONSERVATISMS

Decentralized resources' cost advantage (Figure 13.3) is robust even against implausible improvements in central stations' technology or regulation. For example, if some new sort of fission or fusion reactor could provide free steam to the turbine, the remainder of the central thermal plant would still cost too much to compete. And the cost comparisons shown have two other major conservatisms favoring central plant: they reflect a static snapshot of competitors' costs, not (save one windpower illustration) their continuing rapid decline in real cost; and they count as zero all but 1 (thermal integration) of the 207 "distributed benefits" described in Rocky Mountain Institute's book *Small Is Profitable: The Hidden Economic Benefits of Making Electrical Resources the Right Size*. The market is increasingly counting those benefits which collectively boost value ~10-fold, enough to flip most investment decisions.

This increase in value has three separate causes, excluding such externalities as environmental and social benefits. The most important distributed benefits come from financial economics:

- Small, fast modules incur less financial risk than big, slow projects. In a typical sub-station support application, this can raise the tolerable capital cost of a distributed resource, like PVs, by about 2.7-fold.
- Renewables avoid the financial risk of volatile fuel prices, raising windpower's typical value by about $0.01–0.02/kWh.

These and other financial-economics benefits typically boost decentralized projects' economic value by about an order of magnitude if they're renewable, about three- to five-fold if they're not.

Better known are such electrical engineering benefits as avoided grid costs and losses, increased reliability and resilience, more graceful fault management, free reactive power control (from DC sources inverted to AC), and longer distribution equipment life (by means of reduced heating and tap-changing). Together, these typically increase value by about two- to three-fold—more if the distribution system is congested and new distribution capacity can be deferred or avoided, or if especially reliable or high-quality power is required. Finally, scores of diverse "miscellaneous" benefits typically about redouble economic value—more if "waste" heat can be recaptured.

one, like "negawatts" and micropower, displaces less carbon per dollar spent. This opportunity cost of not following the least-cost investment sequence—the order of economic and environmental priority—complicates climate protection. The indicative costs in Figure 13.3 (neglecting any differences in the energy embodied in manufacturing and supporting the technologies) imply that we could displace coal-fired electricity's carbon emissions by spending $0.10 to deliver any of the following:

- 1.0 kWh of new nuclear electricity at its 2004 US subsidy levels and costs;
- 1.2–1.7 kWh of dispatchable windpower at zero to actual 2004 US subsidies and at 2004–2012 costs;
- 0.9–1.7 kWh of gas-fired industrial cogeneration or ~2.2–6.5kWh of building-scale trigeneration (both adjusted for their carbon emissions), or 2.4–8.9 kWh of waste-heat cogeneration burning no incremental fossil fuel (more if credited for burning less fuel);
- From several to at least 10 kWh of end-use efficiency.

The ratio of net carbon savings per dollar to that of nuclear power is the recip-rocal of their relative cost, corrected for gas-fired CHP's carbon emissions (assumed here to be threefold lower than those of the coal-fired power plant and fossil-fueled boiler displaced). As Bill Keepin and Greg Kats put it in *Energy Policy* (December 1988), based on their still-reasonable estimate that efficient use could save about seven times as much carbon per dollar as nuclear power, "every $100 invested in nuclear power would effectively release an additional ton of carbon into the atmosphere"—so, counting that opportunity cost, "the effective carbon intensity of nuclear power is nearly six times greater than the direct carbon intensity of coal fired power." Whatever the exact ratio, their finding remains qualitatively robust even if nuclear power becomes far cheaper and its competitors don't.

Speed matters too: if nuclear investments are also inherently slower to deploy, as market behavior indicates, then they don't only reduce but also retard carbon displacement. If climate matters, we must invest judiciously, not indiscriminately, to procure the most climate solution per dollar and per year. Empirically, on both criteria, nuclear power seems less effective than other abundant options on offer. The case for new nuclear build as a means of climate protection thus requires re-examination.

Micropower and its natural partner, efficient end use, have surpassed and outpaced central stations despite many obstacles. Being diverse, ubiquitous, plentiful, widely available, largely benign, and popular, they are also hard to stop. To be sure, much work remains to purge the artificial barriers to true competition between all ways to save or produce energy, regardless of which kind they are, what technology or fuel they use, how big they are, or who owns them. But such a free market, for which Kidd rightly calls, seems increasingly unlikely to favor nuclear power. Rather, the economic fundamentals of distrib-uted resources promise an ever-faster shift to very efficient end use combined with diverse generators the right size for their task. That shift could render insufficient or even irrelevant the resolution of the perceived non-economic risks that preoccupy the nuclear industry.

The better the industry and its investors understand this, the more likely they are to fulfil reasonable expectations, apply their talents effectively, and help achieve the global energy, development, and security goals to which we all aspire.

Notes

1   *Editor's note*: A critique by Ian Hore-Lacy of the World Nuclear Association, with Amory's response, is at www.neimagazine.com/story.asp?storyCode=2056395 and www.neimagazine.com/story.asp?storyCode=2056396.

2   *Editor's note*: Amory and many other analysts now distinguish "variable" resources like wind and photovoltaics, the failures of which are predictable, from "intermittent" resources like coal and nuclear, whose failures aren't.

Chapter 14

Nuclear Socialism

2010

Editor's note: *This article is reprinted with permission from* The Weekly Standard, *Vol. 16, No. 6, pp. 15–16, 25 October 2010.*

Given Americans' increasing anxiety over made-in-Washington socialism, it's a wonder that the nuclear power industry has escaped scrutiny for so long. The federal government socializes the risk of investing in nuclear power while privatizing profits. This same formula drove the frenzied speculation that cratered the housing and financial markets. What might it cause with nuclear power?

We got a taste three decades ago. Congress grew infatuated with the promises of nuclear promoters. It overrode the risk assessment of private capital markets, and expanded subsidies for nuclear projects to $0.08 per kilowatt-hour—often more than investors risked or than the power could be sold for. This seduced previously prudent utilities and regulators into a nuclear binge that *Forbes* in 1985 called "the largest managerial disaster in business history" (see "Nuclear Follies," p. 150).

Threefold cost overruns amounted to hundreds of billions of dollars. Three-fifths of the ordered plants were abandoned. Many others proved uncompetitive. Steep debt downgrades hit four in five nuclear utilities. Some went broke. Through 1978, 253 US reactors were ordered (none since). Only 104 survive. Two-fifths of those have failed for a year or more at least once.

New nuclear plants, we're assured, are different—novel enough to merit technology-demonstration subsidies, yet proven enough that investors can rest easy. They're allegedly so much safer than deep-sea oil drilling that we needn't fret, yet so risky that one major nuclear operator insured itself 11 times more against nuclear accidents' consequences than its potential liability to the public. New reactors are supposedly so cheap they crush competitors, yet so costly they need subsidies of 100 percent or more.

That's right: $0.04–$0.06 of new 2005–2007 subsidies, plus $0.01–$0.04 of remaining old subsidies, brings total federal support for new nuclear plants, built by private utility companies, to $0.05–$0.10 for a kilowatt-hour worth $0.06. Some people are outraged that the federal government is subsidizing the new Chevrolet Volt, retailing at $41,000, with a tax credit of $7500. Imagine if the tax credit were $50,000! If new reactors can produce competitive power, they don't need subsidies; if not, they don't deserve subsidies.

Yet nuclear subsidies to some of the world's largest corporations have become shockingly large. A Maryland reactor's developer reckoned just its requested federal loan guarantee would transfer $14.8 billion of net present value, comparable to its construction cost, from American taxpayers to the project's 50/50 owners—Électricité de France (EDF), 84 percent owned by the French government, and a private utility 9.5 percent owned by EDF. The project's builder, AREVA, is 93 percent owned by the French state, yet has been promised a $2 billion US loan guarantee for a fuel plant competing with an American one. EDF just booked a billion-euro loss provision, mainly over the Maryland plant's deteriorating prospects. AREVA's construction fiascoes in Finland and France have "seriously shaken" confidence, says EDF's ex-chairman, and four nations' safety regulators have criticized the design. Meanwhile, the chairman of Exelon, the top US nuclear operator, says cheap natural gas will postpone new nuclear plants for a decade or two. Slack electricity demand and unpriced carbon emissions further weaken the nuclear case. Markets would therefore charge a risk premium. But US nuclear power evades market discipline—or did until 8 October 2010, when the Maryland promoter shelved the project because, for its $7.5 billion federal loan guarantee, it would have to have paid an "unworkable" $0.88 billion fee, or 11.6 percent, to cover the default risk to taxpayers.

Another $8.3 billion of the $18.5 billion nuclear loan guarantees authorized in 2007 was provisionally issued in February 2010 to two Georgia reactors. Taxpayers will be on the hook for about $100 per American family. To offset that risk, the Department of Energy proposed to charge a default fee that's only a small fraction of the likely loss rate that the Congressional Budget Office and Government Accountability Office have estimated. In bankruptcy, taxpayers wouldn't even recover before private lenders—not that there are any private lenders. The Treasury's Federal Financing Bank, financed by new Treasury debt, would issue the DOE-guaranteed loan. Failure would cost taxpayers $8.2 billion net. The developer keeps any upside.

The Georgia project's loan-guarantee default fee is much lower than the Maryland plant's, partly because the Georgia developers have already shifted more of their remaining risks to ratepayers [the US term for electricity customers]. Their project is 54 percent owned by municipal utilities and rural co-ops with access to cheaper financing than private utilities, including subsidized stimulus bonds. Some of these munis and co-ops signed 50-year contracts with the nuclear operators that would put them and their customers on the hook even for power not needed or wanted. In 1982–1983, the analogously financed five-reactor WPPSS ("Whoops") project in the Northwest defaulted on municipal bonds, vaporizing $3–$4 billion in today's dollars.

Moreover, a few southeastern states now make utility customers finance new reactors in advance—often whatever they cost, whether they ever run, no questions asked, plus a return to the utilities for risks that they no longer bear. This scraps all five bedrock principles of utility regulation: payment only for service delivered and only for used and useful assets; accountability for cost and prudence; return matching risk; and no commission able to bind its successors. Such laws recreate for nuclear power the same moral hazard that just shredded America's financial sector.

With such juicy incentives, why won't private investors finance reactors? In 2005–2008, with the strongest subsidies, capital markets, and nuclear politics in history, why couldn't 34 proposed reactors raise any private capital? Because there's no business case. As a recent study by Citibank UK is titled, "New Nuclear—the Economics Say No." That's why central planners bought all 61 reactors now under construction worldwide. None were free-market transactions. Subsidies can't reverse bleak fundamentals. A defibrillated corpse will jump but won't revive.

American taxpayers already reimburse nuclear power developers for legal and regulatory delays. A unique law caps liability for accidents at a present value only one-third that of BP's $20 billion trust fund for oil-spill costs; any bigger damages fall on citizens. Yet the *competitive* risks facing new reactors are uninsured, high, and escalating.

Since 2000, as nuclear power's cost projections have more than tripled, its share of global electricity generation has fallen from 17 percent to 13 percent. That of cogeneration (making electricity together with useful heat in factories or buildings [i.e. CHP]) and renewables (excluding big hydropower projects) rose from 13 percent to 18 percent.

These bite-sized, modular, quickly built projects—with financial risks, costs, and subsidies generally below nuclear's and declining—now dominate global power investments. Last year, renewables (wind, water, solar, geothermal), excluding large hydroelectric dams, attracted $131 billion of private capital and added 52 billion watts. Global nuclear output fell for the past three years, capacity for two.

This market shift helps protect the climate. Renewables, cogeneration, and efficiency can displace 2–20 times more carbon per dollar, 20–40 times faster, than new nuclear power—saving trillions of dollars over decades and avoiding vast financial risks.

Still uncompetitive despite 60 years of handouts, nuclear developers clamor for ever-greater subsidies. The White House, Senate, and House all propose expanded federal loan guarantees ($36 billion was the White House figure); developers demand at least $100 billion. The Clean Energy Deployment Administration endorsed by both houses of Congress could issue *unlimited* loan guarantees without congressional oversight. It would probably fund nuclear and renewable energy like the recipe for elephant-and-rabbit stew—one elephant, one rabbit.

Bureaucrats, not credit markets, would evaluate risks and pick winners. Taxpayers would become America's main energy financiers and almost exclusive nuclear risk-takers. America's once market-based electricity investments would work like China's, Russia's, and France's nuclear command economies. This is bipartisan folly.

As nuclear subsidies spiral toward fiscal ruin, brave voices protest from a handful of think tanks: the Heritage Foundation, the Cato Institute, the George C. Marshall Institute, the American Enterprise Institute, the Competitive Enterprise Institute, the National Taxpayers Union, Taxpayers for Common Sense. Yet most congressional budget hawks—supposedly sages of circumspection and defenders of free markets—urge more nuclear socialism.

Here's a principled alternative. Reverse the energy-subsidy arms race. Don't add subsidies; subtract them. Take markets seriously. Not just for nuclear and fossil fuels but for all so-called "clean" technologies, head toward *zero* energy subsidies, free enterprise, risk-based credit pricing, competition on merit, cheaper energy services, greater energy security, and dwindling [national budget] deficits.

Who wouldn't like that? Why don't we find out?

Section 4

Vehicles and Oil: Goodbye Crude World

Vehicles, in their current state of evolution, and oil are inextricably linked—discussions about one inevitably involve the other. Amory's work on vehicles began in 1990, when he was invited to co-keynote a 1991 National Academy of Sciences meeting on efficient cars and future car technology. It quickly became apparent to him that the vehicle itself needed fundamental redesign. Thus began a nearly two-decade-long exploration of how we design, build, operate, and simply think about cars—although he'd already done extensive study of the oil problem. That exploration culminated in his and his colleagues' design for a new vehicle, the "Hypercar®," which eventually led to the creation of a private-sector firm pushing the limits of advanced composites for automobile structures and other applications. Fiberforge (www.fiberforge.com), as it's now named, has developed economically interesting ways to ultralight everything from sports equipment and military helmets to aerospace and automotive structures.

In the first of these essays, "Energy: The Avoidable Oil Crisis," Amory explains how the United States' historic approaches to oil—drilling, protectionism, and trade—skew oil prices and exacerbate supply problems. They are, he explains, ultimately no match for reducing oil use through increased efficiency or substitution of alternative liquid fuels, or both. "Reinventing the Wheels," which first appeared in *Atlantic Monthly*, was one of his earliest lay syntheses about how vehicles, buildings, and electric grids would come together, offering solutions in all three realms. That developments like the "smart grid" and plug-in vehicles are now becoming commonplace proves the prescience of the 16-year-old ideas within. Years later, in 2000, the second Bush Administration began pushing again—it happens whenever oil prices spike or Alaska needs revenue—for drilling in the Arctic National Wildlife Refuge (ANWR). In mid-2000, the House of Representatives voted to allow drilling. Amory's response was to pen a piece pointing out why conservatives should find this notion particularly offensive and counterproductive. The simple fact that replacing our tires with ones as efficient as the originals would negate the need for "several Refuges'" worth of oil should have been enough to convince any US citizen that the pro-drilling logic was flawed—but the debate lingers on. Amory wrote several pieces on ANWR, the latest being "Drilling in All the Wrong Places," which Rocky Mountain Institute published in 2008 and is included here. For a more detailed but older description of his take on Refuge oil, see "Fool's Gold in Alaska" in the July/August 2001 edition of *Foreign Affairs*. Unfortunately, two months before the 9/11 attacks, the editor changed its title from "The Alaskan Threat to National Energy Security" and required that its alarming analysis of the vulnerability of the Trans-Alaska pipeline to terrorism be condensed to a paragraph, though it remains in the annotated and unabridged version at www.rmi.org/rmi/Library/E01-04_FoolsGoldAlaskaAnnotated. And for a full account, cosponsored by the Pentagon, of how to get the United States completely off oil by the 2040s at an average cost of $15 per barrel, led by business for profit, read Amory's 2004 synthesis *Winning the Oil Endgame* (www.oilendgame.com), to be updated in 2011 by RMI's *Reinventing Fire* (www.reinventingfire.com).

Chapter 15

Energy:
The Avoidable Oil Crisis

1987

Editor's note: *Coauthored with L. Hunter Lovins, "Energy: The Avoidable Oil Crisis" first appeared in the December 1987* Atlantic Monthly.

The United States, having pumped more oil for longer than any other country, has largely depleted its cheapest oil. More oil can be found, but only at higher cost and in more remote and fragile places. Foreign oil now costs less to find and extract than ours, and despite American technological prowess, the cost gap will gradually widen. Only three responses to this trend seem to be available at present: protectionism, trade, and substitution.

The Protectionist Option

Many US oil companies, like companies in other industries whose products can no longer compete in global markets, want tariffs that will make imported oil look as costly as domestic, or a restoration of recently reduced government subsidies that made domestic oil look as cheap as imported, or both. Tariffs would encourage, and subsidies discourage, the more efficient use of oil. Either move would stimulate domestic exploration and extraction of oil, but with side-effects. Either move would make Americans pay more for oil than others pay, making the US economy less competitive. Worse, by making new domestic oil look cheaper than it really is—at least, relative to foreign oil—either move would speed up the very depletion that was supposedly such a worry in the first place.

A more thoughtful variation on the protectionist theme would be to raise the taxes on gasoline and other oil products to discourage consumption. This wouldn't affect oil companies' choices between drilling for oil at home and importing it from abroad: they would do whatever was least expensive—namely, import. It could, however, keep domestic oil companies in business longer, because reduced consumption would slow oil depletion. Unfortunately, though tax increases would spur oil savings by those who could afford to buy more-efficient cars (half our oil is used on the highway), they would burden those who can barely afford the cars they have. More generally, any tax on final energy products is disproportionately hard on people with low incomes, because they spend a larger fraction of their income on fuel. An oil tax would also further distort investment and purchasing choices between oil and other fuels. Both these problems could be avoided by uniformly taxing all depletable fuels as they come out of the ground or into the country. That might be a good idea, and it could greatly enrich the Treasury, but it's an oblique, long-term response to the depletion of low-cost US oil.

The Trade Option

The free-trade alternative to protectionism is to buy the cheapest oil, even if it's foreign. Americans are doing just that. Last year net imports rose to 33 percent of all oil used in the United States—less than throughout 1973–1981, much less than the all-time high of 46 percent, which occurred in 1977, but a bit above the recent low of 27 percent, achieved in 1985. The halving of world oil prices last year, brought about largely by the previous decade's US oil savings, prompted a temporary 3.5 percent boost in domestic oil use while discouraging costly domestic output. If these two trends were to continue (unlikely, since oil prices have about doubled again), they could drive imports above 50 percent of the oil we will use in the 1990s.

Of course, the United States already imports many commodities that others produce better or more cheaply than we do: in 1986, for example, we imported 75 percent of the nickel we used, 92 percent of the bauxite, 70 percent of the tungsten, and 83 percent of the tin. We import coffee and cattle, fish and cheese, perfume and beer, cars and televisions. To pay for these or any other imports, we must export something else that others prefer to buy from us. As Japan has demonstrated, a major industrial power can import nearly all its oil, but if we did that we would have to match Japan's export success as well. To be sure, the potential balance-of-trade burden is easily exaggerated: the US trade deficit for energy, having peaked at $75 billion in 1980, had fallen to $29 billion last year. This was a striking reduction, but those gains were more than offset by the year's $110 billion deficit on non-energy imports. Nonetheless, if oil again cost $24 a barrel, as US oil did in 1980, and if we imported as much of it as we did at the 1977 peak, the dollar outflow would match that of 1980.

A deeper fear is that foreign oil can be cut off by war or politics, much as the United States has embargoed wheat and soybean exports to previously trusting trading partners. For many Americans, the possibility of oil cutoffs suggests not just the inconvenience of gas lines but a threat to this nation's military power, although the latter idea is probably an exaggeration, since the Department of Defense uses less than 3 percent of the nation's oil and is so unconcerned about oil cutoffs that it is depleting its Naval Petroleum Reserve.

National security is too important to be cheapened by invoking it for special pleading. Those who say that national security requires the substitution of costlier domestic oil for foreign oil are glossing over three sets of basic facts.

First, conditions today bear little relation to those of 1973. OPEC now provides only 30 percent of the world's oil output, not 56 percent, and the Persian Gulf only 19 percent, not 37 percent. Oil is plentiful, not in short supply; the oil market favors buyers, not sellers. Once-rich oil exporters, now struggling with budget deficits, can hardly sacrifice revenues, let alone destroy the value of the Western assets that harbor their shrinking cash reserves.

Supplies, stocks, and transportation and marketing arrangements have also become enormously more diverse and flexible than they were 14 years ago. Overland routes to Red Sea and Mediterranean ports now exist, and other parts of the world (Venezuela, Mexico, Nigeria, Indonesia) have two to three million barrels a day of spare output capacity. Five million barrels of oil a day could be immediately forthcoming if needed. In the first half of this year only about seven million barrels a day came through the Strait of Hormuz—roughly half the early-1980s level.

Second, four specific precautions or countermeasures against oil cutoffs are now available: friendly relations, diversification, stockpiling, and military intervention.

The most effective approach would be simply to behave so that others want to continue doing business with us—specifically, those others with whom we have interests in common. In the 1990s, when most US oil imports will probably come from Mexico, Venezuela, and Canada, we may wish we had devoted to those countries' prosperity, stability, and friendship a tenth of the attention we're now lavishing on arguably less vital relationships in the Persian Gulf. Instead, our policies on such issues as immigration, debt, trade, Nicaragua, and acid rain are souring relations in the Western Hemisphere for decades to come.

The United States has already diversified its oil sources. More than half of our net oil imports last year came from the Western Hemisphere and Britain. Of all oil used by the United States in 1986, just 17 percent came from OPEC (including such countries as Nigeria, Indonesia, and Venezuela), 7 percent from Arab countries, and less than 6 percent from the Persian Gulf.

Another basic precaution—stockpiling, in the 530-million-barrel Strategic Petroleum Reserve and in private reserves—has already been taken, and not just by the United States. Japan, for example, has about 150 million barrels of crude oil in anchored tankers—a month's worth of oil, for all uses, for the country. Government stockpiles among 21 advanced nations now contain about 800 million barrels—more than four times the 1979 level. This very large reserve, bought at high cost, can make up for more than a year the net deficit that might be caused by a sudden cutoff of shipments through the Strait of Hormuz. A year is long enough for fuel switching and the reactivation of shut-in wells to fill the gap: the non-Communist world's spare oil-extracting capacity on such a timescale is about 10 million barrels a day, or more than a fifth of the same countries' total oil demand.

Still remaining is the option (assuming it is considered moral, effective, and safe) of threatening to use or using force to maintain access to foreign oil. This card, however, has already been overplayed, and the stakes are high. Earl Ravenal, of the Georgetown University School of Foreign Service, found that in fiscal year 1985 alone, before the [USS] *Stark* attack, the United States spent $47 billion projecting power into the Persian Gulf—$468 per barrel imported from the Gulf in that year, or 18 times the $27 or so that we paid for the oil itself.

Of course, more is at stake in the Gulf than simply the flow of oil to the United States. We are, however, paying a heavy price to ensure that oil is shipped—from a war zone partly of our own making—to ourselves (we receive about 10 percent of the Gulf's oil) and to our business competitors (about 90 percent). What's more, we're paying the price in money borrowed from those competitors and from the oil exporters themselves.

Persian Gulf oil, whose total purchase-plus-military cost in fiscal year 1985 was $495 a barrel plus interest, is hardly a competitive fuel for the American economy. Today some 25,000 members of the US military are in the Gulf region. The costs of that expanded presence and its military risks, even spread over more barrels imported from the Gulf, still amount to hundreds of dollars per barrel. To paraphrase a cartoon by Dan Wasserman, we're spending money we don't have, to defend ships that aren't ours, to ship oil we don't use, for allies who won't pay, in pursuit of a policy we haven't formulated.

Third, the premise underlying the national-security argument—that foreign oil is less secure than domestic—is not necessarily valid. Six years ago our study for the Pentagon (published as *Brittle Power: Energy Strategy for National Security*, summarized in *The Atlantic* of November 1983) found that a handful of people could cut off three-fourths of the oil and gas supply to the eastern states—so efficiently that it would take upwards of a year to restore it—in one evening's work, without even leaving Louisiana. That remains true. 23 percent of all crude oil extracted and 16 percent of all crude oil used in the United States flows through the Trans-Alaska Pipeline System—two and a half times as much as we're importing from the Persian Gulf. Yet the pipeline has already been repeatedly, if incompetently, attacked, and the Army has declared it indefensible. The pipeline is far easier to disrupt and harder to mend than Middle Eastern oil facilities and tanker shipments. We know of many alternative routes for Middle Eastern oil—the Saudis, for example, are completing their second pipeline to the Red Sea, avoiding the Gulf altogether—but none for Alaskan oil. Far more of our oil supply, therefore, is now unavoidably at risk from a single, simple, unattributable act by a lone saboteur in Alaska than could possibly be cut off by an all-out war in the Strait of Hormuz. Seeking additional oil in the Arctic National Wildlife Refuge, where the odds of cost-effectively finding any are at best one in five, therefore would be not just uneconomic; it would also perpetuate one of the gravest threats to US energy security.

The Substitution Option

The third option, though largely ignored, works better and costs less. It avoids all the problems of the first two options. It increases security instead of risks, saves money instead of spending it, and avoids the damage to our economy and environment that would come from rapidly depleting our domestic oil reserves. This option is to avoid using oil in the first place—that is, to reduce oil use through increases in efficiency, or to substitute alternative liquid fuels, or both.

The lower 49 states have two supergiant oil fields, each bigger than the biggest in Saudi Arabia, both nearly as economical (only a few dollars a barrel) and both about four-fifths untapped. They are the "weatherization oil field" in our attics and the "accelerated-scrappage-of-gas-guzzlers oil field" in Detroit. By saving oil, or natural gas that can replace oil, we could eliminate US oil imports. We could do so before any new power plant or synfuel plant ordered now could be built and before production from any new Arctic oil field could begin—and at a fivefold to tenfold lower cost. In fact, if we spent as much to make buildings heat-tight as we now spend in one year on the military forces meant to protect the Middle Eastern oil fields, we could eliminate the need to import any oil from the Middle East.

(An impractical kind of oil saving is sometimes proposed instead: building more coal or nuclear power stations. Since less than 5 percent of our electricity is made from oil, and less than 5 percent of our oil is used to make electricity, the two have almost no connection. Power plants are virtually irrelevant to the oil problem—except that the huge expense involved in building new ones would draw money away from investment in effective oil savings. The modest amounts of oil and gas still burned in power plants—and, for that matter, most of the coal and all of the uranium, too—can be cost-effectively displaced by superefficient new lights, motors, appliances, and building components.)

Saving oil isn't just theoretical. From 1977 to 1985, real US GNP grew 21 percent, the number of registered vehicles grew 20 percent, but total oil use fell 15 percent. The oil saving in 1985 equaled three times our 1986 imports from the Persian Gulf.

Americans now use 38 percent less oil and gas to produce a dollar of GNP than they did in 1973—and they achieved that saving mainly with caulk guns, duct tape, and slightly more fuel-efficient cars, not with the powerful new technologies that can now save even more energy at even lower cost. For example, full use of American-made superwindows, which insulate two to four times better than triple glazing, could save the nation more oil and gas than Alaska now supplies. Widespread use of these efficiency measures would cost less, protect the environment, and deplete no critical resource.

Last year the 13-year-old "energy-efficiency industry" produced, in effect, two-fifths more energy than the century-old oil industry. We're getting less domestic oil at higher costs each year, but more efficiency at lower costs. Reserves of oil are dwindling, but reserves of efficiency are expanding. Why, then, does federal policy emphasize depleting oil quickly and saving it slowly? The 1986 rollback of new-car efficiency standards from 27.5 to 26 miles a gallon is wasting more oil than the areas currently off limits in Alaska and offshore California might yield.

Conversely, improvements in the efficiency of the car fleet in use between 1973 and 1986 (from 13.1 to only about 18 miles per gallon) saved over twice as much oil last year as we imported from the Persian Gulf, or slightly more than Alaska's total output. We can do much better. The most efficient four-to-five-passenger cars in 1985 were getting more than 55 miles per gallon in commercial models and 70 to 100 mpg in prototypes.

After two previous oil crises, in 1973–1974 and 1979–1980, the United States tried ignoring efficiency and boosting supply. The result was overbuilt and insolvent supply industries that couldn't respond to the gush of energy savings produced in the marketplace. Today, with the potential for savings bigger than ever, the Reagan Administration seems determined to make the same mistake. When Donald Hodel, now Secretary of the Interior, was head of the Bonneville Power Administration, he proclaimed imminent electricity shortages in the Pacific Northwest and promoted the now notorious nuclear project WPPSS. Instead of the shortage prophesied, the northwestern states found themselves with a seemingly permanent surplus, triggering a $7 billion default. Now Hodel wants to inflict the same genius on the struggling oil industry.

"Drilling" for oil in our inefficient cars and buildings isn't instant or free. But it's faster and much cheaper, safer and far surer, than drilling anywhere else. Energy savings have already cut the national energy bill by some $150 billion a year. That's an average of more than $1700 a year cash savings for each household in the United States—tax-free extra income that largely recirculates in our local economics, supporting local jobs and local multipliers.

But this achievement represents a mere fraction of the amount of energy efficiency available and worth buying. If Americans were now as efficient as our Japanese and Western European competitors are—and even they have a long way to go—we'd save an additional $200 billion a year, which is more than last year's federal budget deficit. Buying the economically optimal amount of energy efficiency for the rest of this century could lead to net savings of several trillion dollars—enough, in principle, to pay off the entire national debt.

Energy inefficiency costs American jobs in world markets. Japan's higher energy efficiency, for example, gives all its exports an automatic cost advantage over ours, averaging about 5 percent—much more for energy-intensive products. Conversely, whether measured per unit of energy saved or per dollar invested, buying energy efficiency creates several times as many American jobs as supplying more energy: not jobs in boom-and-bust frontier towns, but jobs right in the communities of the people who need them.

The efficient use of oil can also buy time for the decades-long switch to the renewable sources that, one way or another, we'll adopt as oil becomes too costly. This transition won't be quick or cheap, but that's all the more reason for getting started now—before the cheap oil and the cheap money made from it are gone. Already, American oil is becoming costlier than imported oil, and the faster oil is used, the sooner other oil-supplying nations will find their oil becoming costlier than OPEC's huge reserves. The problem that we have now, others will have later, though Saudi Arabia (according to our present knowledge of petroleum geology) will have it last of all.

The short-term oil savings and diversification in our sources of oil extraction that have resulted from the past two oil shocks now offer a unique opportunity: roughly a two-decade-long respite (longer if the exploration of new areas is unexpectedly successful, shorter if federal policy continues to stifle gains in efficiency) from Middle Eastern dominance of the global oil supply. If this interval is frittered away, it could end with the United States, its alternative options expired, needing Middle Eastern oil more than ever. If, instead, we increase our oil efficiency and make sensible use of diverse alternative fuels, this grace period could expire on a United States that no longer substantially depends on oil from the Middle East or anywhere else outside our borders. Without efficient cars, no liquid-fuel future makes sense for long. With efficient cars, alcohols and other liquid fuels made from natural gas and sustainably grown biofuels—abundant or even inexhaustible resources, whose use poses little or no risk to the world's climate—can meet our energy needs at reasonable cost. Efficiency and alternative fuels are natural partners. With both, we can with confidence buy American.

Chapter 16

Reinventing the Wheels

1995

Editor's note: *This article was first published in the* Atlantic Monthly, *January 1995.*

AUTHOR'S NOTE: Since this article was written (coauthored with L. Hunter Lovins) in late 1993, many things have changed. The simple GM mathematical model we'd used to simulate car efficiency turned out to overstate supereffi-cient cars' efficiency by about twofold, but plug-in hybrids, entering major automakers' showrooms in 2010–2011, roughly offset that error in fuel consumption, and the underlying technological data proved valid. (Thus the ambitious efficiency targets stated in the article are overstated in terms of total propulsive energy, but are pretty close in terms of liquid fuel requirements—using the best 2010 technologies.) In the seven years after I open-sourced the Hypercar® concept in 1993, the industry committed roughly $10 billion to this line of development, making hybrids now ubiquitous and lightweighting (led by Ford, Nissan, Audi, and the Chinese industry) a rapidly spreading strategic focus. In 1999, RMI spun off Hypercar, Inc., which in 2000 designed with European partners an uncompromised 67-mile/USgal midsize SUV (114 mpg with hydrogen), respectively 3.6 times and 6.3 times normal efficiency. The capital market collapsed just as we sought production funds, but this highly integrative design (www.rmi.org/rmi/Library/T04-01_HypercarsHydrogen AutomotiveTransition) remains influential, and in 2004 (www.oilend game.com) we showed a one-year payback for its extra cost (which is only because it's a hybrid—the ultralighting was indeed free). In 2007, Toyota showed the first major-automaker concept hypercar; the carbon-fiber 1/X had the interior volume of a Prius, half its fuel use, and one-third its weight (926 lb with extra batteries to make it a plug-in hybrid, or, matching my early-90s predictions, 880 lb without them). The previous day, top carbon-fiber maker Toray announced a factory to "mass-produce carbon-fiber car parts for Toyota." Honda and Nissan did similar deals in 2008 and Mercedes in 2010, when Toray, Toyota, and Mitsubishi Rayon announced they're commercializing the process. In the mid-2000s, Hypercar, Inc. had morphed into Fiberforge Corporation (www.fiberforge.com) and pioneered rapid manufacturing technology for cost-effective carbon-fiber-composite structures. In 2011–2011, BMW, VW, and Audi announced 2012–2013 volume production of electrified carbon-fiber cars—VW's a two-seat hypercar at 230 mpg. In 2009, another RMI spinoff (www.brightautomotive.com) showed a driving prototype of a 3–12 times more fuel-efficient aluminum-intensive commercial van that, unlike other plug-in hybrids, needs no subsidy because its reduced weight and drag eliminated most of its costly batteries. Thus by 2010 the leapfrog we'd envisaged in 1993–1995 was finally off and running. Regrettably, as we'd warned, two of the Big Three didn't adapt in time, and even today, government policy is only starting to catch up.

New ways to design, manufacture, and sell cars can make them ten times more fuel efficient, and at the same time safer, sportier, more beautiful and comfortable, far more durable, and probably cheaper. Here comes the biggest change in industrial structure since the microchip.

On 29 September 1993, the unthinkable happened. After decades of adversarial posturing, and months of intensive negotiations with Vice President Al Gore, the heads of the Big Three automakers accepted President Bill Clinton's challenge to collaborate. They committed their best efforts, with the help of government technologies and funding, to developing a tripled-efficiency "clean car" within a decade, and a year later they reported encouraging progress. Like President John F. Kennedy's goal of putting people on the moon, the Partnership for a New Generation of Vehicles (PNGV) aims to create a leapfrog mentality—this time in Detroit. However, the PNGV's goal is both easier to attain and more important than that of the Apollo program. It could even become the core of a green industrial renaissance—instigating a profound change not only in what and how much we drive but in how our whole economy works.

The fuel efficiency of cars has been stagnant for the past decade. Yet the seemingly ambitious goal of tripling it in the next decade can be far surpassed. Well before 2003, competition, not government mandates, may bring to market cars efficient enough to carry a family coast to coast on one tank of fuel, more safely and comfortably than they can travel now, and more cleanly than they would with a battery-electric car plus the power plants needed to recharge it.

To understand what a profound shift in thinking this represents, imagine that one seventh of America's gross national product is derived from the Big Three typewriter makers (and their suppliers, distributors, dealers, and other attendant businesses). Over decades they've progressed from manual to electric to type-ball designs. Now they're developing tiny refinements for the forthcoming Selectric XVII. They profitably sell around 10 million excellent typewriters a year. But a problem emerges: the competition is developing wireless subnotebook computers.

That's the Big Three automakers today. With more skill than vision, they've been painstakingly pursuing incremental refinements on the way to an America where foreign cars fueled with foreign oil cross crumbling bridges. Modern cars are an extraordinarily sophisticated engineering achievement—the highest expression of the Iron Age. But they are obsolete, and the time for incrementalism is over. Striking innovations have occurred in advanced materials, software, motors, power electronics, microelectronics, electricity storage devices, small engines, fuel cells, and computer-aided design and manufacturing.

Artfully integrated, they can yield safe, affordable, and otherwise superior family cars getting hundreds of miles per gallon—roughly ten times the 30 mpg of new cars today and several times the 80-odd mpg sought by the PNGV. Achieving this will require a completely new car design—the ultralight hybrid, or "Hypercar" (a term we now prefer to our earlier term "supercar," because that also refers to ultrapowerful cars that get a couple of hundred miles per hour rather than per gallon). The Hypercar's key technologies already exist. Many firms around the world are starting to build prototypes. The United States is best positioned to bring the concept to market—and had better do so, before others do. Hypercars, not imported luxury sedans, are the biggest threat to Detroit. But they are also its hope of salvation.

The Ultralight Strategy

Decades of dedicated effort to improve engines and powertrains have reduced to only about 80–85 percent the portion of cars' fuel energy that is lost before it gets to the wheels. (About 95 percent of the resulting wheelpower hauls the car itself, so that less than 1 percent of the fuel energy actually ends up hauling the driver.)

This appalling waste has a simple main cause: cars are made of steel, and steel is heavy, so powerful engines are required to accelerate them. Only about one-sixth of the average engine's power is typically needed for highway driving, and only about one-twentieth for city driving. Such gross oversizing halves the engine's average efficiency and complicates efforts to cut pollution. And the problem is getting worse: half the efficiency gains since 1985 have been squandered on making engines even more powerful.

Every year automakers add more gadgets to compensate a bit more for the huge powertrain losses inherent in propelling steel behemoths. But a really efficient car can't be made of steel, for the same reason that a successful airplane can't be made of cast iron. We need to design cars less like tanks and more like airplanes. When we do, magical things start to happen, thanks to the basic physics of cars.

Because about five to seven units of fuel are needed to deliver one unit of energy to the wheels, saving energy at the wheels offers immensely amplified savings in fuel. Wheelpower is lost in three ways. In city driving on level roads about a third of the wheelpower is used to accelerate the car, and hence ends up heating the brakes when the car stops. Another third (rising to 60–70 percent at highway speeds) heats the air the car pushes aside. The last third heats the tires and the road.

The key to a superefficient car is to cut all three losses by making the car very light and aerodynamically slippery, and then recovering most of its braking energy. Such a design could:

- cut weight (hence the force required for acceleration) by 65–75 percent through the use of advanced materials, chiefly synthetic composites, while improving safety through greater strength and sophisticated design;
- cut aerodynamic drag by 60–80 percent through sleeker streamlining and more compact packaging; and
- cut tire and road energy loss by 65–80 percent through the combination of better tires and lighter weight.

Once this "ultralight strategy" has largely eliminated the losses of energy that can't be recovered, the only other place the wheelpower can go is into braking. And if the wheels are driven by special electric motors that can also operate as electronic brakes, they can convert unwanted motion back into useful electricity.

However, a Hypercar isn't an ordinary electric car, running on batteries that are recharged by being plugged into utility power. Despite impressive recent progress, such cars still can't carry very much or go very far without needing heavy batteries that suffer from relatively high cost and short life. Since gasoline and other liquid fuels store a hundred times as much useful energy per pound as batteries do, a long driving range is best achieved by carrying energy in the form of fuel, not batteries, and then burning that fuel as needed in a tiny onboard engine to make the electricity to run the wheel motors. A few batteries (or, soon, a carbon-fiber "superflywheel") can temporarily store the braking energy recovered from those wheel motors and reuse at least 70 percent of it for hill climbing and acceleration. With its power so augmented, the engine needs to handle only the average load, not the peak load, so it can shrink to about a tenth the current normal size. It will run at or very near its optimal point, doubling efficiency, and turn off whenever it's not needed.

This arrangement is called a "hybrid-electric drive," because it uses electric wheel motors but makes the electricity onboard from fuel. Such a propulsion system weighs only about a fourth as much as that of a battery-electric car, which must haul a half-ton of batteries down to the store to buy a six-pack. Hybrids thus offer the advantages of electric propulsion without the disadvantages of batteries.

One Plus Two Equals Ten

Automakers and independent designers have already built experimental cars that are ultralight or hybrid-electric but seldom both. Yet combining these approaches yields extraordinary, and until recently little-appreciated, synergies. Adding hybrid-electric drive to an ordinary car increases its efficiency by about a third to a half. Making an ordinary car ultralight but not hybrid approximately doubles its efficiency. Doing both can boost a car's efficiency by about tenfold [with the "plug-in hybrids" entering the market in 2010–2011].

This surprise has two main causes. First, as already explained, the ultralight loses very little energy irrecoverably to air and road friction, and the hybrid-electric drive recovers most of the rest from the braking energy. Second, saved weight compounds. When you make a heavy car one pound lighter, you in effect make it about a pound and a half lighter, because it needs a lighter structure and suspension, a smaller engine, less fuel, and so forth to haul that weight around. But in an ultralight, saving a pound may save more like five pounds, partly because power steering, power brakes, engine cooling, and many other normal systems become unnecessary. The design becomes radically simpler. Indirect weight savings snowball faster in ultralights than in heavy cars, faster in hybrids than in non-hybrids, and fastest of all in optimized combinations of the two.

All the ingredients needed to capture these synergies are known and available. As far back as 1921, German automakers demonstrated cars that were about one-third more slippery aerodynamically than today's cars are. Most of the drag reduction can come from such simple means as making the car's underside as smooth as its top. Today's best experimental family cars are 35 percent more slippery still. At the same time, ultrastrong new materials make the car's shell lighter. A lighter car needs a smaller engine, and stronger walls can be thin; both changes can make the car bigger inside but smaller outside. The smaller frontal area combines with the sleeker profile to cut through the air with about one-third the resistance of today's cars. Advanced aerodynamic techniques may be able to double this saving.

Modern radial tires, too, waste only half as much energy as 1970s bias-ply models, and the best 1990 radials roughly halve the remaining loss. "Rolling resistance" drops further in proportion to weight. The result is a 65–80 percent decrease in losses to rolling resistance, which heats the tires and the road.

Suitable small gasoline engines, of the size found in outboard motors and scooters, can already be more than 30 percent efficient, diesels 40–50 percent (56 percent in lab experiments). Emerging technologies also look promising, including miniature gas turbines and fuel cells—solid-state, no-moving-parts devices that silently and very efficiently turn fuel into electricity, carbon dioxide, water, and a greatly reduced amount of waste heat.

In today's cars, accessories—power steering, heating, air conditioning, ventilation, lights, and entertainment systems—use about a tenth of the engine's power. But a Hypercar would use scarcely more energy than that for all purposes, by saving most of the wheelpower and most of the accessory loads. Ultralights not only handle more nimbly, even without power steering, but also get all-wheel anti-lock braking and anti-slip traction from their special wheel motors. New kinds of headlights and taillights shine brighter on a third the energy, and can save even more weight by using fiber optics to distribute a single pea-sized lamp's light throughout the car. Air conditioning would need perhaps a tenth the energy used by today's car air conditioners, which are big enough for an Atlanta house. Special paints, vented double-skinned roofs, visually clear but heat-reflecting windows, solar-powered vent fans, and so forth can exclude unwanted heat; innovative cooling systems, run not directly by the engine but by its otherwise wasted by-product heat, can handle the rest.

Perhaps the most striking and important savings would come in weight. In the mid-1980s many automakers demonstrated "concept cars" that would carry four or five passengers but weighed as little as 1000 pounds (as compared with today's average of about 3200).

Conventionally powered by internal combustion, they were two to four times as efficient as today's average new car. Those cars, however, used mainly light metals like aluminum and magnesium, and lightweight plastics. The same thing can be done better today with composites made by embedding glass, carbon, polyaramid, and other ultrastrong fibers in special moldable plastics— much as wood embeds cellulose fibers in lignin.

In Switzerland, where more than 2000 lightweight battery-electric cars (a third of the world's total) are already on the road, the latest roomy two-seaters weigh as little as 575 pounds without their batteries. Equivalent four-seaters would weigh less than 650 pounds, or less than 850 including a whole hybrid propulsion system. Yet crash tests prove that such an ultralight can be at least as safe as today's heavy steel cars, even if it collides head-on with a steel car at high speed. That's because the composites are extraordinarily strong and bouncy, and can absorb far more energy per pound than metal can. Materials and design are much more important to safety than mere mass, and the special structures needed to protect people don't weigh much. (For example, about ten pounds of hollow, crushable carbon-fiber-and-plastic cones can absorb all the crash energy of a 1200-pound car hitting a wall at 50 mph.) Millions have watched on TV as Indianapolis 500 race cars crashed into walls at speeds around 230 mph: parts of the cars buckled or broke away in a controlled, energy-absorbing fashion, but despite per-pound crash energies many times those of highway collisions, the cars' structure and the drivers' protective devices prevented serious injury. Those were carbon-fiber cars.

In 1991, fifty General Motors experts built an encouraging example of ultralight composite construction, the sleek and sporty four-seat, four-airbag Ultralite, which packs the interior space of a Chevrolet Corsica into the exterior size of a Mazda Miata. The Ultralite should be both safer and far cleaner than today's cars. Although it has only a 111-horsepower engine, smaller than a Honda Civic's, its light weight (1400 pounds) and low air drag, both less than half of normal, give it a top speed of 135 mph and a 0-to-60 acceleration of 7.8 seconds—comparable to a BMW 750iL with a huge V-12 engine. But the Ultralite is more than four times as efficient as the BMW, averaging 62 mpg—twice today's norm. At 50 mph it cruises at 100 mpg on only 4.3 horsepower, a mere fifth of the wheelpower normally needed.

If equipped with hybrid drive, this 1991 prototype, built in only a hundred days, would be three to six times as efficient as today's cars. Analysts at Rocky Mountain Institute have simulated 300–400-mpg four-seaters with widely available technology, and cars getting more than 600 mpg with the best ideas now in the lab. Last November a four-seater, 1500-pound Swiss prototype was reported to achieve 90 mpg cruising on the highway; at urban speeds, powered by its 573 pounds of batteries, it got the equivalent of 235 mpg.

Similar possibilities apply to larger vehicles, from pickup trucks to 18-wheelers. A small Florida firm has tested composite delivery vans that weigh less loaded than normal steel vans weigh empty, and has designed a halved-weight bus. Other firms are experimenting with streamlined composite designs for big trucks. All these achieve roughly twice normal efficiency with conventional drivelines, and could redouble that with hybrids.

Hypercars are also favorable to—though they don't require—ultraclean alternative fuels. Even a small, light, cheap fuel tank could store enough compressed natural gas or hydrogen for long-range driving, and the high cost of hydrogen would become unimportant if only a tenth as much of it were needed as would be to power cars like today's. Liquid fuels converted from sustainable farm and forestry wastes, too, would be ample to run such efficient vehicles without needing special crops or fossil hydrocarbons. Alternatively, solar cells on a Hypercar's body could recharge its onboard energy storage about enough to power a standard southern-California commuting cycle without turning on the engine.

Even if a Hypercar used conventional fuel and no solar boost, its tailpipe could emit less pollution than would the power plants needed to recharge a battery-electric car. Being therefore cleaner, even in the Los Angeles airshed, than so-called zero-emission vehicles (actually "elsewhere-emission," mainly from dirty coal-fired power plants out in the desert), ultralight hybrids should qualify as ZEVs, and probably will. Last May the California Air Resources Board reaffirmed its controversial 1990 requirement—which some northeastern states want to adopt as well—that 2 percent of new-car sales in 1998, rising to 10 percent in 2003, be ZEVs. Previously this was deemed to mean battery-powered electric cars exclusively. But, mindful of Hypercars' promise, the CARB staff is considering and in April 1995 finally announced an intention of broadening the ZEV definition to include anything cleaner. This alternative compliance path could be a big boost both for Hypercar entrepreneurs and for clean air: each car will be cleaner, and far more Hypercars than battery cars are likely to be bought. By providing a large payload, unlimited range, and high performance even at low temperatures, Hypercars vault beyond battery cars' niche-market limitations.

This result brings full circle the irony of California's ZEV mandate. Originally it drew howls of anguish from automakers worried that people would not buy enough of the costlier, limited-range cars it obliges them to sell.

The business press ridiculed California for trying to prescribe an impractical direction of technological development. Yet that visionary mandate is creating the solution to the problems. Like the aerospace, microchip, and computer industries, Hypercars will be the offspring of a technology-forcing government effort to steer the immense power of Yankee ingenuity. For it is precisely the California ZEV mandate that radically advanced electric-propulsion technology—thereby setting the stage for the happy combination with ultralight construction which we call the Hypercar.

Beyond the Iron Age

The moldable synthetic materials in the GM and Swiss prototypes have fundamental advantages over the metals that now dominate automaking. The modern steel car, which costs less per pound than a McDonald's quarter-pound hamburger, skillfully satisfies often conflicting demands (to be efficient yet safe, powerful yet clean): steel is ubiquitous and familiar, and its fabrication is exquisitely evolved. Yet this standard material could be quickly displaced—as has happened before. In the 1920s the wooden framing of US car bodies was rapidly displaced by steel. Today composites dominate boatbuilding and are rapidly taking over aerospace construction. Logically, cars are next.

Driving this transition are the huge capital costs of designing, tooling, manufacturing, and finishing steel cars. For a new model, a thousand engineers spend a year designing and a year making half a billion dollars' worth of car-sized steel dies, the costs of which can take many years to be recovered. This inflexible tooling in turn demands huge production runs, maroons company-busting investments if products flop, and magnifies financial risks by making product cycles go further into the future than markets can be forecast. That this process works is an astonishing accomplishment, but it's technically baroque and economically perilous.

Moldable composites must be designed in utterly different shapes. But their fibers can be aligned to resist stress and interwoven to distribute it, much as a cabinetmaker works with the grain of wood. Carbon fiber can achieve the same strength as steel at half to a third of the weight, and for many uses other fibers, such as glass and polyaramid, are as good as or better than steel and 50–85 percent cheaper than carbon fiber. But composites' biggest advantages emerge in manufacturing.

Only 15 percent of the cost of a typical steel car part is for the steel; the other 85 percent pays for pounding, welding, and smoothing it. But composites and other molded synthetics emerge from the mold already in virtually the required shape and finish. And large, complex units can be molded in one piece, cutting the parts count to about 1 percent of what is now normal, and the assembly labor and space to roughly 10 percent. The lightweight, easy-to-handle parts fit together precisely. Painting—the hardest, most polluting, and costliest step in automaking, accounting for nearly half the cost of painted steel body parts—can be eliminated by laid-in-the-mold color. Unless recycled, composites last virtually forever: they don't dent, rust, or chip. They also permit advantageous car design, including frameless monocoque bodies (like an egg, the body *is* the structure), whose extreme stiffness improves handling and safety.

Composites are formed to the desired shape not by multiple strikes with tool-steel stamping dies but in single molding dies made of coated epoxy. These dies wear out much faster than tool-steel dies, but they're so cheap that their lack of durability doesn't matter. Total tooling cost per model is about half to a tenth that of steel, because far fewer parts are needed; because only one die set per part is needed, rather than three to seven for successive hits; and because the die materials and fabrication are much cheaper. Stereolithography—a three-dimensional process that molds the designer's computer images directly into complex solid objects overnight—can dramatically shrink tooling time. Indeed, the shorter life of epoxy tools is a fundamental strategic advantage, because it permits the rapid model changes and continuous improvement that product differentiation and market nimbleness demand—a strategy of small design teams, small production runs, a time to market of only weeks or months, rapid experimentation, maximum flexibility, and minimum financial risk.

Together these advantages cancel or overturn the apparent cost disadvantage of the composites. Carbon fiber recently cost around 40 times as much per pound as sheet steel, though increased production is leading manufacturers to quote carbon prices half to a quarter of that. Yet the cost of a mass-produced composite car is probably comparable to or less than that of a steel car, at both low production volumes (like Porsche's) and high ones (like Ford's). What matters is not cost per pound but cost per *car*: costlier fiber is offset by cheaper, more agile manufacturing.

Shifting Gears in Competitive Strategy

Ultralight hybrids are not just another kind of car. They will probably be made and sold in completely new ways. In industrial and market structure they will be as different from today's cars as computers are from typewriters, fax machines from telexes, and satellite pagers from the Pony Express.

Many people and firms in several countries are starting to realize what Hypercars mean; at least a dozen capable entities, including automakers, want to sell them. This implies rapid change on an unprecedented scale. If ignored or treated as a threat rather than grasped as an opportunity, the Hypercar revolution could cost the United States millions of jobs and thousands of companies. Automaking and associated businesses employ one-seventh of US workers (and close to two-fifths of workers in some European countries). Cars represent a tenth of America's consumer spending, and use nearly 70 percent of the nation's lead, about 60 percent of its rubber, carpeting, and malleable iron, 40 percent of its machine tools, 15 percent of its aluminum, glass, and semiconductors, and 13 percent of its steel. David Morris, a cofounder of the Institute for Local Self-Reliance, observes:

The production of automobiles is the world's number-one indus-
try. The number-two industry supplies their fuel. Six of America's
ten largest industrial corporations are either oil or auto compa-
nies. ... A recent British estimate concludes that half of the
world's earnings may be auto- or truck-related.

Whether the prospect of Hypercars is terrifying or exhilarating thus depends on how well we grasp and exploit their implications.

The distribution of Hypercars could be as revolutionary as their manufacture. On average, today's cars are marked up about 50 percent from production costs (which include profit, plant costs, and warrantied repairs). But cheap tooling might greatly reduce the optimal production scale for Hypercars. Cars could be ordered directly from the local factory, made to order, and delivered to one's door in a day or two. (Toyota now takes only a few days longer than that with its steel cars in Japan.) Being radically simplified and ultrareliable, they could be maintained by technicians who come to one's home or office (Ford does this in Britain today), aided by plug-into-the-phone remote diagnostics. If all this makes sense for a $1500 mail-order personal computer, why not for a $15,000 car? Such just-in-time manufacturing would eliminate inventory, its carrying and selling costs, and the discounts and rebates needed to move existing stock that is mismatched to demand. The present markup could largely vanish, so that Hypercars would be profitably deliverable at or below today's prices even if they cost considerably more to make, which they probably wouldn't.

America leads—for now—both in start-up-business dynamism and in all the required technical capabilities. After all, Hypercars are much more like computers with wheels than they are like cars with chips: they are more a software than a hardware problem, and competition will favor the innovative, not the big. Comparative advantage lies not with the most efficient steel-stampers but with the fastest-learning systems integrators—with innovative manufacturers like Hewlett-Packard and Compaq, and strategic-element makers like Microsoft and Intel, more than with Chrysler or Matsushita. But even big and able firms may be in for a rough ride: the barriers to market entry (and exit) should be far lower for Hypercars than for steel cars. Much as in existing high-tech industries, the winners might be some smart, hungry, unknown aerospace engineers tinkering in a garage right now—founders of the next Apple or Xerox.

All this is alien to the thinking of most (though not all) automakers today. Theirs is not a composite-molding/electronics/software culture but a diemaking/steelstamping/mechanical culture. Their fealty is to heavy metal, not light synthetics; to mass, not information. Their organizations are dedicated, extremely capable, and often socially aware, but have become prisoners of past expenditures. They treat those historical investments as unamortized assets, substituting accounting for economic principles and throwing good money after bad. They have tens of billions of dollars, and untold psychological investments, committed to stamping steel. They know steel, think steel, and have a presumption in favor of steel. They design cars as abstract art and then figure out the least unsatisfactory way to make them, rather than seeking the best ways to manufacture with strategically advantageous materials and then designing cars to exploit those manufacturing methods.

The wreckage of the mainframe-computer industry should have taught us that one has to replace one's own products with better new products before someone else does. Until recently, few automakers appreciated the starkness of the threat. Their strategy seemed to be to milk old tools and skills for decades, watch costs creep up and market share down, postpone any basic innovation until after all the executives' planned retirement dates—and hope that none of their competitors was faster. That's a bet-the-company strategy, because even one superior competitor can put a company out of business, and the company may not even know who the competitor is until too late. The PNGV is stimulating instead a winning, risk-managed strategy: leapfrogging to ultralight hybrids.

It is encouraging that some automakers now show signs of understanding the problem. In recent months the PNGV has sparked new thinking in Detroit. The industry's more imaginative engineers are discovering that the next gains in car efficiency should be easier than the last ones were, because they will come not from sweating off fat ounce by ounce but from escaping an evolutionary trap. Although good ultralight hybrids need elegantly simple engineering, which is difficult, one can more easily boost efficiency tenfold with Hypercars than threefold with today's cars.

Little of this ferment is visible from the outside, because automakers have learned reticence the hard way. A long and unhappy history of being required to do (or exceed) whatever they admit they can do has left them understandably bashful about revealing capabilities, especially to Congress. And firms with innovative ambitions will hardly be eager to telegraph them to competitors. Corporations share a natural desire to extract any possible business and political concessions, and to hold back from extending to traditional adversaries (such as the media, politicians, and environmentalists) any trust that could prove costly if abused or not reciprocated. Thus automakers are more likely to understate than to trumpet progress. Also, the Big Three are progressing unevenly, both internally and comparatively: their opacity conceals a rapidly changing mixture of exciting advances and inertia. Only some executives appreciate that Hypercars fit the compelling strategic logic in favor of changing how their companies do business, especially by radically reducing cycle times, capital costs, and financial risks. It is difficult but vital for harried managers to focus on these goals through the distracting fog of fixing flaws in their short-term operations. But signs of rapid cultural change are looming, such as General Motors' announcement, last 3 February, that its corporate policy now includes the CERES (Coalition for Environmentally Responsible Economies) Principles, formerly known as the Valdez Principles—a touchstone of environmentalists.

The Cost of Inaction

The potential public benefits of Hypercars are enormous—in oil displacement, energy security, international stability, forgone military costs, balance of trade, climatic protection, clean air, health and safety, noise reduction, and quality of urban life. Promptly and skillfully exploited, Hypercars could also propel an industrial renewal. They're good news for industries (many of them now demilitarizing) such as electronics, systems integration, aerospace, software, petrochemicals, and even textiles (which offer automated fiber-weaving techniques). The talent needed to guide the transition is abundant in American labor, management, government, and think tanks, but it's not yet mobilized. The costs of that complacency may be high.

Cars and light trucks use about 37 percent of the nation's oil, about half of which is imported at a cost of around $50 billion a year. We Americans recently put our sons and daughters in 0.56-mile-per-US-gallon tanks and 17-feet-per-gallon aircraft carriers because we hadn't put them in 32-mpg cars—sufficient, even if we'd done nothing else, to have eliminated the need for American oil imports from the Persian Gulf. Of course, more than just oil was at stake in the Gulf War, but we would not have sent half a million troops there if Kuwait simply grew broccoli. Even in peacetime, the direct cost to the nation of Persian Gulf oil—mostly paid not at the pump but in taxes for some $50 billion a year in military readiness to intervene in the Gulf—totals nearly $100 a barrel of crude, making it surely the costliest oil in the world.

Had we simply kept on saving oil as effectively after 1985 as we had saved it for the previous nine years, we wouldn't have needed a drop of oil from the Persian Gulf since then. But we didn't—and it cost us $23 billion for extra imports in 1993 alone. Gulf imports were cut by about 90 percent from 1977 to 1985 (chiefly by federal standards that largely or wholly caused new-car efficiency to double from 1973 to 1986). Yet they are now reapproaching a historic high—the direct result of 12 years of a national oil policy consisting mainly of weakened efficiency standards, lavish subsidies, and US military forces.

The national stakes therefore remain large. And even though the PNGV is starting to recreate Detroit's sense of adventure, Hypercars still face formidable obstacles, both culturally within the auto industry and institutionally in the marketplace. Whether or not their advantages make their ultimate adoption certain, the transition could be either unnecessarily disruptive, shattering industrial regions and job markets, or unnecessarily slow and erratic in capturing the strategic benefits of saving oil and rejuvenating the economy. Automakers should be given strong incentives to pursue the leapfrog strategy boldly, and customers should be encouraged to overcome their well-known lack of interest in buying fuel-thrifty cars in a nation that insists on gasoline cheaper than bottled water.

Market Conditioning and Public Policy

The usual prescription of economists, environmentalists, and the Big Three—though, it seems, a politically suicidal one—is stiff gasoline taxes. After painful debate Congress recently raised the gasoline tax by 4.3 cents a gallon, leaving the price, corrected for inflation, the lowest both in the industrial world and in US history. But in Western Europe and Japan, taxes that raise the price of motor fuel to two or four times that in the United States have long been in place, with unspectacular results. Gasoline costing two to five dollars a gallon has modestly reduced distances driven but has had less of an effect on the efficiency of new cars bought. New German and Japanese cars are probably less efficient than American ones, especially when performance, size, and features are taken into account. Costlier fuel is a feeble incentive to buy an efficient car, because the fuel-price signal is diluted (in the United States today, by seven to one) by the other costs of owning and running a car. It is, as well, weakened by high consumer discount rates over a brief expected ownership, and often vitiated by company-owned cars and other distortions that shield many drivers from their cars' costs. This market failure could be corrected by strengthening government efficiency standards.

But standards, though effective and a valuable backstop, are not easy to administer, can be evaded, and are technologically static: they offer no incentive to keep doing better. Happily, at least one market-oriented alternative is available: the "feebate."

Under the feebate system, when you buy a new car, you pay a fee or get a rebate. Which and how big depends on how efficient your new car is. Year by year the fees pay for the rebates. (This is not a new tax. In 1990 the California legislature agreed, approving a "Drive+" feebate bill by a seven-to-one margin, although outgoing Governor George Deukmejian vetoed it.) Better still, the rebate for an efficient new car could be based on how much more efficient it is than an old car that's scrapped (not traded in). A rebate of several thousand dollars for each 0.01-gallon-per-mile difference would pay about $5000 to $15,000 of the cost of an efficient new car. That would rapidly get efficient, clean cars on the road and inefficient, dirty cars off the road (a fifth of the car fleet produces perhaps three-fifths of its air pollution). The many variants of such "accelerated-scrappage" incentives would encourage competition, reward Detroit for bringing efficient cars to market, and open a market niche in which to sell them. Feebates might even break the political logjam that has long trapped the United States in a sterile debate over higher gasoline taxes versus stricter fuel-efficiency standards—as though those were the only policy options and small, slow, incremental improvements were the only possible technical ones.

Perhaps people would buy Hypercars, just as they switched from vinyl records to compact discs, simply because they're a superior product: cars that could make today's most sophisticated steel cars seem clunky and antiquarian by comparison. If that occurred, gasoline prices would become uninteresting. Scholastic debates about how many price elasticities can dance on the head of a pin would die away. The world oil price would permanently crash as super-efficient vehicles saved as much oil as OPEC now extracts. Feebates would remain helpful in emboldening and rewarding Detroit for quick adaptation, but perhaps would not be essential. The ultralight hybrid would sweep the market. What then?

Then we would discover that Hypercars cannot solve the problem of too many people driving too many miles in too many cars; indeed, they could intensify it, by making driving even more attractive, cheaper, and nearly free per extra mile driven. Having clean, roomy, safe, recyclable, renewably fueled 300 mpg cars doesn't mean that eight million New Yorkers or a billion still-carless Chinese can drive them. Drivers would no longer run out of oil or air but would surely run out of roads, time, and patience. Avoiding the constraint *du jour* requires not only having great cars but also being able to leave them at home most of the time. This in turn requires real competition among all modes of access, including those that displace physical mobility, such as telecommunications. The best of them is already being where we want to be—achievable only through sensible land use.

Such competition requires a level playing field with honest pricing, so that drivers (and everyone else) will both get what they pay for and pay for what they get. But least-cost choices are inhibited today by central planning and socialized financing of car-based infrastructure, such as roads and parking, while alternative modes must largely pay their own way. Happily, emerging policy instruments could foster and monetize fair competition among all modes of access. Some could even make markets in "negamiles" and "negatrips," wherein we could discover what it's worth paying people to stay off the roads so that we needn't build and mend them so much and suffer their delays and pollution. Congestion pricing, zoning reforms, parking feebates, pay-at-the-pump car insurance, commuting-efficient mortgages, and a host of other innovations beckon state, local, and corporate experimenters. Yet unless basic and comprehensive transport and land-use reforms emerge in parallel with Hypercars, cars may become apparently benign before we've gotten good enough at not needing to drive them—and may thus derail the reformers.

If the technical and market logic sketched here is anywhere near right, we are all about to embark on one of the greatest adventures in industrial history. Whether we will also have the wisdom to build a society worth driving in—one built around people, not cars—remains a greater challenge. As T. S. Eliot warned, "A thousand policemen directing the traffic / Cannot tell you why you come or where you go."

Chapter 17

Drilling in All the Wrong Places

2008

Editor's note: *This article is reprinted from Rocky Mountain Institute's* Solutions Journal, *Vol. 2, No. 2, Fall/Winter 2008.*

Drilling for oil in the Arctic National Wildlife Refuge should offend conservatives because it's insecure, unimportant, unprofitable, and uncompetitive.

Oklahoman ex-CIA Director R. James Woolsey testified against drilling because its

> *real show-stopper is national security. Delivering that oil by its only route, the 800-mile-long Trans-Alaska Pipeline System (TAPS), would make TAPS the fattest energy-terrorist target in the country—Uncle Sam's "Kick Me" sign. ... Doubling and prolonging dependence on TAPS ... imperils [national] security. ...*
>
> *TAPS is frighteningly insecure. It's largely accessible to attackers, but often unrepairable in winter. If key pumping stations or facilities at either end were disabled, at least the above-ground half of 9 million barrels of hot oil could congeal in one winter week into the world's biggest ChapStick®. The Army has found TAPS indefensible. It has already been sabotaged, incompetently bombed twice, and shot at more than 50 times[;] a drunk shut it down with one rifle shot. In 1999, a disgruntled engineer's sophisticated plot to blow up three critical points with 14 bombs, then profit from oil futures trading, was thwarted by luck. He was an amiable bungler compared with the [9/11] attackers.*

Importance? The Energy Information Administration (EIA) says the Refuge's limited and scattered oil—its biggest field is one-tenth of a Prudhoe Bay—could start flowing around 2018, peak in 2027 at 3 percent of US use, and temporarily cut oil import dependence by 2 percentage points and 2025 oil prices by 2 cents a gallon.

Profitability? EIA in May 2008 found today's quintupled oil prices won't yield more or earlier Refuge oil, because drilling costs have soared even higher: Alaskan onshore drilling costs rose 564 percent during 2000–2005, then *really* stood up on end. Today's soaring capital costs for frontier hydrocarbon projects strain even the biggest oil companies' exploration budgets. In 2001, Refuge oil's costs and risks were among the highest in the industry's global portfolios. Today's higher oil prices don't change prospects' *relative* merits, better technologies tend to advantage other prospects more, and volatile oil prices raise financial risks in a sour capital market, so Refuge oil still lacks a sound business case.

Competitiveness? My team's Pentagon-cosponsored 2004 study *Winning the Oil Endgame* (free at www.oilendgame.com) road-mapped eliminating US oil use by the 2040s, led by business for profit, at an average cost of $15 per barrel—lucrative at $26/bbl, far more so with today's far higher prices. Refuge oil would be costlier and slower than those efficiency and supply-side competitors, and they're getting cheaper.

So why press for a project that would create a new and even more vulnerable Strait of Hormuz, depend for decades on a geriatric pipeline (corroding, maintenance-challenged, already past its 30-year design life), yield little oil slowly and riskily, and lose money? Perhaps advocates simply misunderstand the nature of America's oil problem.

The US has lifted oil faster and longer than any other country, so it's more depleted, and the next barrel costs more at home than abroad. A market economy offers only three solutions: protectionism, trade, and substitution.

Protectionism distorts relative prices by taxing foreign oil (violating free-trade rules) or subsidizing domestic oil (suppressing efficient use). Both approaches weaken competitiveness. Both illogically suppose the solution to domestic depletion is to deplete faster—or, as David Brower said, "strength through exhaustion." Oil-less countries like Japan and Germany trade—buying oil from the cheapest sources (diversified and buffered by stockpiles), earning the money to pay for it, and maintaining good relations with exporters. The US buys copiously but lags in earnings and friendships.

By substituting resources that do oil's tasks better and cheaper, the US can lead the world beyond oil. Face facts: America's oil output peaked in 1970 and Texas is now a net importer of oil. Let's get on with what we *can* do together, better than anyone: saving oil quickly and depleting it slowly.

If the US had kept saving oil as fast as it did during 1976–1985, we wouldn't have needed any Persian Gulf oil ever since.

But now wildcatters are finding new gushers of savings: more than 8 million barrels per day (nearly a Saudi Arabia's worth) in the Detroit Formation, 0.9 in the Seattle Formation—in all, over 14 million barrels per day of "negabarrels" (saved oil) that is all-American and inexhaustible, climate-safe and secure, costing an average of $12 a barrel.

If oil companies went to the ends of the Earth drilling for very expensive oil that might not even be there, while innovators and entrepreneurs found all those negabarrels under Detroit, wouldn't the old-fashioned drillers be embarrassed, even bankrupt?

Smart developers drill the most prospective plays first. We should all be able to agree about that. If we do it, then the oil we don't agree about—at least 50 times smaller and several times costlier—will become superfluous, America will be richer and stronger, and the world will be cooler and safer.

Section 5

Our Human Environment: Building Better, Building Smarter

In 1981, Amory and his then wife L. Hunter Lovins moved from England back to the United States. They spent a year in Los Angeles and began thinking about conceptual design for a house they wanted to build in western Colorado, as well as launching their own organization to incubate inside it, which they did in April 1982: Rocky Mountain Institute (RMI). They felt that the building's physical form should reflect both their values and the important work they were doing. So, with architect Steve Conger, they created a design that included a research center at one end, their living space at the other end, and a jungle in between. Owner-built over the course of a year and a half with over 100 volunteers guided by professionals, the 372-square-meter building is about 99 percent passive-solar, superinsulated, and earth-sheltered. It has no heating system in the usual sense, but is kept comfortable even down to −44°C by passive-solar gain through the superwindows. To this day, tropical fruit trees and fish flourish in the greenhouse. Savings of 99 percent in space- and water-heating energy and 90 percent in household electricity repaid their initial cost in 10 months. After several further upgrades, the building is now more efficient and comfortable than ever, and Amory and his wife Judy Hill Lovins recently harvested the 36th passive-solar banana crop.

Amory added this experience to his intellectual portfolio and began doing work in the then nascent "green-building" industry (an unheard-of term three decades ago). By 1991, green building was enough of a concern that Amory and his colleague Bill Browning formed a consulting practice within RMI dedicated to green development. (Both were later elected honorary members of the American Institute of Architects.) Today, green building is a sizable and quickly expanding sector unto itself. Yet, as Amory repeatedly points out, most of the barriers to better building design are either ingrown institutional or chronic cultural habits—we can't seem to get over ourselves. "Japan has always been a source of technical innovation and thinking in different ways," Amory noted after a 2001 trip there. "When [the] field staff of certain utilities come back at the end of a day to catch up on their paperwork, for example, they all sit at one end of the room. That way, they can run the lighting and cooling in one spot and avoid wastefully lighting and cooling the entire room." So it's no surprise Amory shares, when applicable, lessons from other cultures and points out opportunities to improve our own. Indeed, the first essay in this selection is about the cultural anthropology of air conditioning, which can include your clothing, chair, and coworkers—but the most important thing is your perception. The second condenses his Centenary Address to the American Society of Heating, Refrigeration, and Air-Conditioning Engineers, featuring a commercial and residential example of superefficient designs he co-led. Finally, returning abroad, is an essay contrasting US with Japanese design practices and their underlying cultural contexts.

Chapter 18

Air Conditioning Comfort: Behavioral and Cultural Issues

1992

Editor's note: *In 1986, Rocky Mountain Institute founded Competitek, which provided its clients with detailed technical information on electricity-saving technologies and ways to implement them. Operating as an entrepreneurial program within RMI, Competitek grew to serve more than 200 client organizations in 15 countries and generated a path-breaking, world-renowned series of documents that defined the state of the art in electric end-use efficiency. In 1992, it was spun out as a for-profit subsidiary of RMI, moved to Boulder, Colorado, then renamed E SOURCE. Amory continued to work with his colleagues at E SOURCE and in 1992 authored two influential papers on air conditioning from a social science perspective and on the institutional barriers to energy efficiency in buildings. The first of those papers is excerpted here, focusing first on human comfort sensations and then on the equally fascinating interaction of users with air-conditioner controls. Why do such things matter? Roughly half the recent growth in Chinese electricity demand was to run air conditioners—and, as noted eslewhere, poor building designs and air-conditioning assumptions have misallocated about $1 trillion of US capital to air conditioners and their power supplies.*

Executive Summary

Better understanding of what determines people's perceptions of thermal comfort under hot and/or humid conditions could lead to fundamental changes in the design and use of air-conditioning systems, reducing peak loads on electric utility systems and saving billions of dollars in investment in cooling equipment and the power plants to run it. The energy savings could be significant: an increase of just 1 C° in design room temperature in a large commercial building typically reduces cooling energy by several percent. Conventional "comfort engineering" has overlooked factors such as acclimatization, depend-ence, and physiological variations among individuals. Even the basic precept of the engineering paradigm—that people prefer a constant temperature—now appears faulty. When choosing comfort conditions in everyday life, people show a far wider range of preferences than they do in laboratory experiments. Existing comfort standards are thus more stringent than can be justified physiologically or economically. Design approaches based on an anthropological understanding of how people operate air conditioners are likely to save energy and yield other benefits, particularly in the rapidly developing tropical nations of the Pacific Rim. Just as task lighting can improve visual comfort and performance while reducing lighting energy, so "task conditioning" may eventually achieve similar comfort and productivity benefits—office workers stay healthier and perform better given more control over their workplace environment. Studies of how people actually operate window air conditioners also suggest opportunities for better control systems that could both improve comfort and save energy.

1. Four Comfort Paradigms

People are complex. Human behavior generally reflects a variety of motives, and varies between individuals, cultures, and times. In particular, behavior in using space-cooling and air-handling equipment is far more intricate and mysterious than meets the eye. To understand why and how people use space-conditioning energy requires less the insights of utility practice, comfort engineering, and economics than of anthropology, ethnology, sociology, and psychology. Effectively implementing more efficient air conditioning requires a mix of all these perspectives, with special emphasis on the social sciences.

In the *electric-utility paradigm*, air conditioning is an unpredictable, inter-mittent, inefficient, and expensive load to serve: one anonymous utility executive calls it "the load from hell." People use air conditioning chiefly on hot, muggy, still afternoons and evenings, and especially after a succession of several such days. Air conditioning has peak loads, energy intensities, ramp rates, load factors, power factors, diversity factors, coincidence factors, and correlation coefficients. It depends on weather, demographics, architecture, saturations, market segments, time of the day and week, efficiency and load-management programs, and tariff design.

In the *engineering paradigm*, people use air conditioning because their body experiences conditions above a "comfort range" that is claimed to apply to all human beings everywhere. This "comfort range" is defined by labora-tory tests involving four physically measurable parameters—air temperature, mean radiant temperature, humidity, and airspeed—and two personally estimable parameters—clothing and metabolic rate (which depends on activity level). People are seen as ambulatory, semipermeable, watery sacks with a typical mass of 70 kg,[1] length of 1.73 m, and surface area of 1.83 m², endowed with limited thermoregulatory capabilities and with choices of clothing and activity level, expressed respectively in units of the *clo* (0.155 m^{2}C°/W) and the *met* (58.2 W/m^2, or ~106 W for an average sedentary man; modern office work entails ~1.2–1.3 *met* of effort). In this view, it is not necessary for people to have brains or souls, and it may be inconvenient if they do. Rather, they have an Index of Skin Wettedness, a Heat Stress Index, a Predicted Mean Vote. They are deemed to experience Mean Radiant Temperatures and Corrected Effective Temperatures. They feel comfortable when a "comfort equation," taking three lines to print, is satisfied. They have radiative, absorptivity, evaporative, and convective heat transfer coefficients, vasomotor regulation, and neurophysiological mechanisms—including sensations of discomfort in environments too far from a "neutral" temperature at which the heat flows in and out of the body are equal.[2]

In the *economic paradigm*, people use air conditioning because they derive utility from it. They buy as much of it as they want (in comparison to other things with different relative prices and utilities) and can afford. People are seen as rational consumers balancing alternative marginal investment and consumption choices. People have incomes, price and income elasticities, opportunity costs, discount rates, utility functions, marginal cost–benefit calculi, and hedonic indices.[3]

In the *social-science paradigm*, people use air conditioning for complex physiological, psychological, and cultural reasons. In this view, people have values, wishes, desires, needs, ideals, conflicts, hopes, fears, feelings, peer pressures, self-images, myths, and beliefs, all embedded in cultural and historical contexts. People are social as well as individual. They differ not only in their perceptions of comfort but also in their individual and cultural need for (or aversion to) artificial cooling and what it implies about harmony with or separation from the natural world. Potent influences on people's feelings about air conditioning "range from acquiring modern status in a tropical country to the social acceptability of perspiration. ... With air conditioning, more than most energy-using technologies, culture and behavior can dominate the [utility] load characteristics."[4]

These four paradigms are all useful for different purposes and in differing degrees. They are more useful when integrated than when set in opposition to each other. None is complete by itself, and since all are only mental models, none can perfectly predict human choices: the map is not the territory. The engineering paradigm is especially valuable in emphasizing

> *that it is the combined thermal effect of all physical factors which is of prime importance for man's thermal state and comfort. It is impossible to consider the effect of any of the physical factors influencing thermal comfort independently, as the effect of each of them depends on the level of the other factors. Therefore, emphasis [has to be placed] on the combined influence and the interactions between the effects of the various factors.*[5]

But as will be seen, the social sciences offer especially fruitful practical insights that cannot be gained in other ways. Careful field experiments confirm that the engineering and economic paradigms are not only incomplete but often seriously misleading predictors of space-conditioning behavior, so relying excessively on them can lead to serious implementation problems. Some of the best of this evidence is conveniently collected in a Special Issue of *Energy and Buildings*.[6] The following discussion draws heavily on that important contribution to understanding how people choose, use, and experience air conditioning.

2. Who Wants How Much Space Cooling and Why?

The crux of the engineering definition of comfort is that people in a thermal environment (air and radiant temperature, humidity, and air movement) that places them outside the officially defined "comfort envelope" for their activity and clothing levels will be uncomfortable and will want, or in extreme conditions require, artificial cooling. In this view, "all persons are most comfortable at certain temperatures [and] air conditioning should be thermostatically controlled because people want a constant temperature."[7] This theory rests on decades of laboratory measurements in which volunteers (usually paid) sit quietly in carefully controlled climate chambers in a laboratory and vote periodically on their degree of thermal comfort. Statistical fits to their quite variable responses are then used to define conditions in which, theoretically, 80 percent of people will feel comfortable.[8] These conditions are defined by "comfort equations," mainly based on the one published in 1970 by P. Ole Fanger at the Technical University of Denmark and resting on both the statistical data fits and engineering principles of heat transfer between the human body and its environment. Fanger's and subsequently refined comfort equations underlie current international standards, notably ASHRAE 55-81 and ISO 7730.[9] The limitations of this model, however, should be clearly understood:

> The comfort equation concerns itself solely with the average optimum temperature of a group of people. It does not say anything about the sensation experienced at non-optimum temperatures, or what deviation of temperature is acceptable for an individual or group. This requires some quantifiable estimate of either sensation or discomfort; the two are qualitatively different, although often confounded in practice.[10]

Moreover, the curve fits are very fuzzy: on a seven-point balloting scale, the range normally assumed to connote the "comfort range" (2.5–5.5) is as large as 9 C° wide for the average individual, and individual variations can add another ±7 C° or so to that range.[11] And although this model of comfort has proven highly successful in narrow terms and for certain uses, evidence for its general and uniform validity is at best equivocal, for the reasons explored next.

2.1 Physiological diversity

First, different people differ physiologically even within samples relatively homogeneous as to race, culture, age, gender, health, and other attributes. Fanger's classic treatise[12] shows without comment two remarkable graphs illustrating this. In the first (Figure 18.1), different American and Danish subjects' skin temperature at a given ratio of metabolic rate to body surface area varied within a range of ~3 C° (a range which if applied to internal temperature would correspond to the difference between normal and a dangerously high fever). Even more impressively, Figure 18.2 shows that their evaporative heat loss—a measure of propensity to sweat and to lose moisture through insensible skin diffusion—*varied by ~4–6 times* at each activity level throughout a wide range. Figure 18.2 also shows that from one individual to the next outwardly similar individual at the same typical sedentary activity level, *the amount of body heat dissipated evaporatively (y-axis) can vary from about one-eighth to more than three-fourths of total body heat loss (x-axis).* At higher activity levels, this evaporated-away fraction ranges between individuals from about one-fifth to two-thirds.[13] No wonder people can't all agree on what's comfortable! And of course people who differ that much in sweatiness will have profoundly different sensitivity to humidity and air movement too, not just to air temperature.

Within the more heterogeneous US culture, there is also impressive individual variation. In a small New Jersey sample of people for whom air conditioning

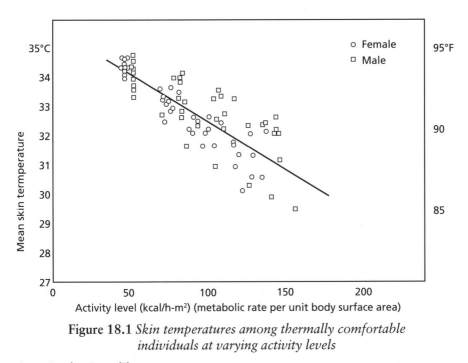

Figure 18.1 *Skin temperatures among thermally comfortable individuals at varying activity levels*

Source: Data from Fanger [2]

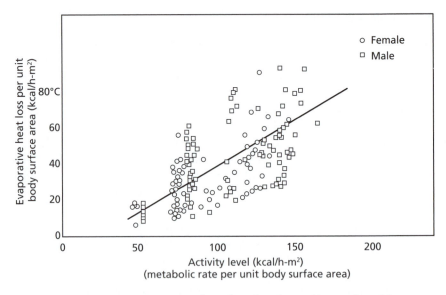

Figure 18.2 *Evaporative heat loss for thermally comfortable individuals at varying activity levels*

Source: Data from Fanger [2]

cost nothing, the steady-state summer comfort temperatures they chose, consistently valid for each individual throughout a cooling season, ranged from 22.2±1.1 to 28.4±0.6°C—a 6.2 C° range, *fivefold* larger than the ±1.2 C° range (one standard deviation each way) of "interindividual variability in preferred temperatures ... that might be expected from laboratory studies."[14] Similarly, daytime temperatures in Japanese households were found to be ~3.1 C° (~5.6 F°) above the normal upper comfort limits, i.e. up to ~30°C.[15] Moreover, international comparative "studies of households with similar incomes and energy prices show that they choose [winter] indoor temperatures ranging from 14 to 21°C (57 to 70°F)."[16] That is, when choosing their comfort conditions in everyday life, people show a far wider range of preferences than they do in carefully controlled laboratory experiments in climate chambers. This presumably reflects some combination of physiological diversity with the diverse expectations, motivations, and beliefs discussed below.

Comfort analysts often erroneously assume that clothing and other conditions affecting comfort are essentially homogeneous within a space. On the contrary, one can commonly observe, especially in summer, male office workers wearing an ensemble in the ~1.0–1.2 *clo* range while adjacent female office workers are wearing as little as 0.5 *clo*—a level unattainable by males within most formal office dress codes, since these are typically written by men and express preferences unrelated to thermal comfort.

This difference, too, may be only the beginning of a major comfort disparity. For example, if a male executive works in a corner office with large areas of sunlit heat-absorbing glass, the resulting high drybulb and radiant temperature may combine with his heavier clothing to make him demand a blast of cold air that more lightly clad females find intolerable. Adjacent office workers can thus experience conditions whose comfort envelopes do not overlap at all, even if men and women have no physiological differences in their comfort sensations.

2.2 Psychological influences

The importance of mind–body relationships that can distort laboratory measurements of perceived comfort was once shown persuasively, if inadvertently, when a reconditioned meat-cooling locker, with white enamel finish and pull-latch door, was temporarily used as a second climatic chamber for thermal comfort experiments. Subjects using this chamber recorded the same comfort sensations and skin temperatures as in a conventional chamber under identical conditions, but consistently balloted a 1.3 C° (2.4 F°) cooler temperature sensation than in the standard chamber. This purely psychological effect of the decor—indistinguishable from physiological coolness by all the standard engineering indicators of comfort—disappeared when the meat locker was redecorated with carpeting, wood paneling, and an acoustic-tile ceiling.[17] Such an experimental result might have been expected to stimulate questions about the soundness of comfort theory's foundations, but apparently this did not occur.

Behavioral feedback and motivational reinforcement can also influence perceptions of comfort, or at least the acceptability of what a laboratory-oriented comfort scientist might consider presumptive discomfort. In 1976, Princeton University researchers showed that summer electricity consumption in a group of townhouses was reduced 10.5 percent by almost daily feedback about consumption, 13.0 percent with frequent feedback plus exhortation to achieve a difficult "conservation" (i.e. curtailment) goal, and 15.7 percent by simply using a signal light to tell people when they could cool their house by opening windows instead of turning on the air conditioner.[18] Confirming that air conditioning is a behaviorally dominated system—one in which behavior and the states of mind it reflects are more important than technical factors—a technical retrofit that saved 20–25 percent of the energy used by identical townhouses yielded scarcely any change in the variance of energy use between those houses, nor in their rank order, because their occupants kept on behaving just as they had done before the retrofit.[19]

An even more impressive experiment encouraged reductions in household energy use by educating people about space conditioning through group discussions or videotapes, giving them feedback on their energy use, and suggesting a schedule of gradual thermostat resets.[20] In Stern's summary:

> *In winter, some experimental groups reduced temperatures by 1.6–1.9°C in waking hours and 2.7°[C] at night compared with controls, with no significant difference in comfort (but some*

added clothing). In summer, the most successful experimental
group increased indoor temperatures about 1.5°C compared with
controls, with minimal change in comfort or clothing levels.
Although these changes seem small, the program cut overall
household electricity demand by 20% in some experimental
groups.[21]

Indeed, this experiment cut central air-conditioner use by an average of 34
percent "with *no* changes in perceived comfort or clothing worn, and with
minimal temperature change in the home"—a result wholly inconsistent with
engineering-based comfort theory.[22]

Perceptions of thermal comfort can also be strongly influenced by such
transient factors as recent meals—especially of the spicy, sweat-inducing foods
so common in tropical countries (and, increasingly, in temperate industrial
countries too)—and by peripheral vasodilators or vasoconstrictors such as
alcohol and nicotine.[23]

2.3 Addiction and acclimatization

Some laboratory-oriented comfort researchers[24] consider trivially small the "at
most only a couple of [Celsius] degrees' variance unexplained by Fanger's
[climate-chamber-experiment-based] comfort theory"—the kind of variance
often observed in different climatic zones and cultures and usually attributed
to acclimatization.[25] From these researchers' perspective of fuzzy curve fits in
theoretical models, it may well be. But such a small difference can have
valuable effects on air-conditioning requirements and hence on both capital
costs and energy use. Each 1 C° increase in room temperature in a large
commercial building typically saves ~2–4 percent of space-cooling energy—
3 percent is a reasonable and possibly conservative estimate—and a roughly
comparable fraction of space-cooling/air-handling capital cost.[26] In New Jersey
households, this coefficient was measured at ~22 percent;[27] in the southeastern
US, where latent loads are higher, at ~14 percent.[28] It is therefore important to
explore in detail whether people all over the world experience thermal comfort
identically, and whether people in a given climate can adjust to feel as comfortable in warmer surroundings.

Thermal experience affects perceived comfort on both short and long
timescales. Recent exposure to air conditioning changes perceptions and expectations of relative temperature by making unconditioned spaces seem hotter:

It seems clear that as air conditioning lowers interior tempera-
tures, it also alters persons' perceptions of "hot" and "cold." ...
The widespread use of mechanical space cooling may have
altered our social understanding of temperature—in ways that
have increased energy consumption. By creating islands of cold
(offices, shops, cars, homes) in warmer environments, air condi-
tioning creates contexts for the routine definition of life in

unconditioned environments as "hot"—hot in contrast to air-conditioned space. Each excursion into the outside environment thrusts a cooled body into a now relatively hotter space. The common summer-time observation "it's sure hot out" is most often overheard when persons enter or leave cooled spaces. ... In [the] words of one resident: "We don't use the air conditioner because it makes it too hot outside."[29]

Over weeks, living in a hot, humid climate without air conditioning causes a more lasting physiological acclimatization to heat—"a well established phenomenon" which "depends first and foremost on an increased ability for maximal sweating."[30] But as to "whether the acclimatization will influence the comfort conditions" and "mean that acclimatized persons will prefer a warmer environment for comfort," Fanger wrote in 1970 that "the present knowledge concerning these questions is inadequate," so "more research is needed."

That research has since been done, and its results, though complex in detail, can be summarized simply: even in hot climates, people who are used to air conditioning experience comfort approximately the same way temperate-climate subjects do—the way assumed as the basis for international design standards based on the comfort equation. Yet in hot climates, people *not* used to air conditioning say they feel comfortable in spaces several Celsius degrees hotter than their cooling-conditioned but otherwise identical colleagues prefer. These field experiments thus show substantial acclimatization, although nobody knows how much of the effect is in the body and how much in the mind, and we may never fully know this because mind and body are connected.

Recent field demonstrations of acclimatization

For example, Lawrence Berkeley Laboratory's John Busch[31] used standard ASHRAE protocols and definitions to compare more than 1100 Bangkok office workers in two groups,

one acclimatized to air-conditioned offices and the other to naturally ventilated offices. Those who are accustomed to working in non-compressor-cooled offices are, by their own ratings, comfortable at higher temperatures than would be predicted by standard thermal comfort studies. They are also comfortable at higher temperature and humidity levels than their compressor-cooled countrymen.[32]

Specifically,

Following the criteria used in developing a widely adopted thermal comfort standard [ASHRAE], it was found that the upper temperature bound for a Thai comfort standard, instead of being the currently accepted level of 26.1°C [79.0°F], should be as high as 31°C [87.8°F] for office workers accustomed to naturally ventilated spaces, and as high as 28°C [82.4°F] for those accustomed to air conditioning. Comparing the responses from the naturally ventilated buildings with those from the air-conditioned buildings and from studies conducted in the temperate regions provides convincing evidence of acclimatization. These and other endings of this study suggest that interior spaces in Thailand can be cooled to a far lesser degree without sacrificing comfort.[33]

Moreover, in the Bangkok offices, *maximum* comfortable temperatures are also higher than in Western theory:

The upper bound of acceptable effective temperature, defined identically to the ASHRAE Standard 55-81 (comfort standard), is 28°C (82.4°F) in the [air-conditioned] buildings and 31°C (87.8°F) in [the naturally ventilated] buildings, both of which are significantly higher than the comfort standard's value of 26.1°C (79.0°F). Translating this finding into practice would result in abundant savings to the building sector. Like most other field studies conducted in the tropics, the present study from Thailand confirms heat tolerances higher than those found in temperate regions or in chamber studies.

These data are summarized in Busch's measured thermal balloting of thermal acceptability in air-conditioned and naturally ventilated offices in Thailand (Figure 18.3).

This is among the latest of many experiments showing that, as de Dear *et al.* conclude, climate-chamber and field studies

have often yielded contradictory results—analyses of the neutral temperatures observed in dozens of field studies have led many researchers to the conclusion that populations are capable of adjusting their comfort temperatures in the direction of the prevailing climatic conditions.[34]

Comprehensive reviews and statistical analyses of the results of over 50 individual field studies on thermal comfort from various global climatic regimes have [found] strong positive correlations between the observed comfort temperature and the mean temperatures prevailing both indoors and outdoors during the field studies.[35]

Advocates of climate-chamber-based comfort theory often claim that these differences are due to uncontrolled-for differences in humidity, clothing, and

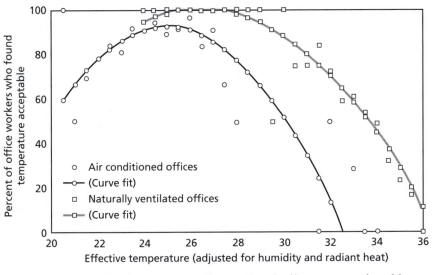

Figure 18.3 *Workers in naturally ventilated offices are comfortable at higher temperatures*

Source: Data from Busch [31]

airspeed, not to acclimatization.[36] With earlier studies, that was probably often true, but not in Busch's work. Even after allowing for differences in clothing and air movement between the air-conditioned and naturally ventilated Thai offices,[37] important differences persisted, apparently reflecting different body sizes (Thais have ~15 percent less surface area and a significantly higher surface/volume ratio than Americans and Danes), cultural norms, and psycho-physiological adaptation to the respective conditioned or unconditioned environments. Specifically, the Thai subjects' comfort ballots revealed "a striking similarity of voting *patterns* between the [air-conditioned and naturally ventilated] buildings, but with a temperature 'phase shift' upward of several" C°: the ASHRAE comfort standard, satisfying 80 percent of building occupants, *had an upper bound ~3 C° higher in the naturally ventilated than the air-conditioned offices.*[38] The Thais' comfort voting on both the seven-point ASHRAE Thermal Sensation Scale and the three-point McIntyre Scale even yielded better statistical explanatory power than Fanger's original work.[39]

A noted Pacific team, too, concluded that similar Australian data for naturally ventilated offices in Brisbane and Melbourne "raise doubts about the relevance of climate-chamber experiments and the comfort models based on them to occupants of naturally ventilated buildings."[40] Previous field research also found hot-climate neutral temperatures ranging from 26.0 to 32.5°C, although some of that work was less carefully controlled or described.[41] In essence, the neutral temperature observed in >35 field studies, far from remaining constant across the range of climates as climate-chamber theory would predict, rose virtually linearly from ~15°C in a 15°C climate to 32.5°C in a

35°C climate—three times the range that could be explained simply by differences in clothing, and indeed inexplicable unless the "respondents who considered themselves comfortable at air temperatures of 32°C [were] in a different physiological state from their European counterparts who were comfortable at 20°C or below."[42] The tropical field studies in non-air-conditioned buildings differ from those in tropical air-conditioned buildings (which are much closer to temperate-climate field and laboratory results) by an amount ranging from ~2–3 C° up to an astounding, though sometimes less reliably measured, ~11 C° (18 F°). Those differences range from ~2–3 times to ~9 times the one-standard-deviation interpersonal variation admitted by the laboratory-based comfort theory used worldwide. Clearly something is amiss.

In final confirmation, another careful field study recently found exactly the same thing in naturally ventilated Singapore apartments and air-conditioned Singapore offices[43] that Busch found in Bangkok. The authors considered three related hypotheses to explain the difference between climate-chamber and field observations: non-physiological influences on perceived comfort, or the subjects' expectations as conditioned by memory and cognition, or that "perception is relevant to and appropriate for the environmental context in which it occurs." Their results were consistent with all three explanations: people, they found, expected air-conditioned offices to be cool and naturally ventilated apartments to be hot, and adjusted their perceptions of what is comfortable according to which kind of building they were in. Hence (emphasis added):

> Perception of indoor climate ... appears not to be strictly determined by the indoor climatic and behavioral (clothing and metabolic) variables that control bodily heat balance. Instead, these data suggest that **thermal perceptions are significantly attenuated by expectations.** Given that climate-controlled office buildings have such constant temperatures and that thermal regimes in Singapore's naturally ventilated public housing are also relatively homogeneous, Singaporeans no doubt have well-established expectations of their indoor climates, and **it is these expectations that appear to form the benchmark for their thermal perceptions.**

This does not mean that HVAC (heating, ventilation, and air conditioning) engineers could be "permitted by building occupants to shift indoor climatic design criteria arbitrarily" and thereby simply establish new expectations for what is comfortable. It does, however, firmly establish that

> expectations are context specific, so ... conditions that were demonstrated to be [perceived as] acceptable in naturally ventilated apartments in Singapore would have inevitably been met with widespread condemnation were they encountered in an [air-conditioned] office building in the same city.

The authors conclude with an epitaph for the undiscriminating use of the pure engineering model of comfort:

> *These two Singapore studies indicate a discrepancy between thermal perception in naturally ventilated apartments and air-conditioned offices of approximately 3 [C°] which cannot be accounted for in terms of the basic heat balance variables (air and radiant temperatures, humidity, air velocity, clothing insulation, and metabolic rate). This discrepancy does, however, seem consistent with a psycho-physiological model of thermal perception in which building occupants' indoor climatic expectations vary from one context to another.*

Implications of acclimatization

Thus acclimatization of some kind is now clearly established, no longer just as an anecdotal report by travelers between hot and temperate climates but as a measured reality. The latest and best field experiments suggest that many factors, some essentially unmeasurable, influence what we call comfort. As McIntyre[44] states, "The human organism is adaptable, and ... if people are given a temperature that is compatible with the local culture and climate, they will adjust dress and behavior to fit." The problem now seems to be not inaccuracies in the climate-chamber experiments, but rather their limited relevance to the complexities of human beings, who have minds as well as bodies. Climate-chamber measurements still dominate in setting building standards only because of an accidental circumstance: that "researchers from engineering and physical science backgrounds rather than psychology have dominated thermal comfort research,"[45] leading to more reductionism than the subject actually warrants.

These findings raise two important practical issues:

1 Thermal comfort research done in and with subjects from temperate climates—i.e. most such research—may be invalid in the tropical climates to which they are often extended without regard to Fanger's little-noticed caveat.[46] The ability of major psycho-physiological adaptation to hot, humid climates (and, conversely, to air conditioning once widely introduced there) to shift preferred comfort conditions within tropical cultures remains controversial[47] but important, because as Kempton and Lutzenhiser point out, it implies that

> *imposing comfort standards developed in temperate regions may result in an unnecessary waste of energy and resources in southeast Asian climates. But this leads to a more fundamental question: If air conditioning is not actually required in the hot and humid urban contexts such as Bangkok, why is it used heavily in more temperate parts of the US?[48]*

2   Space cooling can now be considered addictive or at least habituative—as Schiller[49] suggests when concluding, from US and Australian office studies, that "the prevailing thermal environments in the buildings affect workers' expectations and preferences and [hence] influence their degree of [expressed] discomfort as conditions deviate from these preferred conditions." Based on extensive correlations[50] consistent with their own findings in Singapore and Busch's in Bangkok, de Dear *et al.* likewise conclude that "neutral temperatures shift towards the prevailing level of warmth in buildings" and that the same is true of acclimatization to outdoor temperatures.[51] Thus at least this form of acclimatization is undeniably real, even if its mind/body causality remains unknown and perhaps unknowable.

These discrepancies between theory and field data have also focused more attention on difficulties inherent in defining and measuring comfort:

> *Because comfort is not directly measurable, researchers have struggled with various techniques for estimating this elusive quantity, often through indirect means or inference. Different affective terms have come to characterize the search, like ... "acceptability," "neutrality," "preference," "expectation," or "tolerance." To some degree, the debate pivots on subtle differences in meaning among these terms. [For example,] 51% of the respondents in [the air-conditioned Thai] buildings indicated "no change" [was desired on the three-step McIntyre Scale, but] a much higher 88% voted within the ["acceptable"] central three categories on the ASHRAE Scale. In essence, the percentage of the sample who found the [naturally ventilated] environment acceptable was two-thirds as large as the percentage who found the [air-conditioned environment] acceptable, whereas only half as many respondents (in percentage terms) preferred the [naturally ventilated] environment as compared to the [air-conditioned] environment. "Preference" is clearly a stricter standard than "acceptability".[52]*

Worse yet, de Dear *et al.* have found experimental support for McIntyre's hypothesis that people in a warm climate tend to prefer slightly cooler-than-neutral temperatures and vice versa, and that this ~1 C° "semantic offset" could account for "as much as half" of the ~2 C° "discrepancy between comfort model prediction and field study neutral temperature."[53] Thus a simple semantic difference in how people interpret "neutral" or "preference" or "just right" may have undermined a central premise of comfort theory—that "the zero thermal sensation ('neutral') [state as reported on a seven-point voting scale] was ... the preferred [thermal] state," meaning that people would rather be neither slightly warm nor slightly cool but right in between.

Clearly, Busch concludes, these results require designers to consider:

- whether people live and work well enough in a merely "acceptable" thermal environment that a "preferred" environment cannot justify its extra economic and energy cost;
- how "acceptable" (currently, in theory at least, to 80 percent of a building's occupants) is acceptable enough; and
- whether field studies should supplant laboratory climate-chamber results.

Answering these questions in ways consistent with Busch's findings for the Bangkok offices, and accordingly relaxing "the criteria ... to meet the thermal needs of even a slightly smaller fraction of the building occupant population," could harness "tremendous savings potential." The alternative to this sensible course is unappealing. Kempton and Lutzenhiser[54] remark in their Introduction: "We find the prospect of cooling the anticipated growth in tropical offices and homes to ASHRAE standards ludicrous from a power supply perspective, and terrifying from an environmental one." Now there is a real question whether that prospect is a sensible goal from the perspective of the comfort science that has been used to justify it.

Of the three issues raised by Busch, perhaps the most fundamental is whether to believe climate-chamber or field balloting when they conflict. Common sense would suggest that if people say they are comfortable in particular conditions in real buildings, that is what matters most. To discard their ballots in favor of laboratory results is to rely on a long and tenuous chain of correlations, each with high variance and somewhat limited statistical explanatory power, rather than on a direct measurement of the quantity of interest. That is because in order to apply climate-chamber data (and the comfort equation) outside the laboratory, one must successively "correct" for mean radiant temperature, airspeed, humidity, and estimated clothing and activity levels. This is so difficult that even under well-controlled conditions, Schiller recently found spreads of ~2.4 C° in the neutral temperatures found by different, but all well-established, ways of analyzing the *same* large set of experimental data.[55] That scatter between statistical models is comparable to the debated single-model differences between field and laboratory data for tropical climates. While this doesn't mean that the difference isn't real, it does suggest the wisdom of shortening the logical chain linking observer to conclusion.

Thus, while much of the laboratory work underlying decades of climate-chamber studies is elegant and informative, it seems neither necessary nor desirable to build on its foundation an increasingly tall and unstable edifice of subsequent correlations and corrections when *direct evidence* of expressed comfort perceptions is available instead. Since the objective is to make people feel comfortable in the actual buildings and under the actual conditions where they are polled, not while wearing standard uniforms in climate chambers, *it is simplest just to ask them how they feel and whether they would like the conditions changed.*[56]

Experiments now under way at the University of California at Berkeley and elsewhere with individual-workstation comfort control, both in the field and in climate chambers,[57] may soon yield far better understanding of what people want and of how widely it varies between individuals and times. Such findings about human diversity could in turn lead to an individualized design approach to space conditioning. Just as task lighting can save tens of percent of lighting energy and improve visual comfort and performance by matching the amount and quality of light to the task a particular person is doing at a particular time and place, so "task conditioning" may achieve similar benefits in providing localized, customized, immediately controllable comfort conditions. Commercially available task-conditioning technologies are described in *The State of the Art: Space Cooling and Air Handling*, §5.2.4.[58] Though not yet technically mature, they show promise of providing in offices the same degree of individual control that many people choose to exercise in their own homes.

More broadly, although much debate focuses on the minutiae of comfort balloting in and around the ASHRAE comfort zone, it is important to understand that the body's thermoregulatory zone is vastly wider than the comfort zone. ASHRAE shows[59] that for a typical sedentary subject in low airspeed and 0.6 *clo*, the cooler half of the ASHRAE comfort zone is in fact inside the zone of net cooling (i.e. net loss of body heat); its midline is meant to coincide with, and for some people will coincide with, the boundary between net cooling and net heating; and its warmer half extends only about one-third of the way to the zone of "uncomfort" and only an eighth of the way to the typical limit of steady-state cooling by normal evaporation.

In this light, the reality of acclimatization chiefly via increased evaporation takes on great importance: any individual would need only a very small increase in habitual evaporative cooling—an increase far smaller than normal variation *between* individuals—to extend the upper edge of the comfort envelope very substantially. That is indeed how the Bangkok and Singapore subjects remain able to report comfort in non-air-conditioned buildings ~3 C° hotter than climate-chamber comfort tests would suggest. Even greater comfort-range extension is available through other means, such as increased air movement and more appropriate seasonal clothing.[60] Existing comfort standards may therefore turn out, as older lighting standards did on closer examination, to be overengineered—more stringent than is physiologically or economically justifiable.

2.4 Cultural influences

The universality of human comfort criteria (an ideal point of 25.6±1.2°C or 78.1±2.2°F with 0.6 *clo*, 1.0 *met*, 50 percent relative humidity, and ≤0.2 m/s air movement at sea level) is a central doctrine of comfort engineering, supported by extensive worldwide laboratory experiments and enshrined in international standards that often have legal force.[61] Unfortunately for engineering convenience, it may not be true, for two quite different reasons.

The previous section explored the first reason: acclimatization and other sources of shifts in comfort conditions among people within the same culture. In addition, however, field studies also show appreciable *cross*-cultural variation in the use of and the desire for air conditioning, even without invoking plausible but untested extremes, such as the acclimatized preferences of, say, Inuit and highland Tibetans vs. those of Zambians or Kalahari Bushmen (let alone people with yogic control of their physiological state).

The comfort theory itself implies that if people in different cultures, after due correction for differences in physical variables such as clothing and airspeed, prefer different thermal environments, then they must be experiencing comfort differently. This does not necessarily mean that their bodies work differently, although that remains one possible partial interpretation of some of the tropical data just discussed, and has reportedly been found within the United States but not published. But even if it means only that people have different reactions to or expectations about comfort—if what is happening is psychological rather than physiological—that will have the same practical effect: they will want and use different amounts of air conditioning.

In cold climates, space heating to theoretical comfort levels is considered in many cultures to be enervating, unhealthful, profligate, or otherwise undesirable:

> *In winter, households in countries with similar energy prices and average incomes keep quite different indoor temperatures—from about 14°C in Japan to 17°C in Norway and 21°C in Sweden. These differences partly reflect preferences.*[62]

Conversely, in many hot-climate cultures, people traditionally choose to change their behavior (clothing, daily schedule, siestas, architecture, bathing patterns) rather than buy air conditioners. Both hot and cold climates teach a similar lesson:

> *[Such cognitive factors as ownership and expense for space-conditioning were] probably implicated in the surprisingly low average dwelling temperature of 15.8°C observed [in a 1982] nationwide UK field survey of 1000 houses. The average clothing insulation level of 0.83 clo observed in that study was well below the 2 clo level that would have been required [for people to feel comfortable according to comfort theory]. Therefore it seems that the British have a greater tolerance of cold temperatures in their homes than is suggested by comfort standards. Similarly in Singapore, residents of naturally ventilated apartments have a greater tolerance of warm and humid indoor climates than suggested by the standards.*[63]

Moreover, it seems that widespread craving for air conditioning can be culturally conditioned too. In the modern global context, the apparent intense desire for and extensive use of space cooling seems to some observers characteristically if not uniquely North American: as Cambridge don Gwyn Prins[64] acerbically remarks:

> *Americans are more systematically (but unconsciously) divided on the criterion of temperature than any other. A Condi [Prins's cultural label for a person addicted to space conditioning] is cold in summer, hot in winter; others are the reverse. ... Once one's body has become addicted to air-conditioned air, one has extended one's range of basic, physiological human needs beyond food, shelter and warmth to an acquired need: Coolth. The case is then made with genuine conviction that access to Coolth is a basic human right, based on physiological need.[65] Coolth is actually different from warmth because it is not essential to the preservation of life. The fact that ... millions in the tropical poor countries manage to get along without it demonstrates this.[66]*

As more neutrally summarized by Paul Stern, a US National Research Council psychologist:

> *The economic and engineering perspectives both simplify analysis by making [individual] preference exogenous to social action—they free society from responsibility for preferences. [In contrast, Gwyn Prins's] cultural theory considers the social origins of preference. He theorizes that a "condi" culture of North Americans (and, via cultural imperialism, of others as well) desires separation from and superiority over nature, symbolized by such things as sweat, passion, hunger, and weather.[67] "Condis" prefer coolth because it both symbolizes and embodies victory over nature. Once people come to prefer coolth for cultural reasons, they become addicted to it by a physiological process of acclimatization.*

Whether one accepts Prins's cultural and anthropological characterization or not, it is undeniable that some US regions and subcultures display every sign of both individual and cultural addiction to air conditioners—except, perhaps, denial. Consider this 1983 account[68] of Houston life:

> *It's 90 degrees and muggy outside, but inside Beverly and Louis Lerner's spacious home it's so cold "you could hang a side of meat," says Harry Waters, the Lerners' air-conditioning repairman.*

> *The Lerners say they crave cold air and couldn't survive without it, so they've surrounded their two-story, 15-room home with six independently run, central air conditioners. "If one breaks down, the other five can pick up the slack," says Mrs. Lerner, pointing to a unit that isn't working. "We couldn't get along without air conditioning."*
>
> *"Psychologically, we're addicted," says Buck St. Cyr, who runs an air-conditioning concern here. The addiction is understandable. "There is nothing closer to hell in modern America,"* writes historian David McComb in Houston: The Bayou City, *"than to be caught in a rain in a Houston traffic jam in an un-air-conditioned car. It is possible, at that moment, to appreciate the plight of a steamed clam."*
>
> *Houston's dependence on cool air was most apparent last month when hurricane Alicia cut off power to many areas of the city for up to a week. Thousands of residents crammed into their air-conditioned cars or, if they could afford it, stayed at the city's hotels. After two days in his 95-degree home, Cary Ziter and his pregnant wife fled to the downtown Holiday Inn. "When we finally got to the hotel, we just hung over the air conditioner and lapped up that cool air," he says. "It felt so good we didn't move."*
>
> *[In contrast,] Justus Baird, head of the city health department's epidemiology division, has gone without cool air in his home and car for three years. Mr. Baird says he has become accustomed to the heat. "I'm sure some people think I'm nuts, but you really don't need air conditioning," he says.*
>
> *As Mr. McComb, the Houston historian who has since sought refuge in the arid mountains of Colorado, says: "When you get everything air conditioned, it shuts people off from the outside environment. So Houstonians are a little out of touch with fellow men, and it's easier for them to be hostile." Mr. McComb says the hostility shows up in the city's high crime rate, its wild freeways and what he considers a generally uncaring society.*

It is instructive to compare a very different kind of society[69] with a similar summer climate.[70] Classical Japanese culture strongly emphasizes integration and harmony between people and nature. Japanese open-plan house architecture, traditionally wood-and-paper, was oriented to cooling winds, used wide apertures and long eaves to increase cross-ventilation, was well shaded, and blended inside with outside.[71] Japanese people, too, normally provide comfort by cooling (or heating) people, not rooms or air:

> *Japanese do not have a custom of cooling or heating whole*
> *house. They think that it is waste of energy to cool or heat the*
> *rooms not occupied. So, air conditioners which distribute condi-*
> *tioned air to all of the rooms are not required.*[72]

Natural ventilation was the norm until very recently, and hand fans were common many centuries before electric fans became common (gaining ~94 percent saturation in the past few decades).

Modern Japanese urban apartments, in contrast, have interior walls, lack passive cooling, and are often in settings where noise, dirt, proximity, and crime inhibit the opening of windows and doors; "In these circumstances it is so difficult to control the indoor climate that the number of people possessing an air conditioner has been rising."[73] Even when these less agreeable conditions are imposed on a natural-cooling culture, however, technically informed users of sophisticated air conditioners, with controls more flexible and powerful than are available in other countries, overwhelmingly preferred natural to artificial cooling.

An astonishing 90 percent of these Japanese subjects considered natural cooling healthier than air conditioning, 75 percent considered it more economical, and 70 percent preferred it. On the other hand, only 20 percent considered artificial cooling more comfortable, and 95 percent regarded artificial cooling as a necessity (at least in their urban setting, where 85 percent express concerns about leaving their windows open at night—still a common rural practice). Apparently, the respondents prefer to use air conditioning only when they feel it is really necessary. Few consistently use it by choice: when they feel hot, the first thing 80 percent of them do is open a window; only 5 percent turn on the air conditioner first.

Another Tokyo-suburb survey[74] yielded similar results: 76 percent reported hesitating to use an air conditioner, most of all because it is "bad for the health" (64%) rather than "because electric charges run up" (51 percent), "it is wasteful to use an air conditioner for the sake of only a few occupants" (38 percent), "an electric fan dispenses with an air conditioner" (30 percent), or "energy conservation" (26 percent). When asked if "it is bad for the health to be in the air-conditioned room," 57 percent agreed, only 2 percent disagreed, and 40 percent were neutral; this is logical enough, the authors explained, because a standard Japanese room is only 9.7 to 13.0 m² (105 to 139 ft²), making exposure to direct cool drafts from an air conditioner almost unavoidable.

It appears, therefore, that the urban conditions that reduce to only 21 percent of Fujii's sample the fraction that leave windows open at night are a major driver of the use, albeit the very disciplined use, of household air conditioners in Japan. Nighttime window-opening can be effective in the US too: as was mentioned in §2.2, simply suggesting that Americans turn off the central air conditioner and open the window when a blinking blue signal light in the kitchen indicated it was cool outside (<68°F or 20°C) reduced air-conditioning energy use in a 1976 New Jersey experiment by 15.7 percent.[75]

2.5 Is comfort even important to cooling behavior?

Not only are people's comfort needs and perceptions far more malleable than would appear from the laboratory studies and comfort equations that under-pin official standards, but *thermal comfort may not even be the main determinant of how much space cooling people want and use in their own homes*, where the air conditioner's operation is under their close personal control. If this is true—if how much air conditioning people want to use depends on far more factors than their calculated or even reported thermal comfort level—then space conditioning's energy use would depend less on thermostats than on "soft-science" variables wholly absent from the engineering model. The evidence for this is not exhaustive, but so far as it goes, it is certainly compelling.

Elegantly confirming "that one cannot simply study room air conditioners in the lab or in unoccupied apartments if the goal is to understand actual field operation or the determinants of energy use," Kempton *et al.* plotted how much of the time room air conditioners ran as a function of outdoor temperature in a carefully monitored group of New Jersey apartments. With no occupants, the units ran according to their thermostat setting, as one would expect from the building's physics, and outdoor temperature is well correlated with runtime ($r^2 = 0.64$). But with occupants, the pattern was utterly transformed (Figure 18.4):

> At any temperature, the unit may be off. There is a threshold at 20°C (68°F) below which it is not switched on; above that threshold there is no visible relationship between temperature and runtime. ... This pattern is typical of the apartments we studied".[76]

A Japanese study yielded virtually identical results.[77]

US field studies have richly confirmed "that user behavior is a large, perhaps the largest, component of variation in air-conditioner energy use among similar dwellings."[78] In identical California Central Valley apartments and in New Jersey apartments, different people were found to want enormously different amounts of space cooling *even when they did not pay* for the electricity (nor, in the latter case, for the air conditioner). So important were these differences that in similar apartments of the same New Jersey building (though in various rooms and with various orientations), the use of nearly identical room air conditioners ranged from 2.5 to 1557 hours, *and consumed from 1.2 to 1048 kWh*, during the cooling season monitored.[79] These three-order-of-magnitude energy spreads hardly comport with any simple model of physical comfort or hedonic preference.[80]

Informants expressed in ethnological interviews "many non-economic reasons for refraining from air-conditioning use, such as health, comfort, safety, and aversion to waste."[81] Field observations combining detailed physi-

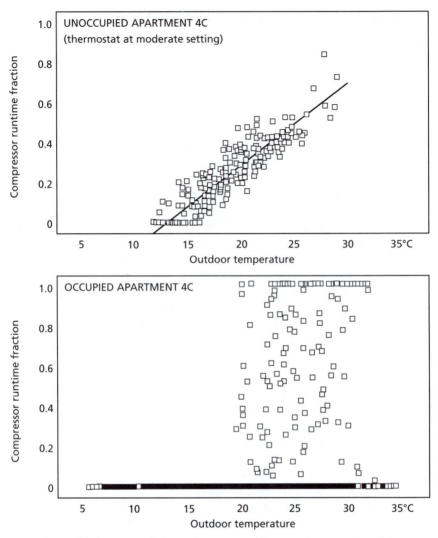

Figure 18.4 *Air-conditioner operation: Not just temperature-driven*

Source: Data from Kempton *et al.* [14]

cal monitoring with ethnographic interviews *specifically refuted both the engineering and the economic models of air-conditioning behavior.* In the Princeton team's[82] classic formulation:

> *These interviews helped to correct our initial simplified stereo-types of how people use air conditioners. We had previously assumed that people turn on an air conditioner when they feel hot and set it to a temperature at which they will be comfortable. We thought that most people limited their use primarily because*

of cost, and thus in our study building where electricity is not billed we expected people to operate their air conditioners whenever the weather was hot.

But just the opposite actually occurred:

Instead we found that operation was rarely thermostatic, and that use was governed by household schedules and by multiple overlapping systems of beliefs and preferences concerning health, thermal comfort, alternative cooling strategies, folk theories about how air conditioners function, and general strategies for dealing with machines. We found a multitude of non-economic factors limiting air-conditioner use.

These reported reasons for using little or no air conditioning even on hot days and at no cost included:

- household schedule—few people left the unit on while away all day;
- health reasons, such as "arthritis," "cramps in shoulders," and "colds;"
- discomfort, such as "the air blowing," "the cold," "cramps," "I get dizzy," or distasteful noise;
- safety (three of eight interviewees had some apparently justified concerns about overloaded wiring, or felt uneasy leaving electrical equipment on at night or while away);
- feelings (expressed by five out of eight interviewees) that it was "incorrect or wanton" to use the air conditioner excessively, even though it and its use were free to them;
- folk physiological theories—consistent in consequence, if not in content, with Figure 18.2—such as "I'm older, less blood to keep cool," or "I am a warm person, maybe I'm a little anemic," or "I can endure heat," or "other people need cooling more than I do because of their body chemistry."

"This rich variety of limits on air-conditioner use," the authors concluded,

helps explain why our field study found a wider range [by fivefold, as discussed above] of actual indoor temperatures than one would expect from climate chamber studies. The latter studies focus exclusively on physical comfort, which our interviews clearly show to be only one of many determinants of indoor temperature choice.[83]

Moreover, most users had a different understanding of how the air conditioner worked than its designers did: they thought, for example, that setting the thermostat to a colder temperature would "make the cold come out faster" and the room cool off faster, as if the thermostat were a sort of valve for coldness, when in fact, owing to icing of the coils, the opposite was often true. This important point, discussed more fully in §3.1, is still another reason field observations diverge from engineering theory, which tacitly supposes that users will employ the control strategies that are rational to the machine's designer.

Of course, people's running air conditioners less (or not at all) may mean not that officially defined comfort parameters were unimportant to them, but that air conditioning didn't much affect their comfort level. Reduced air-conditioner use may scarcely reduce reported comfort, especially where the air conditioner is too big (as it usually is). In a separate empirical study, when the utility's radio load controllers turned off some New Jersey households' air conditioners half the time on the hottest days of the year (625 house-days out of 1455), the participants reporting insufficient cooling rose only from 7% to 15%, and controlled houses had only slightly higher indoor temperatures (by 0.6 F° or 0.3 C°).[84] Households saving more peak demand were no likelier to report warmer indoor temperatures or increased discomfort on the controlled days, nor were those reporting the greatest increases in discomfort the biggest savers. That is, load savings were statistically uncorrelated ($r^2 = 0.02$) with perceived discomfort. This study's further implications for cutting load-management costs are discussed in §7.6.4 of *The State of the Art: Space Cooling and Air Handling*.[85]

This is consistent with Kempton, Feuermann, and McGarity's finding[86] that

> *in this mid-Atlantic climate [New Jersey], air conditioning is far from being a "necessity" to everyone. Energy efficiency analysts shy away from mentioning this for fear of being seen as advocating sacrifice—but this was the view of many (not all!) of our informants. The majority used their air conditioners for a small fraction of the day and lowered room temperature only a few degrees. Residents voluntarily switched off their units while they were still at home even though electricity was free. We surmise that, if not for the multiple non-economic constraints revealed in our interviews, air-conditioner energy use would be dramatically higher than it is today.*

Similarly, a survey of 279 households in identical apartments in hotter Davis, California,[87] found that although one of the two housing complexes studied had been equipped with free air conditioners by the university, in the other complex, 48 households (17 percent) had chosen not to install one; another 10 percent had air conditioners but did not use them; and ~46 percent reported using their air conditioners no more than twice a week. (Both complexes had been converted the previous year from master to individual meters.)

This finding is encouraging for today's electric loads, because presumably even larger fractions might have chosen not to install or use air conditioners if they had to pay for the electricity. But it raises a disturbing possibility for future loads. Air-conditioner usage may be a function more of generational demographics than of climate: as parents and grandparents not raised in air-conditioned homes, and disinclined to use the device casually, are gradually replaced by younger generations that take thermostatically controlled central cooling as much for granted as they do central heating, usage might rise very dramatically even with no change in the capital stock.

This idea also highlights a key difference between the commercial and the residential sectors:

> *When air conditioning first appeared in American commercial establishments, proprietors quickly learned that business suffered in summer the more competitors installed cooling; employers may also have learned that workers are happier and more easily retained in air-conditioned establishments. Competition ratcheted up the standard of coolth, and kept it there.*[88]

In households, similar competitive forces among builders cause the same ratcheting-up in the *installation* of air conditioners:

> *The proportion of new single-family [US] homes equipped with air conditioning grew from about 33% in 1970 to 77% by 1989. Combining new and existing housing, air conditioning saturation has grown from 12% of occupied US housing units in 1960, to 36% in 1970, 55% in 1980, and 64% [in 1987].*[89]

This growth has followed similar patterns in successive US regions: in the South, central air-conditioning saturation in new single-family homes rose earliest and highest, from 36 percent in 1986 to 78 percent in 1976 to 91% in 1986, while in the Northeast, where introduction lagged, the corresponding figures were 7 percent, 13 percent, then a steep rise to 43 percent.[90] Even faster growth in southeast Asia is now adding an estimated 15–25 GW of new regional load *per year*.[91]

But the crucial difference is that in almost all commercial establishments, the people who install, turn on, and control the air conditioner are the managers, not the customers. In households, on the contrary, even though occupants may have little choice of whether an air conditioner is present, they can and do choose how much to use it and how to operate it. Their usage follows personal preference, not construction market forces. If, therefore, householders' patterns of use become less intentional and more incidental (to borrow Lutzenhiser's language from §3.1 below)—as might be expected both with generational change and as more of the workforce, already over half office workers, becomes acclimatized (like Busch's Thais) to air-conditioned buildings—then what will mainly determine air-conditioning energy use will be not the saturation of this appliance but rather the extent and patterns of its use. These variables are cultural; they are not technical nor, probably, substantially economic.

Successful foresight in air-conditioning loads may therefore need not so much econometricians as cultural anthropologists. Indeed, in the 1976 Princeton experiments reported in §2.1, 55 percent of the variance in summer electricity usage between identical townhouses was explained by *purely attitudinal* variables determined by questionnaires—"in psychological research ... a strikingly high attitude–behavior correlation."[92] Only the social sciences offer insight into these variables—e.g. whether people believe air conditioning is important to their health and comfort.

Already, a significant market segment of US household air-conditioner users choose to run their units only rarely—almost exclusively at times of utility-system peak load. They "report that they are fairly indifferent to owning an air conditioner," but their occasional use impels huge utility investments with little compensating revenue. Such users appear to present a promising load-management opportunity,[93] but only if captured while they still retain their present indifference. To the extent that indifference will diminish over the years through generational change and cultural conditioning, it represents a lost-opportunity resource for utility planners:

In summary:

- People's perceptions of summer comfort vary widely and are subject to major cultural conditioning.
- Whether air conditioners get installed often depends on marketing considerations that have little to do with actual demand for comfort.
- How much air conditioners, once installed, are actually used may have only a tenuous relationship to the users' imputed or perceived needs for comfort.
- Rather, their usage depends strongly on complex and often "unscientific" relationships between comfort and numerous systems of belief, expectation, and preference.[94]

Just as people are often observed to choose in their own homes levels of illuminance an order of magnitude below those that engineers presume they will want and need in their offices, so the emergence of personal space-conditioning control and delivery options in the workplace may soon reveal that engineers are designing commercial HVAC systems to provide environmental conditions that many people do not really need or want. The overwhelming dominance of complaints that space-conditioned offices are too hot or too cold suggests that what people really want is often rather far from what comfort engineers think they want.[95]

3. How Do People Want to Control Cooling?

People want to feel in control of their environment. They do not want to feel alienated from machines, cut off from nature, or frustrated by incomprehensible controls. Extensive European and North American data[96] confirm that office workers tend to have lower sick leave and absenteeism and better productivity the more control they have over their workplace environment (temperature, lighting, etc.). Similar gains are sometimes observed if people *believe* they have control, whether they actually do or not: many building managers, unable to make everyone comfortable, simply install dummy thermostats *that are connected to nothing but reduce complaints anyway.*[97] Such managers believe, with good reason, that far more people say they want, and probably do want, a feeling of control over their environment than will actually choose to *exercise* that control once they have it.

In commercial buildings, "thermostat wars" between workers with different preferences, or in areas differently served by a poorly balanced or controlled HVAC system, lead many building managers to lock the thermostat in a sealed box where it can perhaps be seen but not adjusted. This may simply add frustration to discomfort. But if the alternative is deception (dummy controls), the building manager is in a difficult position. That quandary results simply from poor design that makes far more people uncomfortable than the irreducible minimum due to human diversity. It is a sad indictment of the design community that so many building managers must choose daily between being overtly and covertly unresponsive to the needs of those they serve.

Until workers get the individual, workstation-based comfort delivery and control systems whose early (and not very satisfactory) versions are described in §5.2.4 of *The State of the Art: Space Cooling and Air Handling*,[98] it is much easier to understand their interactions with HVAC equipment that is already under their personal control. The simplest, commonest example of this is with the room air conditioner in their households. We therefore return to a fuller discussion of two studies already introduced in §2.5. Chapter 7 of *The State of the Art: Space Cooling and Air Handling*[99] discusses the engineering aspects of controls; here we consider instead their behavioral aspects, which are at least as important and often more so.

3.1 Unexpected operating behavior

People interpret HVAC controls according to their mental models of how machines work, not necessarily according to how machines actually work. In Kempton *et al.*'s pioneering study[100] of New Jersey behavior in operating room air conditioners, most users turned out to believe:

- that if they set the thermostat to a lower temperature, the room would cool faster—the same "valve" theory Kempton had earlier found as a minority view for heating thermostats;[101]
- that colder thermostat settings would cause the air conditioner to consume power at a higher instantaneous (not just monthly) rate; and
- that the air conditioner was either on or off—a binary device with no intermediate conditions (in fact, it often ran the fan without the compressor, but there was little enough difference in noise that it was hard to tell).

Some residents, pursuant to the first of these beliefs, left the thermostat set at its minimum temperature (maximum cooling) at all times, often causing the evaporator coils to ice up. One responded with tedious manual defrosting. Another said "I always turn it on 'super'"—with the thermostat knob set on 9 out of 10—because "It doesn't run real great anyhow." No wonder: the ice was not only inhibiting heat transfer, but also blocking 75–80 percent of the vent area. Apparently the manufacturers had not thought to apply to air conditioners the advanced frost sensors now emerging for residential refrigerator/freezers.[102]

Even more interestingly, none of the eight monitored New Jersey apartments ever changed the thermostat setting.[103] At least one did know how thermostats work in heating systems and said he wanted one on his air conditioner too; he didn't realize he already had it, because it was poorly labeled. In the much larger Davis, California, experiment in 231 student apartments with room air conditioners, 58 percent reported that they manually turned the units on and off without changing the thermostat, 29 percent used the thermostat, and ~13 percent combined the two strategies.[104] Thus in the two most recent, thorough, and informative studies of US household room air-conditioning behavior, *most of the users controlled operation with the on-off switch, not the thermostat.* (A further California study in new, centrally air-conditioned single-family houses[105] found about a third under manual control, a third thermostatic, and a third thermostatic with setbacks.) These differences in behavior are clearly shown in New Jersey data (Figure 18.5) for outdoor temperature (A), a thermostatic air-conditioner operator (B), a manual air-conditioner operator (C), and an infrequent manual operator (D). In plot B, the circled point "A" indicates a change of occupants from one with thermostatic to one with manual operating patterns.

The Davis users, like their New Jersey counterparts, usually reported using the coldest thermostat setting if they ran the unit manually (thus turning the unit's power switch into a "cooling switch"), and the warmest setting if they ran it automatically under thermostatic control. A minority used the opposite settings: automatic operation then caused overcooling, while manual operation yielded only modest air movement that would have been better done by an oscillating portable fan. Some automatic users, too, adjusted the thermostat in a way that suggested they were really using the thermostat as a switch:

> *In what is probably the optimally efficient use of [automatically operated room] air conditioners, for highly-controlled and moderate room cooling (and one certainly not intended by the manufacturer), the user turns on the fan and then rotates the thermostat [knob] until its contacts close. This engages the compressor for a short cooling period, followed by a longer period of air circulation.*[106]

> *When the New Jersey experimenters ... provided more clearly marked, easier-to-use thermostatic controls, people continued to*

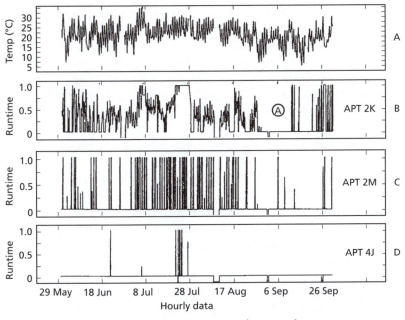

Figure 18.5 *Individual patterns of air conditioner use*

Note: Hourly data for entire cooling season: (A) Outdoor temperature. (B) Runtime fraction for apartment 2K, a thermostatic pattern. (C) Apartment 2M, an on-off pattern. (D) Apartment 4J, an infrequent user with an on-off pattern. The circled A indicates a change of occupants in apartment 2K.
Source: Data from Kempton *et al.* [14]

use them in a manual on–off mode. Manual on–off operation seems to meet people's needs for cooling when and where they want it. Despite the maximum thermostat setting, the empirical result is that manual on–off operation uses less energy.[107]

That's right: the manual operation, though more trouble, was also more energy-saving! The one consistent thermostatic user in the small New Jersey sample used *seven times* the average air-conditioning electricity. In Lutzenhiser's larger, though less directly measured, Davis sample, automatic thermostatic operation was inferred to use ~27 percent more energy than manual operation (102 vs. 80 kWh/month in August). Although neither set of data is definitive, both are highly suggestive. Kempton *et al.*[108] conclude:

Rather than encouraging these perfectly comfortable manual-control users to run their air conditioners continuously under thermostatic control, we would advocate learning from them about how to design better controls. For example, users in this study seem to want air conditioners that will turn off after a set period of time. When used after returning from work, many want it to turn off completely when a desired temperature is reached once. Even on a thermostatically controlled unit, an occupant coming in from the heat might want to press a button for a brief period of maximum continuously-on cooling followed by automatic return to the [higher-temperature] thermostatic level.[109] *Surely such control operations would be even more desirable in the more common situation of users who must pay for their electric use.*

As §3.2 below describes, many of these control options are already available in Japan.

Lutzenhiser explores in detail the interaction of the room air-conditioner controls with their users. The controls no doubt seemed simple to their designers, but to the users the controls often seemed confusing (one said, "It looks like an aircraft instrument panel"). Neither the users nor the building superintendent had an instruction manual, nor were copies available even in microfiche for those 18-year-old models. The proper operation of the controls could not be readily inferred without the instructions. Furthermore, in the somewhat international Davis sample, half, including a substantial number of the American students, said they had never used a residential air conditioner before, and nearly half said that air conditioning was not common in their home region. Yet, interestingly, both manual and thermostatic operation were associated with both correct and incorrect (in engineering terms) theories of how the device worked, and neither was correlated with technical background.

Much of the users' confusion is the fault of the control-panel designers:

The confusion about whether the thermostat dial refers to room air temperature or output air temperature may also reveal an important taken-for-granted technical understanding regarding technical control. Clockwise rotations of controls are, in the US, most often associated with turning something on (car ignitions, keys in locks), or turning something up (increasing the heat of stove burners)—i.e. activating or increasing the output of a mechanical process or system. Analogously, US thermostats and other dials (e.g. the prototypical clock dial or the dials of barometers, speedometers, and a variety of electronic gauges) register higher values to the right (in a clockwise direction) and lower values to the left. The [air-conditioner] control arrangement that we studied uses a thermostat dial ... that registers lower room temperatures with clockwise rotations—exactly the opposite of the culturally-conditioned expectation. The dial strongly embodies a rheostatic character, while signaling "thermostat" in a way that contradicts customary understandings of how a thermostat dial should be laid out (i.e. like a heating thermostat, with lower temperatures to the left).[110]

Lutzenhiser[111] also explains that the manual operation is more rational than might at first appear:

> *Manual control involves the instrumental cooling of people (e.g. "cooling off" or "keeping cool"). It can also involve the instrumental preparation of cool space for persons (e.g. "getting ready for the arrival of X"). Automatic cooling, on the other hand, represents what we might call an incidental approach, one in which cooling occurs regardless of occupancy or activity. The space is cooled incidental to ordinary activity, i.e. space is kept cool in the event that one might want to use it.*

That is, manual operation via the on–off switch means (in Lutzenhiser's plausible interpretation) that the user is trying to cool people near the air conditioner, not simply the room in general: it is a directly task-oriented approach, not an abstract one. This pattern, he remarks, has useful analogies: the Asian tradition in winter is to heat people, not spaces, just as the majority of his largely American student sample cooled people, not spaces; but this is the opposite of many Americans' behavior with respect to air conditioners, which they tend to use "incidentally ('out of habit') rather than intentionally."

3.2 Control design philosophy: A cross-cultural comparison

The completely different control philosophy adopted in Japanese residential air conditioners (typically bidirectional split-system heat pumps as described in *The State of the Art: Space Cooling and Air Handling*, §6.6.2,[112] with mainly wireless handheld remote controls like TV or VCR control units) may introduce different kinds of user confusion. At least, however, it designs controls *from the viewpoint of the user* experiencing the device's operation, not that of the design engineer who understands its inner workings.

In one of the clearest signals of the user orientation, consistent also with Japan's naturalistic cultural tradition, what an American engineer would call "air movement" or "air handling" a Japanese engineer would call "wind," which has "wind velocity" and "wind direction":

> *The "wind" terms follow from an orientation to the system from the point of view of the user, and describe the product of the air conditioner: directionally-controllable moving cool air. The comparable term on US air conditioner controls ("fan speed") refers to the machine rather than its output. "Wind" also communicates the Japanese understanding that these machines are devices that can be made to replicate a naturally-occurring phenomenon ("the wind"), while the analogous English term ("air," e.g. "conditioned air") connotes a static indoor environment.[113]*

Additional major differences between Japanese and American control philosophy include these:

- "Wind velocity" typically has 3–7 steps plus an automatic mode that matches wind velocity to the changing cooling load seen by the variable-speed compressor. However, wind direction can also be automatically deflected by an electronically controlled flap or vane. Its automatic option blows the air out horizontally (from high on the wall) before the machine has run long enough for the air to become cool, then downwards toward the people after the air had become cool. Most remarkably (emphasis added),

  > *In cooling and dehumidifying modes, the flap automatically swings continuously between horizontal and 45 degrees downward, distributing throughout the room and approximating the "gusts" of a breeze. Some newer systems* **can sense the location(s) of persons in the room, directing the wind accordingly.**[114]

- Since 1986, one Japanese manufacturer has offered a "thermal comfort switch" that enables the control system to "learn" from users' evaluations of their comfort on a three-point scale ("I feel cold," "I feel comfortable now," and "I feel hot"), augmented lately by an "I feel sticky" button. No numerical temperature targets are involved; rather, the control system simply adjusts setpoints until the user indicates comfort. Indeed:

  > *In designing a machine that "perceives" the user's psycho-physical states, a new human–machine relation has been constructed. This machine is not merely an instrument of the user, but rather features a "not-very-intelligent" (but "intelligent-enough") memory that does more than just store intentionally programmed user preferences. Instead, the machine accumulates information sensed through its interactions with the user. This makes it a very limited robotic servant who (like servants everywhere) having learned the resident's habits, works diligently to enforce them. By lodging this sort of intelligence in a simple logic circuit, the requirement of premeditated human control is eliminated altogether from the human/machine system.*[115]

It would be valuable to measure the resulting energy use in comparison with the same machine manually controlled in the common American fashion.
- Some systems have control protocols preprogrammed for the seasonally appropriate comfort conditions that the manufacturer considers suitable for the behavior expected in particular kinds of rooms—study, sleeping, elderly person's room, etc. Further, in the children's room/study setting, the

temperature is periodically adjusted by small random intervals, "based upon a theory that beneficial physiological (and presumably mental) effects result from mild environmental stimulation." The belief that variable environments are more healthful and stimulating than static ones has wide anecdotal support, especially in Asian cultures, and merits further research (*The State of the Art: Space Cooling and Air Handling*, §2.2.4).[116]

- In perhaps the most fascinating reflection of the "deep regard for nature ... that [has] produced the Japanese garden, bonsai sculpture, water color artistry, and intricate joinery," embedded in a culture that "is not necessarily designed to wrest materials from a recalcitrant nature or to dominate a hostile natural world," the "wind" produced by many Japanese air conditioners and heat pumps is "a surrogate for natural winds":

  > *One company offers a machine that reproduces the wind patterns recorded at a well-known resort located near Tokyo. Other systems offer wind patterns generated by an algorithm which assumes that natural phenomena vary in a linear relationship of intensity, frequency and duration such that more intense gusts are less frequent and of shorter duration than are gentler gusts. With this logic encoded in control circuitry, the fan speed is varied to produce seemingly natural breezes inside the dwelling. This natural wind emulation is, to our knowledge, found nowhere else in the world.[117]*

- Japanese designers, appreciating users' conflicting desires both for more automaticity and for more opportunities for individual control, tend to make the automatic features primary but to provide (under a small door on the bottom of the handheld remote control unit) a variety of direct control options that override the automatic settings. This reduces the potential both for control conflicts and for user confusion—which occurs in Japan just as in the United States when controls become too complex.[118]

- Most remote controls include countdown and/or preset-time timers similar to those on video recorders and electronically controlled ovens. The timers (emphasis added)

  > *are of value in cooling a room before the resident arrives home from work or shopping. They can also turn the machines off after residents have gone to sleep.[119] This "strategic" use of air conditioning could be contrasted with the "static temperature" approach assumed in U.S. system design—in which users are assumed to desire uniform temperatures at all times, perhaps with a [night] "set-back" (or "set-up"). ... A few Japanese systems are able to "monitor themselves," determining the time when they should begin to heat or cool. They do this by periodically measuring the room temperature and comparing this*

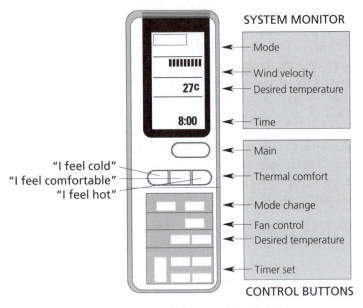

Figure 18.6 *Typical design of hand-held remote control for Japanese air-conditioning unit*

Source: Data from Fujii and Lutzenhiser [68]

*information with user-programmed desired room temperature and usual arrival time data and information on the machine's own cooling performance parameters. **The system is able to calculate the running time required to reach the desired temperature**, and to turn itself on at a time calculated ... so that the room reaches the desired temperature by the time the residents return.*

- Further controls observed in the Akihabara retail electronics district of Tokyo in late 1990 included occupancy sensors (so that coolth follows one around the house), humidistats, and an elementary self-diagnostic device— a red LED indicator of when to clean the coil or change the filter, apparently activated not by a timer but by a sensor of deteriorated heat transfer or airflow.

Such distributed intelligence is characteristic of a society in which microchips are nearly free and are therefore installed in virtually every kind of household device in order to make it more convenient and versatile. But it works only because the designers make a conscientious effort to put themselves inside the heads of the users, so that the air conditioner becomes a tool, not a machine. The contrast with American control designers and their products could not be plainer.

* * *

In conclusion, the kinds of warm-weather comfort conditions people want and how they use air-conditioner controls to achieve them often differ markedly from assumptions based on comfort theory. Understanding these differences is central to the effective provision of comfort with a minimum of energy use and economic cost. Chapter 2 of *The State of the Art: Space Cooling and Air Handling*[120] further explores the surprising flexibility available to designers in defining those conditions that people will experience as comfortable. The options described there exploit ten degrees of freedom in addition to the one most designers use exclusively (air drybulb temperature): namely, variation between individuals, variation over time, allowable excursions beyond the comfort envelope, dynamic rather than static comfort conditions, metabolic rate, clothing, furniture, radiant temperature, air movement, and humidity. Together, these ten additional variables can yield important (~10–30+ percent) savings in space cooling energy without violating international comfort standards, which in turn are often far more conservative than the underlying science would justify. But designers who begin by treating the people they serve as whole human beings, not merely as watery sacks, will do even better in saving energy, saving money, and creating delight.

References and Notes

1   For adult males in dominant European and North American cultures. Other than radiative heat-transfer measurements that depend on skin pigmentation, little human diversity is implicit in normal comfort calculations, although the DuBois formula permits other values to be substituted: surface area in $m^2 = 0.202$ (mass in kg)$^{0.425}$(height in m)$^{0.725}$. Women, who are half of the US office workforce, are somewhat smaller on average and have slightly different physiology. Men normally consider the difference to be small enough to require no special provisions, but in some circumstances this may be incorrect, possibly by an unexpectedly wide margin.

2   More precisely, "thermal neutrality … is defined as the condition in which the subject would prefer neither warmer nor cooler surroundings. Thus thermal neutrality is a necessary condition for thermal comfort, but it need not be a sufficient condition. For instance, a person who is exposed to an extremely asymmetric radiant field [e.g. from an adjacent sunlit window] can very well be in thermal neutrality without being comfortable." P. O. Fanger, *Thermal Comfort: Analysis and Applications in Environmental Engineering*, McGraw-Hill (New York), 1970, at p. 14. An interesting tacit assumption in Fanger's and most other analyses is that perceptions of thermal discomfort are monotonic within the too-hot or too-cold range: if the most people are comfortable at, say, 78°F, then more will say they are too hot at 86°F than at 84°F. However, Fanger's data (at p. 129 and at p. 131, Fig. 26) show this is not the case: there is a pronounced notch in the too-hot side distribution, together with a symmetrical effect (though a plateau, not a notch) on the too-cool side at 70°F. This effect appears to have gone unnoticed, and its statistical significance cannot be deduced from the data provided. Moreover, thermal neutrality is now suspected not to be an accurate indicator of thermal comfort but to be systematically biased by a "semantic offset" (§2.3).

3   For a critique, see L. Lutzenhiser, "Explaining consumption: the promises and limitations of energy and behavior research," *Procs. ACEEE 1990 Summer Study on Energy Efficiency in Buildings*, 2:101–110, and P. Stern, "Blind spots in policy analysis: what economics doesn't say about energy use," *Journal of Policy Analysis & Management* 5(10):200–227 (1986).

4   W. Kempton & L. Lutzenhiser, "Introduction," *Energy and Buildings*, 18(3):171–176 (1992).

5   Fanger [2], at Preface, p. 5.

6   Air Conditioning: The Interplay of Technology, Culture and Comfort, W. Kempton & L. Lutzenhiser, eds., *Energy and Buildings* 18(3) (1992). E SOURCE is grateful to Dr Willett Kempton (then at the Center for Energy and Environmental Studies, Princeton University, now at the Center for Urban Policy Research, University of Delaware) for kindly making available a prepublication proof set and for many helpful comments.

7   Kempton & Lutzenhiser [4], p. 173.

8   Strictly speaking, the ASHRAE comfort zone corresponds to ~90% acceptability, but an additional ~10% will find it unacceptable because of practical local problems with uncomfortable drafts, radiant asymmetry, etc.; B. W. Oleson, discussion in *ASHRAE Transactions* 92(1B):26 (1986).

9   American Society of Heating, Refrigerating and Air-Conditioning Engineers, Inc. (Atlanta), ANSI/ASHRAE Standard 55-1981, "Thermal Environmental Conditions for Human Occupancy," 1981, and International Standardization Organization (Geneva), Standard 7730, "Moderate thermal environments—determination of the PMV and PPD indices and specification of the conditions for thermal comfort," 1984. The latter is widely believed to give a slightly better fit to reported comfort conditions in warmer climates. For a comparison, see K. C. Parsons, "Human response to hot environments: a comparison of ISO and ASHRAE methods of assessment," *ASHRAE Trans.* 93(1):1027–1038 (1987). Standards similar or identical to these are also adopted by such groups as the Conseil Internationale du Bâtiment and the Society of Heating, Air-Conditioning and Sanitation Engineers of Japan (SHASE).

10 D. A. McIntyre, "Chamber studies—reductio ad absurdum?," *Energy and Buildings* 5:89–96 (1982).

11 McIntyre [10], p. 91; each unit on the seven-point scale corresponds to a 3 C° range of drybulb temperature (p. 90).

12 Fanger [2], p. 40. In both of Fanger's graphs, "In order to maintain comfort the ambient temperature was lower the higher the activity level." Note also, in the right-hand graphs at the sedentary activity levels (~70–90 kcal/h·m^2), the systematic gender difference.

13 In confirmation, the evaporative heat loss in recent experiments in Singapore varied by 4.6 times among the 32 Asian subjects, especially among women: R. J. de Dear, K. G. Leow, & A. Ameen, "Thermal comfort in the humid tropics—Part I: Climate chamber experiments on temperature differences in Singapore," *ASHRAE Trans.* 97(1):874–879 (1991).

14 W. Kempton, D. Feuermann, & A. E. McGarity, "'I always turn it on super': user decisions about when and how to operate room air conditioners," *Energy and Buildings* 18(3):177–191 (1992). In addition, of course, these New Jersey subjects freely chose their own real-life clothing and activity levels, rather than sitting quietly in a laboratory climate chamber wearing a standard uniform.

15   T. Sawachi, Y. Matsuo, K. Hatano, & H. Fukushima, "Characteristics of behavioral responses to indoor climate—A case study of Tokyo suburbs in summer," *Procs. ASHRAE Far East Conf. on Air Cond. in Hot Climates* (Singapore, 3–5 September 1987), pp. 22–34.

16   Kempton & Lutzenhiser [4] citing P. C. Stern, "The preference for coolth," *Energy and Buildings* 18(3):262–263 (1992).

17   F. H. Rohles & W. Wells, "The role of environmental antecedents on subsequent thermal comfort," *ASHRAE Transactions* 83(1):21 (1977). This result appears to supersede Fanger's 1970 conclusion that although some experiments had shown a link between color and perceived temperature (subjects had judged heating coils to be hotter when wrapped in green paper [perhaps nearer the eye's peak photometric sensitivity at 555 nm] than in paper of other colors), while others did not, no experiment had yet indicated a psychological influence of color on perceived thermal comfort at a given temperature. Perhaps decorating offices as meat lockers or walk-in freezers would provide comfort at higher temperatures and lower costs than via conventional air conditioning, albeit unattractively.

18   C. Seligman, J. M. Darley, & L. J. Becker, "Behavioral approaches to residential energy conservation," Ch. 10 (pp. 231–254) in R. H. Socolow, ed., *Saving Energy in the Home: Princeton's Experiments at Twin Rivers, Ballinger* (Cambridge MA), 1978, reprinted as a Special Issue of *Energy and Buildings* 1(3) (April 1978).

19   T. Woteki, "The Princeton omnibus experiment: some effects of retrofits on space heating requirements," CES [now CEES] Report #43, Princeton University (Princeton NJ 08540), 1977.

20   R. A. Winett *et al.*, "The effects of videotape modeling and daily feedback on residential electricity conservation, home temperature and humidity, perceived comfort, and clothing worn: winter and summer," *Journal of Applied Behavioral Analysis* 15(3):381–402 (1982).

21   P. Stern [3].

22   W. Kempton, D. Feuermann, & A. E. McGarity [14].

23   A class of comfort models originated by A. P. Gagge ("A new physiological variable associated with sensible and insensible perspiration," *American Journal of Physiology* 120:277 (1973)) treats the body as ~90% core and ~10% skin mass (the outer 1.6 mm), with blood-borne heat transfer governed by vasodilation controlled in turn by deviations from both portions' comfort setpoints (36.6 and 34.1°C respectively).

24   R. J. de Dear, "In defence of space cooling and the science of thermal comfort," *Energy and Buildings* 18(3):260–262 (1992).

25   However, as pointed out by S. Tanabe & K. Kimura, "Thermal comfort requirements under hot and humid conditions," *Procs. ASHRAE Far East Conf. on Air Conditioning in Hot Climates* (Singapore, 3–5 September 1987), pp. 3–21), this debate is confused on a key point:

> *Sometimes it is misunderstood that agreement of the neutral temperatures among different groups of subjects can prove agreement of all thermal sensations. Linear regression analysis ... normally gives good estimation around the median point, but equal sensation changes on the thermal sensation scale may not always correspond to equal temperature changes for the entire temperature range. Criticizing Fanger's work, [R. P.] Clark and [O. G.] Edholm [Man and His Thermal Environment, Edward Arnold, 1985] suggested that it seems*

unlikely that the influences of exercise, body size, body composition, sex, age and effects of adaptation or acclimatization ... have little or no effect on thermal comfort. It was not clearly understood that Fanger did not statistically check thermal comfort for the entire temperature range but only around the neutral point among different groups of subjects. It is apparent that people living in hot and humid areas are acclimatized and have the increased ability to endure a hot environment [A. R. Frisancho, Human Adaptation, U. of Mich. Press, 1981], which does not mean acclimatized persons would prefer warmer conditions. Since the interindividual differences are rather large ... it is quite difficult to determine statistically the differences in the neutral temperature among different groups of subjects.

26 Lee Eng Lock, personal communication, 20 May 1992. A confirmatory calculation by Rob Bishop of the E SOURCE staff for a typical medium-sized office yielded 3.1% in peak load and more in annual energy. The sensitivity depends on many details of system design and control in this complex and highly interactive nonlinear system. In general, the savings are more valuable in a well-designed, finely tuned system than in a crudely designed, oversized one.

27 That is, during the summer, when 70% of the studied townhouses' electricity consumption was for central air conditioning, each 1 F° increase in thermostat setting would reduce air-conditioning electricity consumption by ~12% and hence total summer electricity consumption by ~8%: R. H. Socolow & R. Sonderegger, "The Twin Rivers program on energy conservation in housing: Four year summary report," CES [now CEES] Report #32, Princeton University (Princeton NJ 08540). The air conditioners were EER≈7, 2-t units with no economizers, sized at ~725 finished ft²/t and operating ~700 h/y. The commercial-sector coefficient is lower than the residential because of important parasitic loads that move water and air (including makeup air) over longer distances, and because of large latent loads that are quite insensitive to the air temperature supplied.

28 S. Chandra, "Fans to reduce cooling costs in the Southeast," FSEC-EN-13-85, Florida Solar Energy Center (Cape Canaveral, FL), 1985.

29 L. Lutzenhiser, "A question of control: alternative patterns of room air conditioner use," *Energy and Buildings* 18(3):193–200 (1992).

30 Fanger [2], p. 83.

31 J. F. Busch, "A tale of two populations: thermal comfort in air-conditioned and naturally ventilated offices in Thailand," *En. Build.* 18(3):235–249 (1992); "Thermal responses to the Thai office environment," *ASHRAE Trans.* 96(1):859–872 (1990).

32 As summarized by Kempton & Lutzenhiser [4].

33 These temperatures are expressed in terms of Effective Temperature (ET*), which is normalized to 50% relative humidity and to mean radiant temperature equal to air temperature.

34 de Dear, Leow, & Ameen [13].

35 R. J. de Dear, K. G. Leow, & S. C. Foo, "Thermal experiments in the humid tropics: field experiments in air conditioned and naturally ventilated buildings in Singapore," *International Journal of Biometeorology* 34:259–265 (1991).

36 Conversely, the apparent care and thoroughness of even the best climate-chamber measurements may be illusory: awkward questions can be raised about turbulent vs. laminar airflow, radiant-temperature symmetry and equilibration, ceiling

heights (which may interact with the human body's convective plume—part of why old, high-ceilinged rooms increase summer comfort), displacement vs. ceiling-louver ventilation, rhythms and transients vs. steady-state effects, thorough characterization of airflow vector fields, etc. It is also possible that some supposedly irrelevant conditions within the chamber, such as flickering or poor-spectrum lighting or video displays, could affect physiology in second-order ways not yet understood.

37   The former averaged 0.56 *clo* (standard deviation 0.12) and 0.13 (0.04) m/s, while the latter averaged 0.49 *clo* (0.10) and 0.33 (0.27) m/s, and these differences accounted for a substantial part of the comfort results. Interestingly, the air-conditioned space and heavier Western clothing were private-sector; natural ventilation and more traditionally Thai, climatically appropriate clothing were public-sector. The correction for the effects of clothing and airspeed differences, though more weakly correlated (r^2= 0.74), was an impressive 4.6 C°, dramatizing "how important clothing conventions and local air-movement are in influencing comfort in indoor climate conditions such as" those in the naturally ventilated offices.

38   Specifically, 2.7 C° using Effective Temperature as the explanatory variable (as ASHRAE Standard 55-81 does), or 1.6 C° using Standard Effective Temperature.

39   The goodness of fit was r^2 = 0.91 with Effective Temperature and 0.96 with Standard Effective Temperature as the explanatory variable, vs. Fanger's 0.57–0.83.

40   R. J. de Dear, K. G. Leow, & A. Ameen, "Thermal comfort in the humid tropics—Part II: Climate chamber experiments on thermal acceptability in Singapore," *ASHRAE Trans.* 97(1): 880–886 (1991).

41   Busch cites others' findings of neutral temperatures of 26.0, 26.1, 26.7, and 27.2°C in earlier Singapore studies, and of hot-season values of 31.1°C in India and 32.5°C in Iran; Fanger [2], p. 83, cites studies finding 27.4°C and 28.5°C in Singapore, 26.0°C in Port Harcourt (Nigeria) and Calcutta, and 25.8°C in Port Moresby (New Guinea).

42   M. A. Humphreys, "Field studies of comfort compared and applied," *Building Services Eng.* 44:5–27 (1976); McIntyre [10], the source of the quotation (p. 93).

43   de Dear, Leow, & Foo [35], who found the same neutral Effective Temperature as in Busch's Bangkok naturally ventilated offices—28.5°C.

44   D. A. McIntyre [10].

45   de Dear, Leow, & Foo [35], p. 95.

46   The Summary of *Thermal Comfort* [2], p. 217, states (emphasis added) that "No significant difference was found in the comfort conditions between American and Danish students, between students and elderly people or between males and females. It is therefore suggested that the comfort equation can be applied *within the temperate climate zones* for adults, independent of the sex." A similar statement appears at p. 81. However, at p. 85, Fanger is more equivocal: "Comparison between results from field studies in the tropics and the comfort equation suggests that the equation can also be used under these conditions."

47   A more restrictive, purely physiological model of acclimatization might have been disproven by recent studies suggesting that "humans cannot be adapted to prefer warmer ambient temperatures," but for the experimental flaw that the Singaporean subjects were students from a fully air-conditioned campus, so their "daily routine includes several hours in temperatures below 25°C and 60%

relative humidity," vs. the Singapore normal range of ~26.6±4.1°C at 84% relative humidity (de Dear, Leow, & Ameen, [40]). The partly air-conditioned exposure of these subjects, therefore, while making them perhaps representative modern Singaporeans, could well have inhibited or perturbed the true acclimatization to be expected from continuous exposure to their native climate.

48   Kempton & Lutzenhiser [4].

49   G. A. Schiller, "A comparison of measured and predicted comfort in office buildings," *ASHRAE Trans.* 96(1):609–622 (1990), at p. 616. This effect has even been quantified: in offices, for example, the neutral temperature in °C is 5.41 + $0.73T_m$, where T_m is the mean indoor air temperature to which people have become acclimatized (A. Auliciems, "Thermobile controls for human comfort," *Heat. Vent. Eng.*, April/May 1984, pp. 31–33). Nor is the use of the term "addictive" a rhetorical excess: Kempton has pointed out (in his 30 September 1992 presentation at the Fifth Annual E SOURCE Members' Forum) that based on the field study data, air conditioning appears to satisfy rigorously the same definition of addiction used by clinicians with respect to drugs.

50   M. A. Humphreys, "Field studies of comfort compared and applied," *Building Services Eng.* 44:5–27 (1976) and "The dependence of comfortable temperatures upon indoor and outdoor climates," in K. Cena & J. A. Clark, eds., *Bioengineering, Thermal Physiology and Comfort*, Elsevier (Amsterdam), 1981, pp. 229–250; A. Auliciems, "Towards a psycho-physiological model of thermal perception," *Int. J. Biometereol.* 25:109–122 (1981) and "Psycho-physiological criteria for global thermal zones of building design," id. 26 (Suppl. 2):69–86 (1983).

51   de Dear, Leow, & Foo [35].

52   Busch [31]. Similarly, Schiller [49], p. 620, notes that

   up to one-third of the people voting within the four extreme categories of the thermal sensation scale (±2, 3) also felt these conditions were comfortable. These results suggest that the concept of "comfort" covered a broader range of thermal sensations than commonly assumed and that people voting within the extreme sensations are not necessarily dissatisfied. This is somewhat in contrast to the assumption commonly used [e.g. by ASHRAE] to relate thermal sensation to comfort. ... These results suggest that current standards and practices for maintaining comfortable thermal environments in office buildings need to be reexamined, supplementing laboratory data with information obtained in field studies.

53   de Dear, Leow, & Ameen, Part I [40]. McIntyre [10] cites similar evidence for asymmetrical semantic effects: "in a British winter, people voting 4 [i.e. neutral on a 7-point scale] may want the temperature warmer, but never cooler, [while] in a hot American summer, people voting 4 may want a lower temperature, but never warmer. These results suggest that culture and climate affect people's description of sensation—though not necessarily the sensation itself." McIntyre cites further experiments demonstrating a poor correlation between a near-neutral thermal sensation (3–5 on a 7-point scale) and reported comfort, depending largely on seasonal shifts on perception: in a New York office experiment, for example, over half the people voting 6–7 (conventionally considered outside the comfort range) also reported that they felt comfortable.

54   Kempton & Lutzenhiser [4].

55   Schiller [49] reported 1987 ASHRAE RP-462 findings in ten San Francisco office buildings, using the Berkeley mobile measurement cart to assess microclimatic

conditions at the workstation. She found that at the observed mean conditions (0.55 *clo*, 1.12 *met*), there were substantial differences between the neutral temperature measured by her own 2342-datapoint regression (22.4°C), the TSENS computer model (23.8°C), Gagge's modification of Fanger's Predicted Mean Vote method (23.9°C), and Fanger's original PMV method (24.8°C). A similar 2.8-C° range was found for the optimal temperatures (those at which the smallest percentage of the people were dissatisfied). The Predicted Percent Dissatisfied at her neutral temperature was 12%, compared with Fanger's predicted optimum of 5%, partly (it seems) because of a twofold-either-way range of *clo* values. In discussion, however, Fanger suggested that the combination of a 0.1–0.2-*clo* chair correction and a ~0.1-*met* correction for the higher stress of modern office work (added to the 1.2 *met* assumed from older research) could account for much of the difference. If so, that is not reassuring, because ~0.1-*clo* and -*met* measurement precision is very difficult, so model results requiring such precision would be a rather insecure basis for marginal comfort investments ultimately totaling hundreds of billions of dollars nationwide.

56   However, Fanger correctly notes that one must poll perceived comfort under conditions actually experienced, not merely ask people the temperature at which they think they would be comfortable: such a thought-experiment, he found yielded ~4–5 C° apparent error, [2] pp. 80–81.

57   This should test the "obvious hypothesis that the [non-air-conditioned] samples are more directly responsible for achieving their own comfort," and that this is responsible for their different comfort response than those who passively experience air conditioning that they do not control: R. J. de Dear, personal communication, 19 May 1992.

58   Lovins, Amory B., David Houghton, Robert Bishop, Bristol L. Stickney, James Newcomb, Michael Shepard, and Bradley Davids, *The State of the Art: Space Cooling and Air Handling*, Rocky Mountain Institute (Snowmass, Colo.), 1992.

59   *1985 Fundamentals Handbook*, at p. 8.8, Fig. 2A.

60   Lovins *et al.* [58], Chapter 2. In a multimethod behavioral study, "Clothing factors in energy conservation," *J. Consumer Studies & Home Ecs.* 11:195–205 (1987), B. A. Sommer, S. B. Kaiser, & R. Sommer note, among other interesting things, that the Carter Administration's indoor temperature regulations apparently caused about two-thirds of US office workers to dress differently; of 28 California firms surveyed, half had relaxed and none had tightened their dress codes in a recent five-year period, often moving away from business suits; most people expect executives of either gender to wear a suit; and there is "scant awareness of the connection between dress codes and energy conservation."

61   ASHRAE 55-81 at p. 4 specifies ≤0.15 m/s (30 fpm) in winter and ≤0.25 m/s (50 fpm) in summer. The former is "important to prevent local draft discomfort" at optimal or cooler air temperatures. The latter, however, can be increased for cooling credit; §2.3.4.

62   Stern, "The preference for coolth," [16]. Robert Bishop reports that the comparable New Zealand value is often as low as 10°C, at least in the mornings. Moreover, emphasizing "the learned character of comfort," US indoor thermal standards for winter comfort have risen from 18°C in 1923 to 24.6°C in 1986; in the 1920s, 90% of workers in a US light industrial plant rated 21–22.5°C "too warm." British household temperatures have risen by 1 C° per decade for the past 30 years. Some of these changes are due to clothing—a wool sweater is

equivalent to ~2 C°—but more to changing perceptions of comfort. (See T. M. Nelson, T. H. Nilsson, & G. W. Hopkins, "Thermal comfort: advantages and deviations," *ASHRAE Trans.* 93(1):1039–1054 (1987).)

63 de Dear, Leow, & Foo [35].

64 G. Prins, "On Condis and Coolth," *Energy and Buildings* 18(3):251–258 (1992); cf. other authors' subsequent commentaries.

65 In the American Refrigeration Institute's *Koldfax*, August 1988, at p. 4, the Managing Editor, after a full-page list of heat-related deaths around the world, says they mean that space-conditioning products "not only are valued, but often are necessities rather than luxuries."

66 Of course, despite the useful field data from the studies cited earlier in Bangkok, Singapore, etc., we still lack truly large-scale efforts to measure comfort, happiness, productivity, etc. of those millions with and without air conditioning.

67 More precisely, Prins states that his semiotic analysis suggests the dualism: "cool/dry/civilized/scented/barren/asexual/black/clean/good/death" vs. "hot/wet/wild/sweaty/fertile/sexual/red/dirty/evil/life" (Prins [63], p. 258).

68 B. Burrough, "In Houston, the ubiquitous air conditioner makes tolerable an otherwise muggy life," *Wall Street Journal*, 21 September 1983, p. 31.

69 H. Fujii & L. Lutzenhiser, "Japanese residential air conditioning: natural cooling and intelligent systems," *Energy and Buildings* 18(3):221–233 (1992).

70 Tokyo has 5 F° lower summer design drybulb temperatures than Houston and a 6° higher latitude, but its design wetbulb temperatures are only 0–1 F° lower, and its summer mean daily range is 4 F° smaller. By any measure, Tokyo is very hot and muggy in the summer. It also has 1–4 F° lower winter design temperatures than Houston.

71 T. Sawachi, Y. Matsuo, K. Hatano, & H. Fukushima, "Characteristics of behavioral responses to indoor climate—A case study of Tokyo suburbs in summer," *Procs. ASHRAE Far East Conf. on Air Cond. in Hot Climates* (Singapore, 3–5 September 1987), pp. 22–34.

72 H. Fujii, "Controls for Japanese residential heating/air conditioning systems," *Procs. 1991 ACEEE Summer Study on Energy Efficient Bldgs.* 2:35–42.

73 Sawachi *et al.* [15]. Of course, such conditions are common in cities and suburbs worldwide. For example, a 1990 survey of 384 East Central Florida tract-development households reported the following reasons for not using natural ventilation: too hot/cold/humid, 63.2%; no adequate breeze, 15.4%; lack of security, 10.5%; allergies, 8.6%; excessive noise from traffic, 1.4%; windows hard to operate/inoperable, 0.5%; dog barking, 0.3%. Yet the same sample could hardly be said to be unwilling to take actions to improve comfort: 52.3% always and 32.5% frequently used fans (averaging 4.46 per household) to reduce air conditioner usage; 89.8% reported using a fan with the air conditioner; and 91.7% reported using manual setback. R. K. Vieria & D. S. Parker, "Energy use in attached and detached residential developments," FSEC-CR-381-91, Florida Solar Energy Center, Cape Canaveral, FL 32920, January 1991.

74 Sawachi *et al.* [15]. This paper gives detailed data on usage patterns in different rooms and during various household activities.

75 Seligman *et al.* [18]. The statistics in the 40-household, 1-month experiment were $F(1,35) = 4.64$, $p<0.04$. The experiment was doubly controlled by using four groups of 10 households: blue light only, feedback that thermostat control is the key to saving central air conditioning energy, both, and neither.

76 Kempton, Feuermann, & McGarity [14].

77 Sawachi et al. [15], p. 32, Fig. 3, and p. 33, Fig. 5.

78 Kempton, Feuermann, & McGarity [14].

79 Ibid.

80 Or two orders of magnitude if comparing apartments rather than air conditioners. The corresponding difference in average indoor temperature was a surprisingly small 2.4°C for the whole summer, or 3.7°C for July only. Moreover, the most energy-intensive apartment was on a mid-floor on the north side, and ought therefore to have used less air conditioning, not more.

81 Kempton & Lutzenhiser [4].

82 Kempton, Feuermann, & McGarity [14], pp. 180–181.

83 Ibid., p. 189.

84 W. Kempton, C. Reynolds, M. Fels, & D. Hull, "Utility Control of Residential Cooling: Resident-Perceived Effects and Potential Program Improvements," *Energy and Buildings* 18(3):201–219 (1992).

85 Lovins *et al.* [58].

86 Kempton, Feuermann, & McGarity [14], at p. 190.

87 Lutzenhiser [29].

88 Kempton & Lutzenhiser [4] citing Stern [16].

89 Kempton & Lutzenhiser [4].

90 US Department of Commerce, Bureau of the Census, Characteristics of New Housing: Construction Reports, Series C25, various issues, compiled in G. R. Amols, A. K. Nicholls, K. B. Howard, and T. D. Guerra, *Residential and Commercial Buildings Data Book*, 3d ed., Battelle Pacific Northwest Laboratory, PNL-6454, February 1988.

91 Lee Eng Lock, personal communication, 21 May 1992. Published regional sales of household air conditioners are about 10 million units/yr. They have a minimum compressor shaft input of 1 hp and are more often 1.5–2 hp. A nominal 2–hp compressor, plus motor losses and auxiliaries, draws ~2.2 undiversified peak kW. We assume here a 15% onpeak grid loss, a ~0.8 nominal onpeak power factor, and onpeak diversity of 0.9 (though the actual figure is probably higher).

92 Seligman *et al.* [18], pp. 232–241; the statistics were $r^2 = 0.553$, $F(8,47) = 7.26$, $p<0.001$.

93 Lovins et al. [58], §7.6.4.

94 This does not mean unfounded—just perhaps surprising to researchers. In the Kempton, Feuermann, & McGarity study [14], for example, all of the users' belief systems rested on objectively valid premises, even though outsiders might not have thought of them or attached so much weight to them.

95 In Schiller's sample [49] of ten San Francisco office buildings, for example, ~40% of the people exposed to ASHRAE comfort conditions were thermally dissatisfied, and even under "perfect" theoretical comfort conditions, a minimum of 12% were dissatisfied, rather than Fanger's theoretical minimum of 5%. This puts building operators in an impossible position. Moreover, as McIntyre [10] points out, it appears that "there is an intrinsic variation of discomfort feelings, and … if an opportunity exists to ascribe the cause of discomfort to an identifiable external environmental cause, people will take this opportunity to externalize their discomfort."

96 *Procs. Fifth Intl. Conf. Indoor Air Quality and Climate* (Toronto), 1990, summarized by O. Seppanen & J. R. Wright, *ASHRAE Journal*, October 1990, p. 16.

97   On 30 September 1992, a substantial fraction—perhaps a third to a half—of the facility managers in the audience at the Fifth Annual E SOURCE Members' Forum raised their hands when collectively asked if they had ever done this. All of those subsequently interviewed reported that their dummy thermostats had been highly successful.

98   Lovins *et al.* [58].

99   Ibid.

100   Kempton, Feuermann, & McGarity [14].

101   Lutzenhiser's [29] informants, too, said they turned the thermostat to colder levels "'If we need the cold air to come quickly,' 'when first starting up we turn it to colder,' 'when it's hotter outside,' 'maximum cool when we first turn it on, then adjust to a more economical level,' 'max at start, then turn down,' 'set it to coldest when we have been away and the apartment is very hot.' These responses suggest that some persons imagine the thermostat to be a sort of 'cold rheostat' … with higher [i.e. colder] settings producing cooler air—an interpretation that is powerfully implied by the 'warmer–cooler–coolest' labelling on the thermostat control."

102   Shepard, M., Lovins, A.B., Neymark, J., Houghton, D.J., & Heede, H.R., *The State of the Art: Appliances*, Rocky Mountain Institute (Snowmass, Colo.), 1990.

103   Except for two who used it to turn on the air conditioner gradually because they believed this would save power or avoid blown fuses—a not wholly fanciful belief in the inadequately wired apartment building, since it would make the fan and compressor startup surges non-coincident. But it does not appear that these two users understood and harnessed their units' thermostatic capability.

104   Lutzenhiser [29].

105   J. Lutz & B. A. Wilcox, "Comparison of self reported and measured thermostat behavior in new California houses," *Procs. 1990 ACEEE Summer Study on Energy Efficiency in Buildings* 2:91–100.

106   Lutzenhiser [29], p. 197.

107   Kempton, Feuermann, & McGarity [14], p. 190.

108   Ibid.

109   Professor Jeff Cook reports (personal communication, 9 June 1992) that this option is being studied at Kansas State University with support from the US Department of Transportation.

110   Lutzenhiser [29], p. 198.

111   Ibid., p. 196.

112   Lovins *et al.* [58]

113   Fujii & Lutzenhiser [69].

114   Ibid.

115   Ibid.

116   Lovins *et al.* [58]. See also Lisa Heschong, *Thermal Delight in Architecture*, MIT Press (Cambridge MA), 1979.

117   Fujii & Lutzenhiser [69].

118   Only one-fourth of Fujii's technically informed sample, all of them male, reported using every control; 30% (five of the six being female) used some in ways they believed to be correct; and 35% (half of each gender), including both the elderly residents, said they would check the instructions before changing the settings.

119   As Fujii [72] remarks, "A timer may … be a useful control for both efficient energy use and comfortable thermal environment setting. Once residents fall asleep and become unaware of thermal comfort, it is not necessary to keep cooling."

120   Lovins *et al.* [58].

Chapter 19

The Superefficient Passive Building Frontier

1995

Editor's note: *This article is reprinted with permission from the* ASHRAE Journal, *June 1995, and summarizes Amory's Centenary address.*

I would like to look ahead to the next century and to suggest that what this century has been preparing us for is not just a linear view of progress, but a cyclic view of progress in which we rediscover much forgotten wisdom. I think in the next century of mechanical design, pressures on capital and energy costs, environmental performance, and operability will rapidly shift designs from active to passive, from formulaic to uniquely optimized, and from complex to simple.

I am going to suggest that integrated whole-building design can yield superior comfort with about three to thirty times less mechanical energy and often with lower capital costs, but that achieving this poses fundamental challenges to professional education and practice and to compensation structure.

The United States has already misallocated something like 200 million tons of cooling capacity and 200 peak gigawatts of power supply to run it, at a total marginal cost approaching $1 trillion, through failure to optimize the buildings that that capacity was installed in. We need to do better.

How can we apply and integrate proven methods that build on millennia of design wisdom? As an example, I would like to talk about a house that has been occupied for the past year in Davis, California, as part of the Pacific Gas and Electric ACT² experiment. It is an ordinary-looking tract house of 1672 ft² (155 m²) that is compliant with the strictest energy code in the country (1993 Title 24). The design temperature is 105°F (41°C), the peak about 113°F (45°C).

A design team at Davis Energy Group was first able to eliminate 7 meters of superfluous perimeter by improving the floor plan. In addition, they put the windows in the right place and designed an engineered wall made of an oriented strandwood product that is a kind of synthetic hardwood. In this way they saved three-quarters of the wood, doubled the insulating value to a true R-27, improved strength, airtightness, durability, stability and speed of construction, and saved $2000.

Altogether their design changes saved in the order of 15 percent of the original energy use and $4000 of construction costs. On the interior the designers did a lot of little things to the appliances, lights, glazings, and hot water system, thereby raising the total savings to about 60 percent of original energy use. Along the way they got rid of the furnace, using instead a hydronic backup to a radiant slab coil fed by the 94-percent-efficient gas water heater that they were paying for anyway.

In other words, by getting about twice as much insulation in the shell and much better glazings, they found they did not need the furnace any more. But they still had left a third of the original 3.5 ton cooling capacity and were up to their cost-effectiveness limit. What to do? Well, they had thoughtfully reserved a special "potential cooling elimination package" into which they had put all other measures considered but rejected because they did not save enough *energy* to pay for themselves—yet they also saved *cooling load*. When seven such measures were added to the design, they more than *eliminated* the remaining air-conditioning needs. They therefore achieved even larger savings at lower capital cost because they saved $1500 on air conditioning and ductwork.

The result, therefore, of putting in these supposedly non-cost-effective measures was to give bigger and cheaper total savings. In fact, the design basis was 80 percent savings on space and water heating, space cooling, refrigeration, and lighting, and it appears it is probably working at least that well. In a mature market, construction cost would be about $1800 below normal and present-valued maintenance cost $1600 below normal.

The Davis house shows in a hot climate, and Rocky Mountain Institute's 99 percent passive-solar banana farm in an 8700-degree-day-Fahrenheit climate showed in 1983, that big savings can be cheaper than small savings if you combine the right ingredients in the right way. We are often seeing this phenomenon in hot and cold climates, big and small buildings, and new and even retrofit buildings. We are also seeing it in many other technical systems: motor and lighting systems, hot water systems, computer design, cars, and almost everywhere else we look.

The magical economics come from single expenditures with multiple benefits. For example, superwindows have ten main benefits. They do not just save HVAC energy. They also provide such superior radiant comfort that they save a lot of energy indirectly through relaxed thermostat setpoints. They also let you downsize, simplify or even eliminate mechanical equipment.

HVAC simplification and the reduction of loads in commercial buildings create many important indirect benefits. You may, for example, go from big rectangular ducts to small round ducts, saving 70 percent of the metal and more on labor, reducing plenum height, getting more stories per unit height, and saving structural loads and plan areas no longer taken up by those big duct sections and wiring closets and mechanical rooms. You can rent out the space next to the mechanical rooms because the equipment becomes very quiet. Altogether, these kinds of indirect savings that pyramid through all aspects of the design may save more capital cost than the reduction in mechanical capacity.

In a cold climate, just superwindows' ability to eliminate perimeter zone heating in a commercial building more than pays their marginal cost, making their nine other benefits free. You save not only the capital cost, but also some floorspace and flexibility in reconfiguring the space in perimeter offices.

In addition there is better UV control from superwindows, better noise suppression and reduced maintenance costs. You facilitate the entry and control of daylighting to displace both lighting and cooling energy and capacity, separating light from unwanted heat with near perfection. You end up creating such superior visual, thermal and acoustic comfort from an integrated design that recent case studies show you may well get labor productivity benefits of 6 to 16 percent. These benefits could be worth an order of magnitude more than the entire energy bill.

This example suggests that if we properly count multiple benefits and take credit for those that are real and measurable in rigorous engineering/economic terms, we will very often find that the way to make a building inexpensive to construct is to make the windows expensive. This is not the usual value-engineering approach of squeezing pennies out of each component separately, but it is investing our money in a highly integrated fashion to put more in some places so we can put a lot less in others.

Let me give a few examples of how this can happen. To create comfort, there are many things we can do to expand the comfort envelope: for example, better mean radiant temperature and less asymmetry in it, or air movement, or ventilative chairs. Then there is load reduction. That is remarkably powerful if we combine systematic reduction of internal gains like lights and plug loads with reduction of external gains through insulation, superwindows, shell albedo, mass, shading, and orientation. Just making the building the right shape and pointing it in the right direction is often good for about a one-third saving in energy use.

It is not unusual in office design to be able to go from 250 or 350 ft^2 (32 m^2) per ton up to 800 or 1000 ft^2 (74 to 92 m^2), and in state-of-the-art designs, 1200 ft^2 (111 m^2) per ton. Obviously, downsizing the mechanicals at $3000 per whole-system ton is good business and, in fact, it will often facilitate the use of passive and alternative cooling. Passive may be ground coupling, ventilative, or radiative. Alternative may be desiccant, absorption, evaporative, and combinations thereof. Then you can cut any remaining mechanical refrigerative system to 0.6 kW/ton or less, including all auxiliaries from supply fan through cooling tower (an indirect-evap-plus variable-speed recip system was recently designed at 0.14 kW/ton); then do better controls; and then perhaps coolth storage. Thus you gradually nibble away at the original energy use with a chain of successive savings until almost nothing is left.

Let us consider a 200,000-ft^2 (18,850-m^2), 20-year-old curtainwall office building with dark glazing units that are starting to fail from old age. Normally you would just replace what is there with more just like it. It turns out at almost the same cost you can save three-quarters of the energy with the retrofit. How do you do that? You reglaze it with superwindows that are twice as good at letting in light as heat, admit six times as much light, and insulate four times as well as the old dark units, yet cost almost the same. You flood the space with deep daylighting, glare-free and nicely distributed. You also put in very efficient lighting and plug loads (totaling 0.5 W/ft^2 as used), and you thereby reduce the design cooling load from 750 tons to under 200 tons.

Now, ordinarily you would have renovated the 750 tons for maybe $800 a ton; that's $600,000. Instead, you can rebuild 200 tons to get not 1.9 but 0.5 system kilowatts per ton. So it is almost 4 times as efficient, and may cost 2.5 times as much per ton, but you have almost 4 times fewer tons. You save $200,000 on the mechanical retrofit, and that is what pays for the lighting and glass retrofit. Calculated payback: minus 5 to plus 9 months. There are over 100,000 big, old curtainwall buildings out there that many of you can retrofit in this fashion.

The general strategy, then, for commercial retrofit, is to rigorously avoid internal heat gains and have an exemplary envelope with tuned superwindows and deep daylighting. Some load reductions you can do only in new buildings, but many you can do in retrofits. Then you have smaller and much better mechanicals to the extent they are still required, and superefficient drivesystems and controls. You typically end up with site energy around 10,000–20,000 site BTU/ft^2·yr—80–90 percent savings—but construction costs go *down* by several percent.

Such good design needs better compensation structure. If design professionals of any kind are compensated for what they spend, not for what they save, they have a perverse incentive that rewards inefficiency and penalizes efficiency.

I think it is important for design professionals, as for other parties in the real-estate process, to be rewarded for energy efficiency: for example, to be allowed to keep as extra compensation a percentage of whatever lifecycle costs they save. That could double or triple a conventional fee. It's a fair reward for the extra work. It certainly gets people's attention. And it would help, I think, to reintegrate the design process and to substitute true engineering optimization for rules of thumb.

The sort of world engineers will face in the next hundred years will require us all to do much more with much less. I am grateful for this opportunity to offer a few perhaps provocative insights into how this is starting to happen and how it can return us again to the existential joys of real engineering.

Chapter 20

Foreword to *Sustainable Design Guide* of the Japan Institute of Architects

1996

When the distinguished Secretary-General of the Japan Institute of Architects, Mr Kōjiro Takano, asked me to help introduce this book to the Japanese profession and public, I was at first a bit puzzled. Why should someone from his ancient culture—one so uniquely sophisticated in its aesthetic refinement, so experienced at avoiding any false distinction between people and the rest of the natural world—want any ideas from a brash and too often brutish young society? How could the culture that developed the wonders of classical Japanese gardens and *ikebana*, painting and tea ceremony, benefit from the one that invented strip malls, Las Vegas, and sterile, sealed office towers? How might the masters of elegant frugality, in everything from brush-strokes to *haiku* and from swords to ceramics, benefit from the almost undisputed leaders (despite fierce competition lately) in crude and ugly wastefulness? I recalled the time Gandhi-ji, from another ancient culture, was asked what he thought of Western civilization—and dryly replied, "I think it would be a very good idea."

A few years ago, on the meditation porch of an exquisite little temple above Kyōto, I was reminded what is really important about Japan. It's not the technology. The important things Japan can teach the world come most of all from its spirituality and its traditional aesthetics. The stunning technical achievements that seem to the superficial eye to define the spirit of modern Japan are in truth the least important part: far less important, for example, than the simple but profound idea that technology is foremost an art-form, to be judged first on aesthetic criteria.

And so I began to wonder whether the ageless roots of Japanese design might not have become somehow disconnected from the overlay of the glittering new technologies that they should be guiding; whether the manipulative, nature-alienating techniques of the past half-century might not be smothering the older wisdom of people's being integral with nature and not apart from it;

whether, in short, a new fusion of Japan's venerable sensibilities with the West's youthful enthusiasm for innovation might perhaps help Japan to rediscover the best in its traditions. And it began to seem possible that "green" architecture—creating serene and beautiful structures that grow organically in and from their place, structures that do not exploit or pollute but rather increase harmony with the whole world around them—might be a good place to start seeking this *kansei* synthesis.

Perhaps a society can become too good at technical artifice; can be tempted to do things because they are doable, not because they are wise. Too many modern commercial buildings have become like that, in East and West alike. They wall people off from the outdoors, immerse them in conditioned air and artificial light, cocoon them in uniform banality that is vacuous and ultimately deadly to the spirit. The air comes from coils and fans, not from trees and gardens; the light from phosphors, not from our sacred neighborhood star; the sounds from humming motors and lighting ballasts, forward-curved centrifugal fans and duct resonances, not from songbirds and waterfalls; the diversions from acrid fumes and eye-hurting glare, not from flowers and butterflies.

However skillful the designers, such buildings cannot provide the conditions in which people evolved, in which the body is healthy and alert and the spirit tranquil and nourished. Yet so pervasive has this travesty of architecture become that few people today have ever experienced real comfort—thermal, visual, or acoustic—and so they don't know what they're missing. When finally people *do* experience a space in which they can feel comfortable, see what they're doing, and hear themselves think, they do far (in eight recent case studies from Rocky Mountain Institute, about 6–16 percent) more and better work; they remain healthier and more alert, friendlier and happier. Businesses typically pay (at least in America) a hundred times as much for people as for energy; so such an improvement in labor productivity is an order of magnitude more valuable than eliminating the entire energy bill! Yet it remains extremely rare.

To provide such genuine comfort seems to require a fusion of Eastern and Western insights. From the East, especially from Japan, comes the idea—utterly contrary to the comfort theory used by Western mechanical engineers, yet utterly obvious to any evolutionary biologist—that people will prefer and will thrive in a subtly dynamic environment rather than a static one. So it is that modern Japanese room air conditioners slightly vary the temperature using a pseudorandom-number generator, and deliver air not in a steady flow but in a series of pseudorandom gusts—sometimes even in a pattern typical of a certain famous resort near Tōkyō. So it is, too, that a Japanese air-handling system might occasionally add subliminal traces of fragrance—sandalwood, say—to stimulate the sensorium. And so it is that great Japanese teacups and handles are pleasurable to touch, yet no two quite alike; that great Japanese food is as beautiful to see as to taste, and expresses the uniqueness of the occasion and of the learned chef's personality.

In contrast, the typical Western mechanical engineer would strive to *eliminate* every such pesky trace of variability with thermostats and humidistats and photosensors, to render the human experience uniform and constant down to the last lux of light and molecule of air—as if people were dead machines, not dynamic organisms. All too often, the architect, who seems to have forgotten where the sun is, designs a box that is randomly oriented, "all glass and no windows," shuttered with blinds and curtains to block any natural light that might somehow find its way in. Disagreeable drafts of hot and cold air are noisily blown at people to try to overcome the radiant and airborne chill or heat of the poorly conceived building envelope. Controlled ventilation replaces operable windows. Glaring downlights replace the luminous sky under which people first evolved. In every case, the artificial substitutes for natural conditions are less and worse: they do not let people do or be or feel their best.

Yet what if the correct Japanese ideal of the subtly dynamic environment were blended with some of the best of Western technology: for example, with "superwindows," not yet available in Japan, that can let in natural light without unwanted heat (to the maximum extent that's theoretically possible—visible transmittance is twice the shading coefficient), and that can insulate as well as twelve sheets of glass and block noise as well as four to six, yet look like two and cost less than three? (Such glazings enable us at Rocky Mountain Institute to grow bananas with no heating system while outdoor temperatures plummet to −44°C, and in Davis, California, to build a new house that provides superior comfort with no air conditioning while outdoor temperatures soar to +46°C—in both cases at lower-than-normal capital cost, because of the eliminated space-conditioning equipment; see Chapter 19). Could not a building designed with the beautiful Japanese idea of blending indoors and outdoors as simply as drawing open the shōji, yet also equipped with the American superwindows, provide the best of both worlds? Could it not automatically help keep us warm in the winter and cool in the summer, while integrating with advanced American daylighting design that floods the ceiling with soft, glare-free light, even tens of meters from the nearest window?

Let us imagine even more. Could not that supplementary uplighting, as the best American design also teaches, be free of glare, flicker, and hum; tuned to the eye's red, green, and blue retinal cones; serene and restful to the eye; and automatically dimmed according to how much natural light is coming in—and so use only a few watts per square meter? Could not the best of both Japanese and American computers, monitors, faxes, and photocopiers then reduce to the same value the heat they release into a fully equipped modern office? Could not chairs with comfortable mesh rather than insulating upholstery, and ceiling fans, and perhaps more sensible clothing, and other simple improvements greatly expand the conditions in which people feel comfortable, even in a Tōkyō summer? And could not the combination of the climatically responsive architecture (built form, massing, shading, orientation, albedo, vegetation, and all the other variables the architect can command) with the superwindows and the efficient lighting and office equipment then reduce normal cooling loads by four- or five-fold? Could not, finally, the lower cost of that smaller cooling and air-handling equipment, if indeed conventional equipment is still required at all, more than pay for making it several times as efficient—thereby *reducing* total construction cost, while saving 80 to 90 percent of the building's normal energy use, and meanwhile greatly improving comfort, amenity, and aesthetics?

Yes, such magical buildings can indeed emerge from a green design that artfully fuses the best of East and West. Some such buildings (not yet enough) already exist. They combine superior human *and* financial performance. They create delight when entered, harmony when occupied, regret when departed. They can be built in practically any desired style or size, program or climate. They simply (or at least it *seems* simple when it's all done) combine a biologically and spiritually informed appreciation of what people are and want, a completely integrated design process, and a toolkit of advanced technologies. They require the art and science of design to yield a result that is simple, not complex; passive, not active; gracefully responsive rather than stubbornly resistant to climate and sunlight and weather; uniquely optimized, not formulaic. And when you are in such a building, you know in your bones, as instinctively as if you were still on the savannah where humankind was born, the time of day and the time of year. You know which way is south. You know if it's snowing, if it's cloudy, whether there's a rainbow, whether the stars are out. You know you are a child of the natural world—not a refugee from nature, not a prisoner walled in a drab cell, cut off from our planet's constant change.

Green design can work its wonders on any scale, from a cottage to a city. In South Amsterdam, it creates the headquarters of the major bank NMB—an anthroposophic 50,000-square-meter complex where changing sun angles reflect off colored metal (part of the highly integrated artwork) to bathe the lower stories in ever-changing colors. (In newer buildings, these might even match the body's daily cycle of "appetite" for light that's now cooler, now warmer.) Every office has natural air and natural light. Bankers in three-piece suits dabble in the water that gurgles and splashes down flow-form sculptures in the bronze handrails along the soaring stairs. Absenteeism is down 15 percent, and working hours far exceed design expectations, because the workers can't bear to go home, and hold all sorts of evening events at their workplace. How did this miracle happen? Simple: the bank's Board of Directors ordered "an organic building that would integrate art, natural materials, sunlight, green plants, energy conservation, quiet, and water"—and that would not cost one guilder more per square meter. They insisted on a highly integrated and transdisciplinary design process, led by an architect, Ton Alberts, who had never before designed a commercial building. And when it was done, the building became the most readily recognized in all Holland, and the bank, with its newly progressive and creative image, soared from fourth to second biggest in the country, changed its name to ING, and bought Baring's.

In Leicester, England, green design uses natural convection and thermal mass to eliminate the cooling and air-handling equipment in DeMontfort University's engineering building: the mechanical engineering students must study diagrams of fans and chillers because the 10,060-square-meter building has none to show them, despite machine-shop loads up to 97 watts per square meter. The electrical engineering students learn about lighting systems in a daylit room with all the lights off. Similar approaches will soon eliminate cooling, air-handling, and perhaps also heating equipment in another university building—in subarctic Bozeman, Montana.

But green design goes far beyond balancing the gains and losses of heat, the natural flows of light and air, the harmonious coupling with the earth beneath and the heaven above. Green design extends infinitely outward in the six directions. It seeks to heal natural and human communities; to do no harm; to regenerate and restore; to *create* an abundance of energy and water, food and health, tranquility and beauty. The design is *with* nature—not a destruction or exploitation covered by the lame excuse that people, being part of nature, could do any ugly thing and still call it somehow "natural." Green design first asks the place, as Wendell Berry adjures us to ask, "What does this place require us to do? What will it allow us to do? What will it help us to do?" For Berry-*sensei*—the poet and farmer who also said, "What I stand for is what I stand on"—the architect is the servant of the place, doing what its spirits want, not the reverse. Green design is sacred architecture: it honors and enriches the Buddha-nature of all beings.

Green design is mindful of where materials come from and where they go. It does not take the rainforest tree from the orangutan to make plywood for a concrete form (let alone throw the wood away afterwards). It does not use up anything it cannot replace, destroy anything it cannot recreate; rather, it creates *more* diverse and abundant life than it borrows. It does not take as much, in quantity or in quality, as it gives back. It does not steal from our children, but leaves them a greener and richer and more peaceful world. And it uses materials that not only regenerate themselves but also protect the health and uplift the spirit of the people who use them. It does not poison the air with noxious fumes or the soul with artificiality. It sustains the best of Japan's rich palette of traditional arts and materials, applied in new settings to create well-being and happiness.

Green design treats water not as cubic meters of throughput but as *habitat*. It uses water to enhance life, to create ever more diverse and beautiful conditions in which life can thrive. It makes water leaving the building and the site cleaner than water arriving. It helps rainwater flow where it has always wanted to flow, through the capillaries and arteries of the earth, rather than diverting it through sewer-like concrete pipes. The money saved by substituting porous paving and natural drainage swales is instead reinvested to create gardens everywhere—as when Village Homes, a 1970s solar housing development (also in Davis, California), diverted ¥80,000 of such drainage infrastructure savings per house into community parks, gardens, and family microfarms. Soon it became the first American subdivision noted for the quality of its organic fruits and vegetables. Much of the landscaping's upkeep cost is paid by profits from the annual almond harvest.

Green design treats nutrients, such as human wastes, as a precious gift that can support still more life—as the basis of an aquacultural garden that creates fish and flowers, fruits and vegetables. It is bursting with the green world, both inside and outside. (I write this as I watch two green iguanas teach their own ancient meditation amidst the bougainvillea in the 85-square-meter atrium in the middle of our 99-percent-passive-solar "greenhome" 2200 meters up in the Rockies. The waterfall splashes, adjusted to harmony by an itinerant Japanese waterfall-tuner informed by his culture's millennia of aesthetic experience. My commute to work is ten meters across the "jungle": why not install vines and swing to work? Later, I smell the night-blooming jasmine, and the miniature hedgehogs run silently about, eating bugs in the moonlight. Something is missing ... ah! The frogs are not yet singing.)

Green design goes further yet. It focuses on how all people—including old and young, poor and infirm—can get access to the places they want to enjoy. It helps people to live, work, shop, and recreate nearby, so that as much as possible they can be *already* near where they want to be, and not need to go somewhere else. It integrates buildings with purposes with people. It favors walking and bicycling over cars and trains. It reduces isolation and fosters neighborliness. It protects street life, the public realm, the neighborhood, the community. And it designs communities around people, not cars. In Village Homes, the houses, some nearly invisible beneath luxuriant grapevines and shrubs, face one another across tree-shaded swales—magical green corridors full of flowers and playing children. Cars are kept discreetly at the rear: their narrow, tree-shaded roads and the separate, heavily used walking and biking paths along the swaled greenways enter the site from opposite directions like interlocking fingers. Such details matter: crime is one-tenth that of comparable nearby projects with normal "dead-worm" street layouts; Village Homes command a ¥25,500/m² premium; and they sell in less than one-third the normal time.

A green building's connections into the community run as deep as any taproot. Its links are not only physical but economic, communitarian, and spiritual. Thus the Inn of the Anasazi, a small up-market hotel in Santa Fe, New Mexico, supplies its famous restaurant with organic food from local traditional farmers who would otherwise lose their land. Its excess food goes to homeless shelters, its table scraps to compost, its kitchen scraps to the organic pig farm. Its workers, drawn from all three local cultures—Native American, traditional Hispanic, and Anglo—serve as a focus for intercultural conflict resolution, and are paid part-time to volunteer with community action groups. The hotel's furniture and decorations are made by local artisans from local materials. Its toiletries are made by a local Indian tribe from traditional medicinal herbs, and are also sold by the hotel. Peeling back layer upon layer reveals ever more intricate connections of the hotel into the community that sustains it and that it in turn helps to sustain. It is not a colonial occupying power; it is fully integrated with its place and its peoples. Why is not every building so organically rooted? Why do so many look as if they just landed by parachute?

The technical opportunities of green design are engaging, exciting, challenging. Energy use can be reduced by around 90 percent, water use and materials flows often by several times, while capital cost typically falls and both functional and financial performance improve. Achieving all this requires a highly integrated design process that brings together architects, engineers, artists, builders, and those who will use and operate and maintain the building. This team must remain cohesive not only throughout design but also through construction, commissioning, training, and initial operation, to make sure the design intent is fully carried out. And to reward the design team for their harder-than-usual work, it is appropriate, indeed essential, that they be compensated for what they *save*, not for what they *spend*: traditional compensation structures, based on the *cost* of the building or equipment, actually have the perverse effect of rewarding inefficiency and penalizing efficiency. No wonder we so often get the very opposite of what we want!

Another challenge of green design is to do not only the right things but in the right order. Here too is an opportunity to blend the best of East and West, so as not to adopt a gadget without the *gestalt*, but rather to integrate the world's best technologies into a seamless whole. Some designers forget that the integration is as important as the parts. Particularly in Japan, one sees solar-powered demonstration houses whose roofs are full of costly solar cells that provide perhaps one-half or one-third of the electricity devoured by a big heat pump striving to maintain comfort despite an inefficient building envelope, glazings, lights, and appliances. Solar cells are a wonderful Japanese technology, but the designers of such houses forgot something even more important: to start by making the *rest* of the house equally cleverly. Suppose the building were first made thermally passive, gains balancing losses automatically, so it didn't need the heat pump. (There are several much simpler ways to handle summer humidity.) Suppose the lights and appliances were then made extremely efficient, with Japanese technologies that can cut the house's total electric load to an average of barely over 100 watts. Then its heating and cooling needs would be roughly zero; its electrical needs could be met by only a few square meters of solar cells; and it would all work better and cost less. (That's how our own building works.) Is it because Japanese designers get so excited about solar cells that some get tempted to focus on energy supply before superefficient energy use, and so lose the rich prize of their proper integration? Is it perhaps because Japan is so used to thinking of itself as a country poor in fuel, when in fact Japan is probably the *richest* of any major industrial country in *energy*—the myriad forms of renewable energy—as such Japanese analysts as Dr Haruki Tsuchiya and Professor Yasuhiro Murota have been persuasively showing for the past 20 years? And if some Japanese designers design houses in this odd way, then what is it, conversely, about so much American energy design that misses the basic lessons, so superbly invented in Japan, of technical aesthetics and the dynamic nature of human comfort? What more can we learn from each other?

Yet to my mind, the technological challenges, fascinating and compelling though they are, are the least part of the green design opportunities that now beckon us. To apply technology wisely, in a mature relationship with our needs and capacities; to create architecture as rich as our cultural heritage; to create, above all, designs that celebrate life over sterility, restraint over extravagance, beauty over tawdriness; these truly define the aesthetic and spiritual challenge that calls architects to draw the best from all societies.

There is a name for this challenge. Years ago, Bill McLarney was inventing some advanced aquaculture in Costa Rica. He was stirring a big tank of algae one day when a brassy lady from North America strode in and demanded, "Why are you standing there stirring that green goop, when what really matters in the world is love?"

Bill replied, "Well, there's *theoretical* love; and then there's *applied* love"— and kept on stirring.

Sustainable design brings to all architects everywhere a remarkable opportunity for applied love. May we all prove worthy of that obligation.

Section 6

Energy Security and the Military: Blood and Treasure (and Opportunities)

Amory started lecturing at military staff colleges and meeting with the civilian and uniformed leadership of the US military in the 1970s. In 1982, his and Hunter Lovins's *Brittle Power* study for the Pentagon—the first and still the definitive unclassified synthesis of what's now called "critical infrastructure" vulnerabilities, and of design principles for energy resilience—turned heads, and by 1995, he and his colleagues at Rocky Mountain Institute were helping the US Navy overhaul how it designs buildings. During 1999–2001 and 2006–2008, Amory served on two Defense Science Board task forces that outlined the problems brought on by the military's colossal energy use—a use that hazards both mission effectiveness and the uniformed men and women themselves. Like much of his work, Amory's writing about the military is interwoven with the subject of energy insecurity, and these writings offer several keen observations. First is the simple notion that our energy systems (power stations, powerlines, pipelines, etc.) are highly vulnerable to attack and accident (among Amory's best accounts of this are *Brittle Power*'s Chapters 7 and 13, about "War and Terrorism" and "Designing for Resilience," respectively; also noteworthy is that the book repeatedly anticipated the use of hijacked aircraft as terrorist weapons). Second, that by creating the systems mentioned above, we are handing our warfighters an assignment they really didn't sign up for, such as guarding pipelines in "Faroffistan." Third, what we really need to do is ask ourselves what real security is, then seek the least-cost, most effective ways to get it. And fourth is that the military, both for its own good reasons and as a new contribution to its national-security mission, should help inspire and lead the civilian sector's shift from brittle, overcentralized systems to resilient, distributed, renewable energy resources. By 2010, the Pentagon had largely adopted this goal.

Summarizing *Brittle Power*, "The Fragility of Domestic Energy" was published in *The Atlantic* in 1983 and points out the causes of and alternatives to domestic energy vulnerability. The ideas in the article were brought to the fore again on 14 August 2003 during the massive Northeast Blackout, which knocked out power to ten million Ontarians and 45 million northeastern Americans; "Towering Design Flaws" appeared in Toronto's *Globe and Mail* a few days after the lights went out. The several other essays in this selection—from *Whole Earth Review* (2002), *Joint Force Quarterly* (the magazine of the Chairman of the US Joint Chiefs of Staff, 2010), and *Foreign Policy* (2010), bring together the pieces of the energy/security puzzle and describe why the military is a key part of the solution to problems that the military itself is often charged to address. In this section you'll also find US Senate testimony Amory was invited to deliver in 2006, explaining what energy security is and how to get it in ways that enhance both prosperity *and* security.

Chapter 21

The Fragility of Domestic Energy

1983

Editor's note: *In 1979, the Pentagon's Defense Civil Preparedness Agency commissioned Amory and L. Hunter Lovins to analyze the vulnerability of American energy systems and suggest remedies. The two energy analysts were surprised by what they uncovered—namely, an energy system highly vulnerable to every kind of accident or malice conceivable. In 1981–1982, they turned their research into the ground-breaking work* Brittle Power: Energy Strategy for National Security. *In November 1983, they published this greatly condensed version of their findings in the Atlantic Monthly. Unfortunately, over the next two decades, severe new cybervulnerabilities were added to the electric grid by routing critical controls via the internet without precautions against malicious intervention. These were not discovered until 2006. After the 9/11 attacks in 2001, and especially after the second Defense Science Board task force on which Amory served reported in 2008, such fatal flaws began to be fixed—but so slowly that he's surprised every morning when the grid still works.*

Orthodox energy policy since 1973 has rested on three assumptions:

1 Imported oil, especially from the Middle East, is unreliable.
2 Domestic energy supplies are secure. (This assumption is used to justify such diverse projects as Arctic gas pipelines, offshore oil drilling, coal leasing, the development of synthetic fuels, and breeder reactors.)
3 Secure energy supplies cost more than imported oil, but are worth it.

The oil shocks of 1973 and 1979 vindicated the first of these assumptions, but at any moment the lesson could become far more emphatic. By attacking the oil terminals at Ras Tanura and Ju'aymah, an Iranian jet or saboteurs in a couple of dinghies could cut off five-sixths of Saudi Arabian oil exports for three years, the time that would be needed to rebuild key components; after that, of course, the attack could be repeated. The Persian Gulf's facilities are extraordinarily centralized: on average, each Saudi Arabian well lifts about a thousand times more oil per year than a typical American well does. The super-giant Ghawar field alone—a strip about a dozen miles wide—has, until recently, lifted oil faster than any country except Saudi Arabia as a whole, the US, and the USSR. The oil from this field is collected at a single site; in 1977, a fire there, of questionable origin but officially described as accidental, cut off ARAMCO's output for ten days. The Middle East's dense webs of pipelines, pumping and compressor stations, refineries, and natural-gas plants have been prime targets in the battles between Iran and Iraq, and have been subject to sporadic attack by various antagonists in the region for more than a decade.

The lumbering supertankers that bring Middle Eastern oil halfway around the world to Western ports are also insecure. Naval planners shudder at the tankers' vulnerability to submarines, but even pirates in small boats manage regularly to board and rob tankers off the coasts of Singapore and Nigeria. Moreover, it is not at all unusual for such ships to sink or blow up without any assistance.

So manifest is the fragility of Middle Eastern oil that the second and third assumptions listed above have been accepted uncritically as corollaries. The supply has been presumed to be unreliable because the oil is foreign, and a more fundamental defect has been ignored: in reality, the Middle East's political volatility matters mainly because of the extreme geographic concentration of the oil drawn from the region.

Domestic energy, of any kind, from any source, and at any price, has seemed the obvious answer. However, most domestic sources, current and potential, are also centralized—and hence vulnerable to different, but equally serious, kinds of disruption. Amid the enthusiasm for all-American energy, this cause for concern passes unnoticed. Analysts who are quick to note that some 500 miles of pipe in eastern Saudi Arabia carry a sixth of the non-Communist world's oil have failed to observe that three-fourths of the oil extracted in the US comes from four states (Texas, Alaska, Louisiana, and California); that more than half the refinery capacity is in three (Texas, Louisiana, and California); that three-fifths of the petrochemical capacity is in one (Texas); and that five-sixths of the interstate natural gas comes from or passes through one (Louisiana)—3.5 percent of this amount being processed by a single plant equivalent in terms of energy output to 20 giant power stations.

Only a tenth of the total energy supply in the US now comes from imported oil. Making cars and buildings more energy efficient could reduce that fraction to zero within a decade. Nearly all of the remaining 90-odd percent is accounted for by domestic sources. This share, however, can be disrupted about as easily as foreign oil, faster, and in larger increments. Moreover, the vulnerability of American supplies to sabotage, natural disaster, or technical accident is increasing year by year, because successive administrations, wrongly assuming that any domestic source must be reliable, have promoted, with subsidies totaling more than $10 billion a year, those energy technologies that are the least secure.

Already, a handful of people in Louisiana could shut off three-fourths of the gas and oil supplies to the eastern states in an evening, leaving them out of action for as long as a year. A small group could black out the electric power to a city, or perhaps a region, for weeks or months. Certain attacks on natural-gas systems could incinerate a city. Low-technology sabotage of certain nuclear facilities could make vast areas uninhabitable. The electromagnetic pulse of a single nuclear explosion high over the Midwest could accomplish most or all of these disruptions simultaneously, by burning out electronic devices throughout the contiguous US that control refineries, gas-processing plants, oil and gas pipelines, power plants, and gas- and electricity-distribution grids.

These loopholes in national security are the work not of enemies abroad but of highly qualified and patriotic engineers who faithfully performed the wrong task: designing a reliable system for supplying cheap energy to a technological paradise, in which everything works according to the blueprints, unsullied by human fallibility or malice.

The result, built up piece by piece without regard to the fragility of the whole, is enormously complex and mostly interdependent networks of aerial wires, shallowly buried pipelines, fuel-storage facilities, gas-processing plants, oil refineries, and billion-dollar power plants—systems that take years or decades to be constructed and whose routine operation very few people understand. The systems depend on split-second computer timing and elaborate, instantaneous communication. Those that govern the distribution of electricity and natural gas—together delivering 41 percent of the energy we use—are tightly coupled and not particularly flexible. Electric-power grids depend on the ability of many large, precise electric generators to rotate in exact synchrony across half the continent. Regional gas grids require the continuous maintenance of minimum pressures. Many of the spare parts for these systems are special-order items that are too expensive to be stockpiled, yet require months or years and unusual skills to be made and installed. Some key components are not built in this country. Moreover, supposedly redundant backup devices can themselves suffer unstoppable, grotesquely cascading failures. (For example, in 1969, technicians at the Oak Ridge Research Reactor tried to deploy a safety system with three redundant channels. The backup devices in each channel suffered seven independent and simultaneous failures, a coincidence that, according to the official postmortem report, was "almost unbelievable.")

When all these traits are considered together, it appears more than conceivable that a well-coordinated attack on electric grids and pipelines, and on the computers that orchestrate their output, could enmesh the US in spreading waves of chaos from which recovery would be difficult at best.

According to the General Accounting Office and private security experts, most energy control and distribution centers are protected by chain link fences or Keep Out signs, or not at all. (An audit done by the GAO in 1979 revealed that only a single locked door protected the computer controlling a pipeline that carries a quarter of the crude oil fed to Midwestern refineries. A public road provided free access to the facility.) Hardening the most vulnerable sites, however, buys little security. Fences, alarms, and guards may discourage the casual or incompetent attacker, but will elicit stronger attacks from serious adversaries, or divert their attention to softer targets. There are far too many soft targets to defend. Vital control centers, pumping and compressor stations, switchyards, and the like—each of which governs huge amounts of energy—number in the thousands. Moreover, energy lifelines are stretched thin over hundreds or thousands of miles. The Trans-Alaska Pipeline System alone spans nearly 800 miles of rough country. America's principal oil and gas pipelines would stretch more than 12 times around the equator; the overhead electrical transmission circuits, more than 15 times. The total length of gas pipelines in the US exceeds a million miles.

The scale of modern energy systems is such that a relay failure in Oregon can cause a blackout in Arizona (and did so last year). Their concentration is such that the loss of three major domestic oil pipelines would interdict nearly 5 million barrels per day—substantially more than all the imported oil now used in the US.

One can hardly imagine a target more ideal than the domestic energy system for easy disruption, widespread catastrophic failure, and slow, difficult recovery. Major failures could produce abrupt backward lurches, of decades if not centuries, in this country's economic progress and in its standard of living. Suddenly, the United States might find itself grappling with problems of daily survival that for years have been confined chiefly to the poorest countries.

When the Pentagon's Defense Civil Preparedness Agency (now incorporated into the Federal Emergency Management Agency) commissioned us in 1979 to analyze the vulnerability of American energy systems and seek remedies, we were unprepared for what we found. Practically no one in government, we discovered, knew or cared about any form of energy insecurity other than interruptions of oil imports. And that problem had been "handled" with the establishment of the Strategic Petroleum Reserve (a huge repository of oil stored by the government in Gulf Coast salt formations), which one person in three nights could render useless. Three pipelines ship the reserve's oil to refineries. Each would have to be cut in only one place for the oil to stop flowing. It could take six months or more to mend a break in pipe running through a swamp or across a river; damage to important pumps or controls could shut off the oil for an even longer period. Pipelines carry three-quarters of the crude oil that US refineries use, and about a third of the refined products they produce. A single pipeline system, the Colonial, whose 4600 miles of main pipe cover 1600 miles of territory stretching from Texas to New Jersey, handles half of the shipping of these refined products.

Our research revealed a comprehensive denial of reality: policymakers tend to be so preoccupied with Persian Gulf oil that they fail to consider the frailty of their favorite alternatives. The Trans-Alaska Pipeline, for example, carries a seventh of all the crude oil fed to American refineries.

Its failure would cost more than $700 per second, and in three winter weeks could turn the line into "the world's biggest Chapstick," as 9 million barrels of hot oil congealed inside. (The pipeline's proprietors believe that the pumps are powerful enough to get the oil moving again, but no one knows for sure.) It would take as long as seven months to replace a large section of the labyrinth of 48-inch pipe at the system's north end. The line has already been bombed twice, incompetently and with only light damage. The most accessible and dispensable of its pumping stations was blown up accidentally, in 1977.

Although the Army has declared the Trans-Alaska Pipeline indefensible, the system's owners seem to perceive no security problem. The Reagan Administration wants to augment the oil pipeline with a still more vulnerable Arctic gas pipeline, which would carry half as much fuel at three and a half times the cost. In fact, the new line's estimated cost is so high that even though Congress has waived all legal impediments, it probably will not be built. If it ever is built, however, it will hardly enhance energy security.

Other federal proposals invite similar concern. Consider, for example, the government's hope that by the year 2000, mining in Wyoming's Powder River Basin will yield 500 million tons of coal a year—three-fifths as much as is now mined each year in the US as a whole. The coal would be carried out of the basin along the Burlington Northern rail lines; its concentration would make it as easy a target as the oil shipped through the Strait of Hormuz, in the Persian Gulf. What if the US did come to depend on all this coal? Coal trains and bridges have been bombed frequently in Appalachia during labor disputes; why not in Wyoming, and by attackers with larger motives?

Consider, too, the Department of Energy's encouragement of the electric utilities to build more than 400 giant coal and nuclear plants. The DOE forecasts that otherwise the current overcapacity of electric power plants will turn to shortage. The electricity generated by the new plants would have to be distributed by longer, higher-voltage lines than are now typical, requiring more-specialized and more-vulnerable switchgear and controls. Investors do not seem to share the DOE's enthusiasm for another trillion dollars' worth of power plants: no nuclear plant has been ordered in the US since 1978, and more than a hundred have been canceled since 1972; no large power plant of any kind has been ordered since 1981. For the DOE's projections to be fulfilled, the plants would have to be ordered at an average rate of one a week, starting now. If built, the plants will make blackouts more likely and bigger than they have been thus far.

Offshore oil is favored by the secretary of the interior, James Watt, as a secure substitute for Persian Gulf oil. The Coast Guard says that in good weather it could put a vessel alongside a threatened platform in the main Gulf of Mexico fields in eight hours. Only an incompetent saboteur could fail to destroy the platform in eight minutes.

Twenty billion dollars in federal subsidies are being offered to yet another component of energy security—giant synthetic-fuel plants. Only two countries have ever substantially relied on such plants: Germany, during World War II, and, more recently, South Africa. The German plants were bombed by Allied aircraft, the South African plants by African National Congress saboteurs. Synthetic-fuel plants, like refineries, are dense clusters of high-pressure tanks and pipes carrying flammable liquids and explosive gases, such as hydrogen, interlaced with sources of heat to ignite any fuels that escape. Their configuration makes the saboteur's task easy. Such plants also require prodigious amounts of water and power—supplies that are not hard to interdict.

In sum, all of the energy sources currently being promoted as the backbone of American energy supplies into the 21st century are precisely those least suited to surviving the uncertainty and violence that seem likely to characterize the future.

These risks are frighteningly real: so real as to make us ask whether they should be studied at all. Might it not be better to hope that they will simply pass unnoticed? Unfortunately, it is already far too late for that, as a glance at the newspapers reveals. Worldwide, significant attacks on energy systems are now occurring about once a week (not counting the ones in El Salvador, which occur more or less daily). In the past decade or so, they have occurred in 26 states in the US and in more than 40 other countries. Such attacks are becoming more frequent, more intense, and more sophisticated.

Since 1972, 117 countries have suffered terrorist attacks of some kind, and 10 of these—all advanced countries—account for more than half of the incidents. So far, the US has been very lucky, but few experts on terrorism expect our luck to hold for long. According to the Federal Bureau of Investigation, from 1972 until 1978, minor facilities of American electric utilities were bombed, on average, every two weeks, mainly by political-protest groups. Someday, however, attacks on the country's utilities may become more common, and they may have consequences more severe than the local and symbolic effects that have so far been typical.

Because the risks are real, we have taken great care that our analysis not provide a cookbook. Formal government classification review and extensive peer review have reinforced our own sense of discretion. Yet we feel that the only thing more dangerous than publicly discussing these previously unrecognized forms of energy vulnerability is not discussing them, for if the vulnerability increases, while remedies languish unused, everyone's security will suffer. Since the record of attacks shows that terrorists already know that energy systems are a soft target, it seems time that political leaders and the public be informed of the hazard, and of what they can do about it.

Our work for the Defense Department led the General Accounting Office last year to reiterate its warnings that the US is poorly prepared to deal with assaults on its electrical-power system. Three congressional hearings have been held, a committee on energy vulnerability has been established at the Georgetown Center for Strategic and International Studies, and further analyses have been undertaken at Los Alamos National Laboratory and by some energy companies. Our work seemed to be welcomed by the military and security communities, but not by the Department of Energy. And our final report to the Federal Emergency Management Agency, delivered in November of 1981, seems to have sunk without trace.

The government's limited interest may be owing, at least in part, to the natural assumption that experts have everything under control. After all, the energy supplies of advanced industrial countries are ordinarily quite reliable. Unfortunately, modern energy systems are so complex that nobody can foresee all the ways in which they might fail, even accidentally. Each major failure has been a surprise to the designers. Further, designing energy systems to be reliable in the face of predictable kinds of technical failures—as the engineers have done with commendable thoroughness—does not provide, and may even discourage, a more vital characteristic: resilience in the face of unpredictable kinds of failures (especially from sabotage). Energy engineers tend to design highly centralized, monolithic systems. On their own, they do not fail often, but when they do fail, they fail big. Thus, if a relay failure blacks out New York, the normal response is to improve the relay—but at the same time, the centralized architecture that caused the cascading grid failure in the first place is not only preserved but expanded. Thus the next crash is not prevented, it is enlarged.

Someone unacquainted with the hundreds of actual incidents that we have compiled might reasonably suppose that the country's energy supply could never be seriously jeopardized—just as a regional blackout was considered implausible until 1965; as the hijacking of three jumbo jets in a day was until 1970; as the seizure of more than 50 embassies was until the 1970s; and as an air-raid on a nuclear reactor was until 1981 (when Israel destroyed one in Iraq). But, given the stakes, nobody would wish to be in the position of the British intelligence officer who, on retiring in 1950, after 47 years of service, reminisced: "Year after year the worriers and fretters would come to me with awful predictions of the outbreak of war. I denied it each time. I was only wrong twice."

Energy vulnerability in its widest sense—the potential for interruptions of any form of energy supply, by any means, on any scale, at any time and place—has serious political implications. The threat of terrorism can fundamentally alter the political balance between large and small groups in society. This may in turn erode the trust and the civil liberties that underpin democratic government.

Modern energy supplies depend on technicians with specialized skills. Strikes that disrupted electric power helped to unseat the ruling party of Britain in 1974 and (with the aid of oil and gas workers) deposed the Shah of Iran in 1978. Similarly, power strikes, or threats of them, have been used as political instruments in Argentina, Australia, Israel, Puerto Rico, and elsewhere. Coordinated attacks on electric-power systems hastened the fall of President Salvador Allende, in Chile; in 1972, they disrupted a presidential inauguration in Portugal and frustrated a coup they were meant to precipitate in El Salvador (where the blackout prevented the plotters from communicating with one another or with the public).

The vulnerability of cities and factories to even simple kinds of energy sabotage invites a particularly debilitating kind of economic warfare, such as is now being waged in El Salvador, Peru, Afghanistan, and South Africa. Key power lines and plants are periodically destroyed in all four countries. (In South Africa, the attacks provoke what appear to be retaliatory raids on oil facilities in Angola and elsewhere.) All over the world, valuable energy targets are typically clustered in exposed positions. The burning of the main oil depot in Salisbury, Rhodesia, in 1978 destroyed nearly half a million barrels of oil, which the embargoed regime had painstakingly accumulated from Iran and from South Africa's synthetic-fuel plants. Thus a raid with simple munitions increased Rhodesia's budget deficit by 18 percent within a few minutes. The value destroyed was at least a million times that expended in rockets and tracer bullets. Attacks on oil depots have succeeded in Italy, have partly succeeded in The Netherlands, West Germany, and the US, and have been narrowly foiled in Chile and Israel.

The expense can entail lives as well as money. Modern liquefied-natural-gas (LNG) facilities often store flammable inventories equivalent in energy to megaton-range strategic warheads. Some such facilities are sited near cities—Boston, London, and Tokyo, among others—even though a major leak could cause a firestorm. Just the radiant heat from a large LNG fire can cause third-degree burns a mile or two away. There have already been near misses: for example, an LNG terminal on the River Thames near London has almost been ignited three times by oil spills and fires. An audit done by the GAO in 1978 found that LNG trucks—each potentially a portable, quarter-kiloton firestorm—had only slightly better security than potato trucks; unauthorized people could readily enter the terminals where the trucks are filled and drive them away. Finally, if a saboteur cut off a city's piped natural-gas supply long enough to extinguish the pilot lights, and then turned the gas on again, any errant spark would ignite a conflagration.

The huge radioactive inventories in nuclear reactors and spent-fuel facilities place the fallout potential of a sizable nuclear arsenal in the hands of anyone with the simple means required to cause a major release. Various attempts at sabotage of nuclear facilities—bombings, arson, and destruction of equipment, rather than the more minor incidents that can take place during antinuclear demonstrations—have already occurred in Spain and France, and less frequently in other European countries, the US, Argentina, and Brazil. By 1980, more than 400 acts or warnings of violence (mostly telephoned bomb threats) had occurred at US nuclear facilities; each year, several more actual incidents and dozens of new threats are reported. Worldwide, more than a hundred attacks or significant breaches of security have occurred at nuclear facilities. Fissionable materials stolen from a nuclear plant can be made, directly or indirectly, into crude nuclear bombs, which might, in turn, be aimed at a nuclear facility. The long-term, long-range radiological consequences of exploding such a sub-kiloton bomb in a public area near a large reactor would probably be similar to those of a one-megaton groundburst.

Because attacks on the energy system can be devastating, yet cheap, pointed, and deniable, they offer an attractive means of clandestine or surrogate warfare, against which a free society has no effective means of defense. It is hard to see why any of the conventional means of overt warfare—costly, indiscriminate, and inviting retaliation—should be preferred to clandestine attacks on a country's energy infrastructure (or, for that matter, on equally vulnerable targets in telecommunications, data processing, food, water, and so forth). From this point of view, the US is spending about $10,000 per second on military defenses for the front door while the back door stands ajar unnoticed.

Deterring attacks by thousands of nuclear missiles does not provide security if a few satchels of high explosives have, in the meantime, upset the national economy by blacking out New York City for upwards of a year. Just as the delivery vehicle of choice for a nuclear warhead may now be a tramp freighter, rental van, or parcel-service truck—modes that can be anonymous and therefore undeterrable—so future strategic attacks may occur in the form of abrupt, complete, but seemingly accidental breakdowns of infrastructure vital to national life. Energy vulnerabilities have broad implications for NATO, too. For example, even if no nuclear bombs were used to counter a Warsaw Pact thrust across the north German plain, collateral damage to the four large reactors now sited there could readily release about as much radioactivity as would issue from the groundburst of many thousands of tactical nuclear warheads.

Military history teaches important lessons about energy vulnerability. Hermann Goering and Albert Speer stated after World War II that the Allied Forces could have saved two years by bombing the Nazis' central power plants early. (The Allies had mistakenly believed that the German electric grid could reroute power flexibly enough that a few plants or switching centers would not be missed.) Japan's power system, however, did successfully withstand heavy assault during the war, because it was decentralized. Its thousands of small and dispersed hydroelectric plants, which were thus all but impregnable militarily, generated 78 percent of Japan's electricity Those plants sustained 0.3 percent of the bombing damage. Japan's large power stations, which supplied only 22 percent of the country's power, sustained 99.7 percent of the damage.

The accidental blackout of virtually all of France in 1978 showed that substituting domestic energy (nuclear power, actually based on imported uranium) for foreign oil does not necessarily protect people from freezing in the dark, and may even cause large numbers to do so at one stroke. This lesson was repeated with the accidental blackout of Israel in 1979, of most of southern Britain in 1981, and of Quebec in 1982 (the most recent of many there). Several countries—notably Sweden, the People's Republic of China, and Israel—are already pursuing energy decentralization as a national-security measure. The Red Army is reportedly anxious to follow suit, but the Politburo forbids this on the grounds that any sort of decentralization would threaten its own authority.

Energy insecurity is not necessary; it is not even economic. Cheaper alternatives exist. Design lessons from biology and from many engineering disciplines—computer theory, aeronautics, naval architecture, nuclear science—can be embodied in practical, available, and cost-effective energy technologies. Systematic use of these principles can make the energy system so resilient that major failures, from any cause, become impossible. Investing in a resilient energy supply would enhance American military preparedness, minimize the threats to be prepared against, and make defense costs less onerous. Best of all, such an investment would be paid back. According to a study released by the Solar Energy Research Institute in 1981, the US could double its energy efficiency and convert at least a third of its energy supply to renewable sources within the next two decades. The institute's data suggest that such a shift could save several trillion dollars, make the energy sector deflationary, and provide as many as a million jobs.

A resilient system is one that has many relatively small, dispersed elements, each having a low cost of failure. These substitutable components are interconnected not at a central hub but by many short, robust links. This configuration is analogous to a tree's many leaves, and each leaf's many veins, which prevent the random nibblings of insects from disrupting the flow of vital nutrients.

Such dispersed, diverse, and redundant systems can yield striking economic savings. For example, when a power engineer in Holyoke, Massachusetts, saw the blackout that struck most of the Northeast in 1965 rolling toward him, he was able to separate the city from the collapsing grid and power it instead with a local gas turbine. Within four hours, the money saved by not having to shut off electricity to the city repaid the cost of building the turbine. In 1978, the residents of Coronado, California, were not aware that the surrounding San Diego grid had been blacked out, because their power was supplied by an independent cogeneration plant. In the bitter winter of early 1977, when Midwestern factories and schools were closed by natural-gas shortages, rural New England was unaffected, because its supply was bottled, rather than delivered through the nexus of pipelines; as a result, systemwide collapses could not occur. In 1980, officials from the Department of Energy were cutting the ribbon on a West Chicago gas station powered by solar cells when a violent thunderstorm blacked out the city. The station's power was not interrupted. Likewise, a Great Plains farmer who uses windpower was once watching the television news and saw a report that his whole area was blacked out. He went outside and, sure enough, all his neighbors' lights were off.

All of these examples illustrate the architectural principle of resilient energy supply. But the last two also show the security advantage of harnessing natural energy, which cannot be depleted or disrupted by wars, strikes, embargoes, sabotage, and the like.

Renewable energy sources are often dismissed as unreliable. Yet several analysts have shown that a variety of renewable sources in combination can be more reliable than nonrenewable sources. Stormy weather, bad for direct solar collection, is generally good for windmills and small hydropower plants; dry, sunny weather, bad for hydropower, is ideal for photovoltaics. A diversity of sources, each serving fewer and nearer users, would also greatly restrict the area blacked out if a grid connecting them failed. And when renewable energy sources do fail, they fail for shorter periods than do large power plants. Windmills in appropriate sites might stand becalmed, at reduced output, for tens of hours; but reactors typically fail for 300 hours, at zero output. It can be cloudy (not a serious impediment to properly designed solar systems) for days or weeks; but a total eclipse lasting several months—the natural analogy to an oil embargo—is most unlikely.

Finally, while the intermittence of renewable supplies is caused by well-understood effects that are fairly predictable (rotation of the Earth, calm, cloudiness, drought), the nonrenewable supplies are intermittent for reasons that are much harder to predict (terrorism, reactor accidents, strikes, and international politics). One can have greater confidence that the sun will rise tomorrow than that no one will blow up Ras Tanura today.

The most resilience per dollar is achieved by the most productive use of energy, whatever the source. For example, if America's car fleet averaged 65 miles per gallon (15 fewer than an advanced diesel Rabbit tested two years ago), the cars could run for hundreds of miles on half-filled tanks. The stocks of oil extracted from the ground but not yet sold would run the fleet for about a year, whereas now, if the pipeline feeding a refinery is cut, the refinery must shut down in a few days, and its customers would run out of products in about a week. Thus, using energy more efficiently diminishes fuel stocks more slowly, buying precious time to mend what is broken or to improvise new supplies.

Fuel-efficient cars are only one of many ways to make the failure of an energy supply less critical. If, for example, the heating system of a superinsulated house in Montana were to fail in midwinter, it would probably take days or weeks—not hours—for the indoor temperature to drift down into the mid-50s. The warmth from people and from sun shining through windows (and, if the electricity were on, from lights and appliances) would make lower temperatures physically impossible. The body heat from a few neighbors, coming in to take refuge from their sievelike houses, would restore the superinsulated house to a comfortable temperature. If they brought along a large dog, the house would overheat unless the windows were opened. Alternatively, the house could be evenly heated by any small, improvised source of heat, such as burning junk mail in a large tin can. If well designed, the house need not have been equipped with a heating system at all.

If electricity were consumed more efficiently, the country's large hydropower plants could team up with smaller units of supply—micro-hydropower plants, industrial cogeneration equipment, windmills—to meet the demand. New technologies for saving electricity can perform the tasks of existing technologies more cheaply. Heat-saving renovations of buildings and passive-solar systems can reduce or even eliminate the demand for electric heating and air conditioning; changes in the configuration of motors can drive industrial machinery twice as efficiently; household appliances can be designed in ways that cut their demand for electricity by at least 75 percent; improved bulbs, fixtures, and electronic controls and the creative use of daylight can reduce light bills by between 60 and 90 percent.

Analyses by an energy study group at the Harvard Business School and by the Solar Energy Research Institute have shown, and market experience is confirming, that renewable sources, chosen carefully and acquired sensibly, are generally cheaper—many by several times—than the centralized, nonrenewable sources that they stand ready to replace. Within decades, resilient energy sources could wholly supplant the vulnerable supplies, both domestic and foreign, on which the US now depends. Using energy in an economically efficient way can buy the country enough time to complete the transition comfortably.

The transition is already well under way. Since 1979, the United States has gotten more than a hundred times as much new energy from savings as from all the expansions of the energy supply combined. Of those expansions, more new energy has come from renewable sources, which account for about 8 percent of the total US supply, than from any or all of the nonrenewables. That is, sun, wind, water, and wood have been adding to the American energy supply at a faster rate than oil, gas, coal, and uranium, singly or combined. Higher energy efficiency is far outpacing them all. The amount of energy needed to make a dollar of GNP has fallen by a fifth in nine years, and is still falling by several percent per year.

These savings reflect only a tiny fraction of those that are technically possible and economically attractive. Energy technology is advancing with extraordinary speed. Many of the most important innovations have been developed within the past two years: windows so heat-conserving that they can capture more solar heat than they lose, even facing north; refrigerators using an eighth as much electricity as a standard model; ships and jetliners twice as fuel-efficient as today's fleet; light bulbs that cost $25 but that yield better light than standard bulbs, use a quarter as much electricity, last ten times as long, and return their cost within a year or two; moderately priced heat exchangers that bring copious amounts of fresh air into tightly built houses, yet capture four-fifths of the heat or coolness from the air they withdraw; ice ponds that cool buildings using a tenth of the energy that air conditioners require; ways to make existing buildings so heat-tight that they don't need heating—the list goes on.

Renewable sources of energy, too, are penetrating the market. Since 1979, more new electric generating capacity has been ordered from small hydropower and windpower plants than from coal or nuclear plants or both. Wood-burning, though seldom done in the best way, now delivers about twice as much energy as nuclear power, despite the government's outlays of more than $40 billion in nonmilitary nuclear subsidies. The rate of practical progress is similarly impressive in Europe, Japan, and a few developing nations.

In 1980, Americans spent some $15 billion on energy-saving devices and renewable sources (not including purchases of fuel-efficient cars)—about a fifth of the total US investment in energy equipment. A manifestly imperfect market is working remarkably well, proving that in a large and diverse society, energy security does not require, and may not even be able to tolerate, central management. The problem of secure and affordable energy supplies is starting to be solved, by individual and community action—from the bottom up, not from the top down. Washington will be the last to know.

More than $400 billion a year is spent nationwide for conventional fuels and power, and this huge sum is a keen spur to local initiative. Typically, between 80 and 90 percent of the dollars spent on energy leave a community, never to return. This drain can be equivalent, in a town of 100,000 people, to the loss of about 10,000 jobs. In contrast, higher energy productivity and the harnessing of renewable energy, using local skills and manufacturing capacity, can keep the money, the jobs, and the economic multipliers at home. Thousands of communities are already moving with remarkable speed to capture these opportunities. They would not act so quickly if resilience cost more than vulnerability. Energy security and price are *inversely* related. The resilient technologies are what a truly free market in energy services would reward, if one existed.

These technologies would enter the market even faster if the severe price distortions created by federal subsidies were removed, and if people were not prevented from responding to price signals by many silly rules and customs left over from the cheap-oil era (3000 obsolete building codes, conflicting incentives for landlords and tenants, restrictive lending and zoning regulations, inequitable access to capital and information, and so forth). But this will require greater willingness than the Reagan Administration has shown to expose all technologies—even synthetic fuels and nuclear power—to free competition.

Since we cannot afford (and do not need) to do everything, we must compare investment alternatives. Consider, for example, five ways to spend $100,000 to save oil:

1   Catalyze a program of door-to-door citizen action to weatherize the worst buildings, as Fitchburg, Massachusetts, did in 1979, and as dozens of towns have done since. Experience shows that over the first ten years, the investment of $100,000 in such a program can save 170,000 barrels of crude oil, at about $0.60 a barrel.

2   Pay the extra cost (at the highest published estimate) of making 44 cars achieve 60 mpg. The first decade's savings: 5800 barrels at about $17 a barrel.

3   Buy about 3000 barrels of foreign oil, put it in a hole in the ground, and call it a "strategic petroleum reserve." After ten years, the oil may be available, but the storage and carrying charges—probably between $50 and $70 a barrel—will be unrecoverable.

4   Buy a small piece of an oil-shale plant. After ten years, it will have produced nothing. After that, if it works, it will produce up to 9000 barrels

of synthetic oil per decade, probably retailing at between $70 and $120 a barrel, in 1982 dollars.

5   Buy a tiny piece of the Clinch River breeder reactor. After ten years, it will still be under construction. After that, if it works, the $100,000 investment will yield up to 500 "barrels" of energy (as electricity) per decade, retailing at over $370 per barrel, in 1982 dollars, and probably uncompetitive even with roof-mounted photovoltaic cells.

By having failed to allow a truly competitive marketplace to operate and make the alternatives clear, the US has pursued these options in reverse order, choosing the worst buys first. Energy security will come slowly until we take economics seriously.

Chapter 22

Military Transformation and the Roots of National Security

2002

Editor's note: *This piece is adapted from Amory's remarks at a January 2002 workshop on "Capstone Concepts for Defense Transformation" at the National Defense University, Fort McNair, Washington, DC, and updated in July 2002.*

On 11 September 2001, the Revolution in Military Affairs shifted into fast forward. The asymmetric warfare we had worried about for decades became a reality. A poorly financed and technologically impoverished antagonist proved it could mount devastating attacks on the United States.

Asymmetric warfare's first major US episode gave over a millionfold economic leverage to the attackers, doing trillions of dollars of direct and indirect damage with about a half-million-dollar budget. What's most surprising (but understandable, given the historically sheltered nature of our society from such events) is how psychologically effective it was, even though the survival rates were quite high—around 90 percent in the World Trade Center, which is quite astonishing, and roughly 99.5 percent in the Pentagon attack.[1]

It's also now very clear that you can't effectively guard an open society, especially one that has inflicted itself with alarming vulnerabilities built up over decades. Vulnerabilities include energy, water, wastewater, telecoms, financial transfers, and transportation. If you destroy some critical bits of infrastructure, you can make a large city uninhabitable pretty quickly. This threat becomes more worrisome as weapons of mass destruction gain more customers.

Telecoms and financial transfer by electronics are particularly vulnerable. *The Los Angeles Times*, *Washington Post*, and *Wall Street Journal* recently reported a greatly increased incidence in recent months of probing cyberattacks from the Middle East on electric grids and other critical infrastructure by computer crackers.

As you look over the list of other issues that erode security—the effect of climate change and conflict on increasing flows of refugees; the risks of famine and war; water problems; disease outbreaks (as simulated by the Army War College); the spread of exotic species and invasive pathogens and genetically modified organisms—it's not a pretty picture for a peaceful world.

Traditional thinking about all these issues has been influenced by the supposition that governments are the axis of power and the locus of action, so we need to focus on governmental and international institutions and instruments. That's the wrong mindset, dangerously incomplete and obsolete, in a world that is now clearly tripolar, with power and action centered not just in governments, but also in the private sector and an internet-empowered civil society. There are complex interactions among these three actors. Increasingly, government is the least effective, most frustrating, and slowest to deal with, so one ought to focus attention on the other two. Also, each of these three has a kind of antiparticle, as in particle physics. You can have rogue governments like the Taliban, rogue businesses like Enron, and rogue nongovernmental organizations like al-Qaeda.

In a tripolar society, power is enlarged and diffused, and everything can happen a lot faster, because there are a lot more ways and channels for it to happen. In the model that we grew up with, governments rule physical territory in which national economies function, and strong economies support hegemonic military power. In the new model, already emerging under our noses, economic decisions don't pay much attention to national sovereignty in a world where more than half of the 100 or 200 largest economic entities are not countries but companies. Governments can no longer control their economies or look after their people. With trillions of dollars of capital sloshing around instantaneously at a whim, you might have more economic growth, but you also have extreme local volatility.

You might suppose that the rise of the private sector enhances the prospect for peace, because war is bad for most businesses, and business could therefore be expected to take steps to reduce conflict. But so far, taking into account all of the ingredients of stability, globalization is clearly making stability deteriorate. This is mainly because the trends of the past decade or two have made losers greatly outnumber winners. The gap between rich and poor has grown, and that unwelcome growth is apparently accelerating. According to the World Bank, of the 6 billion people on Earth, 3 billion live on less than $2 a day, and 1.2 billion live on less than $1 a day, which defines the absolute poverty standard. Access to clean water is denied to 1.5 billion people. Meanwhile, the world's richest 200 people are worth an average of $5 billion each. This naturally increases envy and anger. Typically, Western (and especially American) firms get blamed.

The instability of economies and polities erodes a sense of national or other identity, and therefore decreases stability and makes conditions ripe for nationalism and fundamentalism of all stripes. When nations can't take care of their people, people lose confidence in them and often tend not to vote, because they're not pleased with any of the candidates. Then you get movements backing such candidates as Jean-Marie Le Pen in France, with eerie parallels to the rise of Hitler. The growing influence of extreme right-wing parties, now in or tilting governments in at least eight Western European countries, certainly indicates that the problem is not just limited to poor countries.

Hierarchical government is in many respects losing effectiveness and credence. What needs to emerge, and may be starting to emerge, is networked governance. But that only works if it's really tripolar, engaging all three poles—the public and private sectors, plus nongovernmental organizations (NGOs) or civil society. While that networked governance—the tripolar world—jells, shifting ad hoc coalitions are seeking topical solutions between pairs or occasionally triplets of those three poles. This is a very sharp contrast to our old mental model of negotiations and treaties between sovereign nations. For example, business and civil society are increasingly joining forces to do what government can't or won't do. Civil society can either grant or withhold the legitimacy that gives business its franchise to operate, and by shifting purchasing and investment patterns can profoundly accelerate the revolution already visible in business leadership.

Also, of course, evil globalizes, whether through the spread of weapons of mass destruction (by two or sometimes all three of the poles in interaction) or through globalized crime and drugs. And homogenization—culturally driven, largely by the media—fosters the Jihad vs. McWorld polarity. None of this is welcome, but all of it is being either encouraged or tolerated by US policy—often strongly encouraged, in a way that causes resentment.

In hindsight, it's clearly an error to think of 9/11 as evil in a vacuum. There has been much debate about root causes, trying to figure out why people are so angry with us. A lot has been said about perceptions of humiliation and deculturization, the hypocrisy that weights non-American lives and freedoms less than our own, unfairness, bullying, and so on.

This is not surprising to readers of such works as Jonathan Kwitny's 1984 book *Endless Enemies: The Making of an Unfriendly World* (out of print). A *Wall Street Journal* reporter who lived in dozens of countries, particularly in Africa, Kwitny painted an appalling picture of how thoroughly the US government had destroyed what should have been good commercial and cultural relationships—by messing in other people's affairs, backing the wrong people, not understanding whom we were dealing with, and just being disagreeable. His basic conclusion was that if we want other peoples to think well of us, we should be the kind of folks they'd like to do business with, and should ensure that whoever comes to power there should never have been shot at by an American gun. It seems a very pragmatic and principled approach.

Working in about 50 countries, I've been endlessly impressed with how stupidly our country can behave, even through its experienced diplomatic apparatus (as we saw this spring in Venezuela). We Americans are thoroughly disliked, to a degree much greater than our political leaders seem to realize. That's going to be very hard to turn around even if we start now. In fact, we're going hard in the opposite direction, eroding or undercutting practically every peace-promoting, risk-reducing effort put forward by the international community, appearing hypocritical and unilateral, imposing mass-media culture, and showing little understanding of the values of diversity and tolerance or even, all too often, of the rule of law for which we supposedly stand.

The new American doctrine of exceptionalism (what used to be called "isolationism") is uniting the rest of the world, even our closest allies, against us. I think we will look back on the rapid destruction of treaty regimes that have taken decades to create, and of the credibility we were trying to build, and ask, "What on Earth possessed us to do that?"

Strategies for Security

In a remarkable, almost Churchillian speech on 2 October 2001, Tony Blair said, "We need, above all, justice and prosperity for the poor and dispossessed." Martin Luther King Jr. reminded us that "Peace is not the absence of war, it is the presence of justice." We also, I think, need to remember George Kennan's prescient warning, at the start of the Cold War, that the biggest danger was that we'd become like our enemies. Many elements of the Patriot Act passed by Congress after 9/11—abrogating civil liberties, ignoring the Freedom of Information Act, generally constricting the flow of public information—move us in that direction.

Military superiority won't be enough to win the "War on Terrorism." It is said that the kind of leadership we need on Afghanistan has five dimensions:

1 a political one, in which we enhance stability and marginalize the bad actors, so we don't create more monsters like the Taliban and al-Qaeda;
2 a diplomatic dimension where we try to move potential belligerents into a more sympathetic or at least more tolerant stance;
3 an informational dimension in which we show the region, Islam, and the whole world that we're not blaming but rather trying to help the people;
4 a humanitarian and economic dimension, in which we improve people's lives so the seeds of conflict don't flourish; and
5 a military dimension, in which we bring bad guys to justice, maybe use covert operations and encourage the overthrow of the bad guys, or as a last resort, defeat them in battle.

But it seems to me that what's missing from this five-sided approach is a strategic context. So I'd like to talk a little about what security is, where it comes from, and who's responsible for it, because it's clearer every day that the world's best armed forces, costing $11,000 a second, are not making us secure. That's because—as military professionals have understood for a long time, but not always articulated—*there is no significant military threat to the United States that can be defended against.*

That is, it is not technically possible to defend effectively against ballistic missiles. It is certainly not possible to defend against, say, nuclear warheads or other weapons of mass destruction that are smuggled in without leaving a radar track or other return address. Someone could wrap a warhead in bales of marijuana, put it in a shipping container, bring it aboard a ship into any of our harbors, and nobody would notice.

The point is that anonymous, asymmetric attacks can be quite devastating, but are undeterrable in principle, because you don't know who is responsible for them. That can be especially true with suicidal adversaries. We have already learned that interdiction by prior intelligence can't be relied upon. So the only lastingly effective defense is prevention—not so much at the level of intelligence foresight, which doesn't work reliably, but at the level of root causes, of eliminating the social conditions that feed and motivate the pathology of hatred.

This requires a comprehensive (though not indiscriminate) engagement in a geopolitical and ideological sense that goes far beyond traditional military means and digs down to the foundations of what our society aims to become.

Security has two main elements. The dictionary defines "security" as "freedom from fear of privation or attack." Freedom from fear of privation and freedom from fear of attack are not independent, but are both vital to being and feeling safe.

Can we be and feel safe in ways that work better and cost less than present arrangements? Is there a path to security that is achieved from the bottom up, not from the top down; that is the province of every citizen, not the monopoly of national government; that doesn't rely on the threat or use of violence; that makes others more secure, not less secure, whether on the scale of the village or the globe? Can a new approach to building real security also advance other overarching goals, and, ideally, save enough money to pay for other things we need?

I think we can do that.

Freedom from Fear of Privation

Let's start with freedom from fear of privation, which has many obvious elements: reliable and affordable energy, food, water, shelter, sanitation, health; a sustainable and flexible system of production, transportation, communication, and commerce; universal education, strong innovation, vibrant diversity; a healthful environment; free expression, debate and spirituality; a legitimate and accountable system of self-government at all levels. I would suggest that preserving our security requires all these things for others, too. As Dick Bell of the Worldwatch Institute remarks, weapons and warriors cannot keep us safe "in a world of extreme inequality, injustice, and deprivation for billions of our fellow human beings."

Helping others live decent lives is a worthy mission that our nation has undertaken before. General George Marshall said in 1947 that "there can be no political stability and no assured peace without economic security." He said that US policy must therefore "be directed not against any country or doctrine, but against hunger, poverty, desperation, and chaos." That was right then and it's right now.

You can argue about numbers, and certainly there's plenty of room for innovation in how services are delivered honestly and effectively. But for what it's worth, the UN Development Programme says that, today, every poor person on Earth could have clean water, sanitation, basic health, nutrition, education, and reproductive healthcare for about $40 billion a year. That's a good deal less than we're spending on our anti-terrorist program in the United States. It's less than a quarter of the tax cut that the President and Congress bestowed on us last year.

But where is the determination to build a muscular global coalition to create a safer world in those fundamental ways? Wealthy nations have reduced their foreign aid contributions in recent years. The $11 billion the United States now allots annually to foreign aid amounts to 0.11 percent of the nation's gross domestic product. (Canada and major European countries spend about three times as much of their GDP on aid.) The Bush Administration has announced a major and long-overdue increase in foreign aid. That could be a very good thing. But it's a small part of what's required, and it's not being framed in the sense or with the vision that General Marshall did half a century ago.

Aid from rich countries is often leveraged to elicit certain behaviors from recipient nations. Treasury Secretary O'Neill said in Ghana that American aid will be directed only to those African nations that exhibit good governance and also "encourage economic freedom"—in other words, those that privatize their industries, reduce subsidies, and open their markets to goods from the United States. But in fact the United States, along with other rich nations, continues to move away from a policy of open markets, slapping tariffs on foreign steel and lumber, and instituting an additional $35 billion in annual farm subsidies. This appears to our friends abroad, particularly in Europe, to be pure electoral opportunism, rejecting the very principles of free trade that we have been urging them to adopt, as well as stifling poor countries' exports to the US.

Beyond the simple application of more cash and making trade authentically fair, other routes are available to economic security in the developing world. We wouldn't normally think of a light bulb as an instrument for security, but building real security can be as simple and as grass-roots-based as a compact fluorescent lamp (CFL) costing about $3–12 in competitive markets (see box opposite). There are many more techniques like that.

Freedom from Fear of Attack

The other side of security is freedom from fear of attack. In an RMI book, *Security Without War*, published in 1993 but written several years earlier, Hal Harvey and Mike Shuman nicely lay out a new security triad: (1) conflict avoidance or prevention, (2) conflict resolution, and (3) non-provocative defense.

A BRIGHT AND SIMPLE IDEA

A compact fluorescent lamp saves 75–80 percent of the electricity used by an incandescent bulb, lasts 8–13 times longer, looks similar, fits the same fixtures and, over the course of its life, will save about $30–80 more than it costs. In fact, it's generally cheaper to give away CFLs than it is to run fossil-fueled power plants needed to power incandescent bulbs; that's why Southern California Edison Company gave away more than a million such lamps.

One such CFL, over its life, will avoid putting in the air from a typical coal-fired power plant one ton of carbon dioxide, eight kilograms of sulfur oxides, and four kilograms of nitrogen oxides. If the electricity is generated from oil, the lamp saves a barrel of oil and its attendant emissions. Or, if we're talking about a nuclear power plant, one CFL, over the course of its life, will avoid making two-fifths of a ton TNT-equivalent of plutonium plus half a curie (which is a lot) of strontium-90 and cesium-137.

If widely deployed, CFLs could cut by a fifth the evening peak load that crashes the grid in Bombay. They could raise a North Carolina chicken grower's profits by a fourth. They could raise a Haitian family's disposable income by as much as a third, because so much of the sparse cash economy goes for electricity.

A widely unrecognized advantage of such ways of saving electricity is that manufacturing them takes on the order of a thousand times less capital than expanding the electricity supply. When you invest in CFLs, you also get your money back about ten times faster, so it can be quickly invested again. If we do the cheapest things first, then the power sector, which currently gobbles up about a quarter of global development capital, could become a net exporter of capital to fund other development needs.

Such lamps are also the key to affordable solar power that lets girls learn to read, advancing the role of women and reducing population pressure. Currently half a billion CFLs are manufactured annually; the largest maker is China. CFLs can be bought at the local supermarket, and the average person can install them herself. Most of us would never guess such a simple thing could have such an impact globally. But clearly, if we so choose, we can make the world more prosperous, better educated, less polluted and, of course, safer through shared prosperity and justice—one light bulb at a time.

Conflict avoidance/prevention

Conflict avoidance/prevention, which might be called "presponse," has historically been a low priority, but it ought to be the highest priority. It's by far the most cost-effective way not to be attacked. It comprises elements like justice, hope, transparency, tolerance, and honest government. Many governments are still run by crooks or thugs, but I'm encouraged by the movement within the Organisation for Economic Co-operation and Development and by such groups as Transparency International to expose and stop corruption.

Conflict prevention also includes what Harvey and Shuman call "leader control." They note that it's almost impossible to find instances of wars between two democracies, or between two societies that, whatever their outward form of government, have effective ways to find out what their government is up to and tangibly express their displeasure if they don't like it.

Effective leader control tends to discourage adventures by leaders who are either crazy or wanting to divert attention from domestic difficulties. It's enhanced by speeding up the information revolution, so citizens can communicate with each other and with the outside world by a diversity of means that will be hard to block. In the earliest days of *perestroika*, someone asked Gorbachev's senior adviser on science, energy, education, and arms control—Academician Yevgeny Pavlovich Velikhov—how the then-Soviet government intended to keep control once citizens got access to modems, faxes, copiers, and the like. His prescient reply was: "You don't understand. The information revolution is our secret weapon to ensure that the reforms of *perestroika* are irreversible."

Another critical tool for preventing conflict is advanced resource productivity—getting lots more work out of each unit of energy materials, water, topsoil, and so on. As Paul Hawken, Hunter Lovins, and I describe in our book *Natural Capitalism*, advanced resource productivity can actually prevent conflict in four ways. First, it can make aspirations to a decent life realistic and attainable, for all, for ever. It takes a while, but it's definitely going in the right direction. It removes apparent conflicts between economic advancement and environmental sustainability. You can implement it by any mixture of market and administrative practices you want. It scales fractally from the household to the world. It's adaptable to very diverse conditions and cultures.

Second, resource productivity avoids resource conflicts over things like oil and water. As a result, military professionals can have negamissions. Military intervention in the Gulf becomes Mission Unnecessary because the oil will become irrelevant. Just moving to Hypercars® will ultimately save as much oil in the world as OPEC now sells.[2]

Third, resource productivity can make infrastructure invulnerable by design. That's the argument set out in our Pentagon study from 20 years ago, *Brittle Power: Energy Strategy for National Security* (now reposted at www.rmi.org/rmi/Library/S82-03_BrittlePowerEnergyStrategy). And finally, an argument that's a little more complex: resource productivity can unmask and penalize proliferators of weapons of mass destruction. With the late Lenny Ross, we made that argument in detail with respect to nuclear proliferation in *Foreign Affairs* in summer 1980, in an article entitled "Nuclear Power and Nuclear Bombs." It's enlarged in a book, now out of print, called *Energy/War: Breaking the Nuclear Link*.

The basic argument is that if we use energy in a way that saves money, that is enormously cheaper than building or even just running nuclear plants, so any country that takes economics seriously won't want or have nuclear plants. They're simply a way to waste money. In such a world, the ingredients—the technologies, materials, skills, and equipment—needed to make bombs by any of the 20 or so known methods would no longer be an item of commerce. They wouldn't be impossible to get, but they'd be a lot harder to get, more conspicuous to try to get, and more politically costly for both the recipient and the supplier to be caught trying to get, because for the first time, the reason for wanting them would be unambiguously military. You could no longer claim a peaceful electricity-making venture. It would be clear that you were really out to make bombs. The burden would be on you to show that that's not what you had in mind—to do something so economically irrational.

Interestingly, there is a parallel argument, which hasn't been fully fleshed out yet, for certain chemical weapons. In particular, adopting organic agriculture, which tends to work better and cost less and be better for health and nutrition, and can at least equally well feed the world, means that you don't have organophosphate pesticide plants, which means that you just removed the main "cover story" for nerve gas plants. And there's even a weaker, but not trivial, form of the argument: if you're not using transgenic crops, which you shouldn't be if you understand biology and economics, that will remove an innocent-looking cover for making genetically modified pathogens.

Getting back to the roots of conflict in resource rivalries: the broader case I'm making is that resource conflicts are unnecessary and uneconomic—a problem we don't need to have and it's cheaper not to have. For example, 13 percent of US oil now comes from the Persian Gulf, which is clearly risky. Proposed domestic substitutes, such as drilling in the Arctic National Wildlife Refuge, are at least as risky, and probably more so, because the Trans-Alaska Pipeline is about the fattest energy-related terrorist target there is. And therefore, in promoting expanded drilling in Alaska, the Department of Energy has been undercutting the Department of Defense's (DOD's) mission.

Both these kinds of vulnerability, both oil imports and vulnerable domestic infrastructure, are unnecessary and a waste of money. To displace Persian Gulf imports would take (at historic refinery yields of gasoline) only a 2.7-miles-per-gallon increase in the light vehicle fleet. We used to do that every three years when we were paying attention. Most if not all United States oil use could be profitably displaced within a few decades with current technology. This can happen surprisingly quickly. For example, from 1979 to 1985, GDP increased 16 percent, oil use fell 15 percent, and Gulf imports fell 87 percent. We could do that again in spades. DOD itself owns many billions of dollars a year of oil-saving potential, as laid out through a Defense Science Board panel on which I served (see *More Capable Warfighting through Reduced Fuel Burden*, www.acq.osd.mil/dsb/reports/ADA392666.pdf). Everything you could do to achieve that also improves warfighting capability.

I would call your attention particularly to the second of the October 2001 Shell planning scenarios, *Exploring the Future: Energy Needs, Possibilities and Scenarios*. It lays out a technological discontinuity that leapfrogs to a hydrogen economy led by China. This causes global oil use to be stagnant until 2020 and then go down. I think that's perfectly plausible, and in fact, my colleagues and I are helping it happen.

Conflict resolution

Conflict resolution is the next layer of defense if conflict avoidance or prevention fails. That's the realm of better international laws, norms, and institutions. Given space constraints, I won't elaborate on it here. There's a huge body of literature and practice on those things. Hal Harvey's and Mike Shuman's *Security Without War* is especially good.

Non-provocative defense

If the previous two layers of protection both fail, and conflict occurs, the last layer of defense, and a very powerful one, is "non-provocative defense," which reliably defeats aggression, but without threatening others. The concept was developed in Denmark and Holland by the children of World War II resistance leaders, who wanted to apply the lessons from their parents' experience defending their homelands against a powerful invader.

To date, Sweden has executed the most sophisticated design of military forces for non-provocative defense. Its coastal guns cannot be elevated to fire beyond Swedish coastal waters. It has a capable air force, but with short-range aircraft that can't get very far beyond Sweden. The radio frequencies used by the Swedish military are deliberately incompatible with both NATO and the Warsaw Pact, so Sweden will stay neutral.

In every way, by technical and institutional design, they've sought to make Sweden a country you don't want to attack, but one that is clearly in a defensive posture. This approach can ultimately create a stable mutual defensive superiority—each side's defense is stronger than the other side's offense. Each has, by design, at most a limited capacity to export offense.

The basic point of non-provocative defense is to structure and deploy your forces so your adversaries must consider them mainly defensive. That is, you minimize your capability for preemptive deep strikes, or strategic mobility, and you maximize homeland defense. This means four objectively observable technical attributes: low vulnerability, low concentration of value, short range, and dependence on local support.

Non-provocative defense means layered deployment in non-provocative postures. That's a theory that was well developed, much criticized, and ably defended in Europe in the 1980s. It had to be, because the towns there are only a few kilotons apart. It depends on forces that are at least as robust as the attacker's forces, but with a decentralized architecture that increases their resilience. It doesn't exclude cross-border counterattack, but that would be limited in scope and range. The defensive superiority should reduce the risk and the attraction of adversaries' building and using offensive arms. Of course, non-provocative defense doesn't stop terrorism, any more than National Missile Defense would. But the resilient design helps to disincentivize terrorism by reducing its rewards, just as the full spectrum of nonmilitary engagement undercuts terrorism's ideological and political base.

There seems, however, to be a worrisome contradiction in current strategic doctrine. To combat current threats, the US undoubtedly needs light, agile, deployable, sustainable forces. But those forces don't fit the definition of non-provocative; indeed, their global reach makes them look like just the opposite. In our short-term need, therefore, lie the seeds of long-term danger. We're shifting toward a "global cop" role—and not so much the neighborhood-policing cop on the street befriending everyone and heading off trouble, but the SWAT team that forays out of its fortress only to smite perpetrators. Such force structures and deployments will encourage us to act in ways that use those forces. Worse, they are likely to induce in others the attitudes and behaviors that elicit precisely the asymmetric threats to which the US is most vulnerable.

Since what's viewed by others as provocative depends on observed military facts, not on declared political intentions, there is no obvious solution to this paradox. The nearest I can see is to strive mightily to prevent conflict, merit trust, and try to make the global-cop role temporary and brief by making the world safer.

It's a lot better to prevent conflict from scratch than to combat a broadly based terrorist movement. There are some strong stars we can steer by. Our interests in the Third World would be much better advanced by democratization, anticorruption, sustainable development, resource efficiency, fair trade, demand-side drug policies, pluralism, tolerance, and humility than by most of what we're doing now. Third World security would be better advanced by those elements plus transparency and collective tripolar security arrangements—possibly even including an idea some people have had of some countries' giving up their armed forces and buying a credible kind of insurance from, say, the UN, paying fees for sharing in protective forces. And of course, the non-provocative new triad approach that I outlined can enhance everyone's security, but never at the expense of anyone else's security.

To start rebuilding America's lately tarnished credibility as a partner in that sort of world, we're going to need renewed US leadership in multilateral tasks, whether it's the Non-Proliferation Treaty, plutonium reduction, the Chemical and Biological Weapons Conventions and its enforcement, climate protection, or anti-land-mine efforts. It's a very long list, and right now our government is on the wrong side of every one of those issues.

And, of course, there's America's deepest potential strength: the primacy of underlying moral values and civics, which is much referred to rhetorically, but less honored in practice. This will require us to transform more than the military. Military transformation is only part of the challenge to American idealism and ingenuity to building real security. The foundation, which is a very sound notion from about 1787, is the shared and lived belief that security rests on economic justice, political freedom, respect for law, and a common defense. To make that work, we're going to have to bridge the widening gulf in our society between its civil and military elements. We'll also need to address the problem that military hardware and service vendors in the private sector have an unlimited self-reinforcing feedback loop where they coproduce weapons and fear, and there is no equilibration—no negative feedback to limit the self-reinforcing cycle of supply and demand.

Until now, the weapons vendors have had a radical monopoly, as Ivan Illich describes it, on providing "security services." If the only way we can imagine to get security is by buying more weapons, then the demand for weapons appears to be inelastic, especially if reinforced by the sometimes corrupt political process of buying them. Instead, if we have other ways of providing security, of which weapons are just one and must compete with other modalities, fairly and at honest prices, then we will gain much cheaper ways to provide the security services we want.

In the Cold War, security was viewed as a predominantly military matter. Appended and subordinated to military security were economic, energy, and resource security (consisting, for example, of our Naval fleets in and around the Persian Gulf). Environmental security wasn't even on the agenda; in fact, it was officially viewed as harmful to security and prosperity. But in the post-Cold War view, we need to add back the missing links between all four of these elements, and to turn the wasted resources into prosperity and peace. You can imagine these four elements as vertices of a tetrahedron—an immensely strong structure, especially if it surrounds a kernel of justice, whose presence, as Dr King said, is peace.

Notes

1   Of course, if, as is widely suspected, there were more hijackings planned, perhaps a total of six, it makes you wonder what the other three targets were. If certain of those possibilities had succeeded, we'd have woken up to a very different country.

2   In 2000, a young firm I chair, called Hypercar, Inc. (www.hypercar.com), designed—for a few million dollars in eight months—the direct-hydrogen-fuel-cell, uncompromised, competitively priced, mid-sized SUV that the Administration's FreedomCAR initiative intends to develop over the next 10 or 20 years. This concept car is a quintupled-efficiency mid-sized SUV. It can handle five adults and up to 2 cubic meters of cargo. It hauls half a ton up a 44-percent grade, and weighs half as much as usual because its structure is carbon fiber, not metal. Carbon is so strong that the ultralight SUV is at least as safe as a standard steel one, even if they collide. It goes from zero to 60 miles an hour in 8.2 seconds, gets the equivalent of 99 miles per gallon [later modeled more precisely at 114 mpg with hydrogen or 67 with gasoline—Ed.], and drives 330 miles on 7.5 pounds of safely stored compressed hydrogen. It needs that little fuel because it can cruise at 55 mph on the same energy as a normal car of that class uses just for its air conditioner. The only emission is hot water, so I'm tempted to put a coffee machine in the dashboard. It's a very stiff, sporty car with all-wheel digital traction control. It can be designed for a 200,000-mile warranty. The body doesn't rust or fatigue. It doesn't dent in a 6-mph collision. We think it can be made at a competitive cost, with many times less capital and at least an order of magnitude fewer parts.

Chapter 23

Towering Design Flaws

2003

Editor's note: *As Amory has pointed out since the 1970s, blackouts are the result of poor planning decisions—a brittle, centralized grid and inefficient pricing policies. Then the 14 August 2003 Northeast Blackout suddenly put Amory and Hunter Lovins's two-decade-old thesis about "brittle power" back in the public eye. This piece was penned for Toronto's* Globe and Mail *and appeared a week after the power went out (21 August 2003). The* New York Times *had declined it as superfluous to prior major coverage of conventional solutions (which would tend to worsen blackouts).*

The usual suspects—politicians, regulators, deregulators, utilities, and environmentalists—were promptly rounded up when the 14 August blackout lost 61 billion watts of capacity in nine seconds. Yet the real culprit was none of the above—just as in 1965, 1977, and other regional blackouts that I described in a 1981 report for the Pentagon, *Brittle Power: Energy Strategy for National Security* (see www.rmi.org/rmi/Library/S82-03_BrittlePower EnergyStrategy).

The real cause is the overcentralized power grid. Its giant machines spin in exact synchrony across half a continent, coordinated by frail aerial arteries and continuous, precise technical controls. Usually, it works well. But every few years by mishap, or anytime by malice, it can fail catastrophically.

A fixed-wing aircraft can glide to a safe landing without engines, but without instantaneous active control and a tail rotor, a helicopter drops like a stone. The grid is more like a helicopter. Seeing this demonstrated may inspire terrorists to make it happen more often. After previous major blackouts, more giant power plants were linked by more, longer, and heftier transmission lines. Some of these changes relieved local power shortages, but most were unhelpful. Ontario's latest power woes were prolonged because nuclear plants dislike sudden shutdowns and don't restart gracefully: They're the opposite of a plant providing power at peak times—guaranteed unavailable when most needed.

New power lines, plus wholesale competition, have also spawned huge new long-distance power sales. That much power traveling that far can slosh around uncontrollably if a local mishap roils the flow and circuit breakers don't instantly open. But the unimaginably complex grid's "fault tree" of potential failure modes is growing new branches faster than we lop off old ones. The well-meaning operators are always surprised—but if they keep building the same architecture, it will keep failing for the same basic reason.

Modernizing with fast, solid-state switches and advanced controls may help block blackouts, and often boosts existing lines' capacity. Market structures whose rules require and reward high reliability are thus essential (and missing in much of the US). But as one utility executive notes, the emerging policy consensus—that we need to build more and bigger power lines because usage has outpaced capacity—is as wrong as prescribing bloodletting for a patient with a high fever. It reflects a fundamental misunderstanding of what is amiss.

In fact, more wires may make cascading failures more likely and widespread. And they're almost always slower and costlier than three functionally equivalent alternatives: using electricity efficiently, letting customers choose to tailor their usage to price, and decentralized generation.

The cheapest, fastest way to save energy dollars and pollution is to use energy efficiently. (My household electric bill is $7 [Cdn.] a month for a 372-square-meter living space, before counting my larger solar production, which I sell back to the local electricity power cooperative at the same price—now allowed in 38 states.) Ottawa just earmarked $1 billion for conservation (and to meet Kyoto obligations). Each saved kilowatt-hour (kWh) saves 3 kWh of coal at the power station. In the 1970s, Canada had world-class energy-efficiency programs; now they're unimpressive, any incentives to use energy more efficiently removed by rules that reward distributors for selling more electricity and penalize them for cutting customers' bills. Ontario has corrected this perverse incentive for gas but not electricity distributors, and doesn't let efficiency bid directly against generated electricity. This "inefficiency tax" on every home and business hurts competitiveness.

A second key option, "demand response," signals participating customers (electronically or by price) to avoid unneeded use when power is scarce. This needn't inconvenience anyone: If your electric water heater or air conditioner were off for 15 minutes, you'd never know.

Ontario lets this resource compete conveniently against supply only for big customers (and in pilot-project communities). Yet new "smart meters" now make load management profitable for homes as well as businesses—and, like efficient use of electricity, the strategy frees up transmission capacity without building more. A few hundred megawatts of load management, as well as properly opening breakers, might have averted the blackout.

Demand response also stabilizes electricity prices and markets. If California in 2000 could have dispatched load management equaling just 1 percent of power demand, then when nutty rules led suppliers to withhold supply and boost prices, entrepreneurs could simply have shorted the power market, dispatched their load management, and taken $1 billion (US) of the suppliers' money—enough to deter such antisocial behavior.

In the US, where inefficient gas-fired turbines make nearly all peak power, demand response saving just 5 percent of US peak electric load would save about 9.5 percent of all US natural gas. That could quickly return natural gas prices from around $6 to $7 (US) per gigajoule to just a few dollars for years to come, on both sides of the border—and quickly. Between 1983 and 1985, the 10 million people served by Southern California Edison Company used efficiency and load management to cut the decade-ahead forecast of peak demand by 7 percent of actual load per year, at only 1 percent of the cost of new supply. Today's technologies and delivery methods are far better.

A third option: decentralized (also called "distributed") generation is unaffected by transmission failures. Last week in Ontario, Markham, Hamilton, and Sudbury District Energy, among others, isolated their engine-generators from the collapsing grid; they kept running. So did some New York-area industries with microturbines and homes with solar cells. These islands of light in a sea of darkness were powered by local generators that had been installed mainly to save money, but delivered reliability, too. A megawatt generated where it's needed is far more reliable than a megawatt generated far away—yet Ontario prices them the same, with no credit for reliability.

Throughout electricity's first century, power plants were costlier but less reliable than the grid, so ever-bigger power plants backed each other up through the grid. But new power plants are now cheaper and more reliable than the grid, so delivering reliable and affordable power now means generating it at or near the customers (see: www.smallisprofitable.org).

Central thermal power stations stopped getting more efficient in the 1960s. They stopped getting cheaper in the 1980s, and stopped getting bought in the 1990s: Now utility orders are back to Victorian levels. Yet public policy continues to favor central plants and big transmission lines. Transmission is still centrally planned, and needn't compete fairly with its cheaper alternatives.

Our problem isn't too few power lines; it's obsolete rules, rewarding perpetuation of an inherently vulnerable grid. Letting all options compete fairly—whether they save or produce energy, no matter how big they are, what kind they are, or who owns them—would gradually and profitably build a power system as resilient as the internet.

Then major failures, instead of being inevitable by design, would become impossible by design.

Chapter 24

How Innovative Technologies, Business Strategies, and Policies Can Help the US Achieve Energy Security and Prosperity

2006

Editor's note: *In March 2006, the US Senate Committee on Energy and Natural Resources asked Amory to testify on the topic of Energy Independence, specifically how the United States can achieve energy security and prosperity. This is his prepared testimony.*

Both energy independence and its purpose, energy security, rest on three pillars:

1 Making domestic energy infrastructure, notably electric and gas grids, resilient.
2 Phasing out, not expanding, vulnerable facilities and unreliable fuel sources.
3 Ultimately eliminating reliance on oil from any source.[1]

Listing them in this order emphasizes that achieving the third goal without the first two creates only an illusion of security. Hurricane Katrina might as well have read my 1981 finding[2] for DOD that a handful of people could cut off three-fourths of the Eastern states' oil and gas supplies in one evening without leaving Louisiana.

We should worry not only about already-attacked Saudi oil chokepoints like Abqaiq and Ras Tanura, but also about the all-American Strait of Hormuz proposed in Alaska.[3] DOE policy that didn't undercut DOD's mission would:

- shift from brittle energy architecture that makes major failure inevitable to more efficient, resilient, diverse, dispersed systems that make it impossible;[4]
- avoid electricity investments that are meant to prevent blackouts but instead make them bigger and more frequent;[5]
- stop creating attractive nuisances for terrorists, from vulnerable LNG and nuclear facilities to overcentralized US and Iraqi electric infrastructure;[6]
- acknowledge that nuclear proliferation, correctly identified by the President as the gravest threat to national security, is driven largely by nuclear power.[7]

Each of these self-inflicted security threats can be reversed by cheaper, faster, more abundant, and security-enhancing alternatives, available both from comprehensive energy efficiency and from decentralized supply. For example, nuclear power has already been eclipsed in the global marketplace by resilient, inherently peaceful, lower-cost, and lower-risk micropower.[8] That's a big win for national security and profitable climate protection,[9] and a vindication of competitive markets over central planning.

Energy independence is not only about oil. Many sources of LNG raise similar concerns of security, dependence, site vulnerability, and cost: Iran and Russia won't be more reliable long-run sources of gas than Persian Gulf states are of oil. Fortunately, half of US natural gas can be saved by end-use efficiency and electric demand response with average costs below $1 per million BTU— four times cheaper than LNG[10]—making LNG needless and uncompetitive.

America's oil problem is equally unnecessary and uneconomic. Seventy-seven weeks ago, my team published *Winning the Oil Endgame*—an independent, peer-reviewed, detailed, transparent, and uncontested study cosponsored by the Office of the Secretary of Defense and the Chief of Naval Research.[11] It shows how to *eliminate* US oil use by the 2040s and revitalize the economy, led by business for profit. Welcomed by business and military leaders, our analysis is based on competitive strategy for cars, trucks, planes, and oil, and on military requirements.

Our study shows how the US can redouble the efficiency of using oil at an average cost[12] of $12 per saved barrel, and can substitute saved natural gas and advanced biofuels (chiefly cellulosic ethanol) for the remaining oil at an average cost of $18 per barrel. Thus *eliminating* oil would cost just one-fourth its current market price, conservatively assuming that its externalities are worth zero. Side-benefits would include a free 26-percent reduction in CO_2 emissions, a million new jobs (three-fourths in rural and small-town America), and the opportunity to save a million jobs now at risk. America can either continue importing efficient cars to displace oil, or *make* efficient cars and import neither the cars nor the oil. A million jobs hang in the balance.

The key to wringing twice the work from our oil is tripled-efficiency cars, trucks, and planes. Integrating the best 2004 technologies for ultralight steels or composites, better aerodynamics and tires, and advanced propulsion can do this with two-year paybacks.[13] For example, new low-cost carbon-composite manufacturing techniques can halve cars' weight and fuel use, improving safety, comfort, and performance without raising manufacturing cost.[14]

Oil elimination's compelling business logic would drive its eventual adoption. But supportive public policy could accelerate it without requiring new taxes, subsidies, mandates, or federal laws; this could be done administratively or by the states.

Many innovative policies could also transcend gridlock. Size- and revenue-neutral feebates[15] could speed the adoption of superefficient cars far more effectively than gasoline taxes or efficiency standards, and would make money for both consumers and automakers.[16] Novel policies could also support automotive retooling and retraining, superefficient planes, advanced biofuels, low-income access to affordable personal mobility, and other key policy goals, all at zero net cost to the Treasury.[17]

Early implementation steps are encouraging. Our analysis helped Wal-Mart to launch a plan to double its heavy truck fleet's efficiency and to consider tripled efficiency a realistic goal.[18] The Department of Defense is also recognizing fuel-efficient platforms as a key to military transformation. Military needs for ultralight, strong, cheap materials can transform the civilian car, truck, and plane industries—much as DARPA created the internet, GPS, and the chip and jet-engine industries—and thus lead the nation off oil so we needn't fight over oil: negamissions in the Persian Gulf, Mission Unnecessary.[19]

The surest path to an energy policy that enhances security and prosperity is free-market economics: letting all ways to save or produce energy compete fairly, at honest prices, no matter which kind they are, what technology they use, where they are, how big they are, or who owns them. That would make the energy security, oil, climate, and most proliferation problems fade away, and would make our economy and democracy far stronger.[20]

Notes

1   Since oil is a fungible commodity in a global market, national energy policy correctly recognizes that the problem is oil use, not imports: see n. 13. For example, even if the US imported no oil, it would still be a price-taker in the world market, so its economy, like its trading partners', would still be buffeted by oil-price volatility. Oil infrastructure is also inherently vulnerable even if it is domestic (n. 2).

2   A. B. & L. H. Lovins, *Brittle Power: Energy Strategy for National Security*, Brick House (Andover MA), 1981, and Rocky Mountain Institute, 1989; OCR scan reposted at www.rmi.org/rmi/Library/S82-03_BrittlePowerEnergyStrategy; summarized in Chapter 21 of this book.

3   Former director of Central Intelligence R. James Woolsey, an Oklahoman not per se hostile to petroleum, testified against Arctic National Wildlife Refuge drilling on national-security grounds (Energy Subcommittee of USHR Science Committee, 1 Nov. 2001), and wrote that such drilling's

> *real show-stopper is national security. Delivering that oil by its only route, the 800-mile-long Trans-Alaska Pipeline System (TAPS), would make TAPS the fattest energy-terrorist target in the country—Uncle Sam's 'Kick Me' sign. ...*

TAPS is frighteningly insecure. It's largely accessible to attackers, but often unrepairable in winter. If key pumping stations or facilities at either end were disabled, at least the above-ground half of 9 million barrels of hot oil could congeal in one winter week into the world's biggest ChapStick®. ... The Army has found TAPS indefensible. It has already been sabotaged, incompetently bombed twice, and shot at more than 50 times. Last 4 October [2001], a drunk shut it down with one rifle shot. ... In 1999, a disgruntled engineer's sophisticated plot to blow up three critical points with 14 bombs, then profit from oil futures trading, was thwarted by luck. He was an amiable bungler compared with the 11 September attackers. Connect the dots: doubling and prolonging dependence on TAPS hardly seems a prudent centerpiece for what advocates whimsically called the Homeland Energy Security Bill. ... Reliance both on Mideast oil and on vulnerable domestic energy infrastructure such as TAPS imperils the security of the US and its friends. (R. J. Woolsey, A. B. & L. H. Lovins, "Energy security: It takes more than drilling," *Chr. Sci. Mon.*, 29 Mar. 2002, www.rmi.org/rmi/Library/S02-05_EnergySecurityMoreThanDrilling. For documentation, see hyperlinks to p. 73 in A. B. & L. H. Lovins, *For. Aff.*, pp. 72–85, July/Aug. 2001, www.rmi.org/rmi/Library/E01-03_FoolsGoldAlaska, and later supplementary references at www.rmi.org/sitepages/pid171.php# E01-04.)

4   "Surprises and Resilience," *RMI Solutions*, pp. 1ff, spring 2006, www.rmi.org/Content/Files/RMI_SolutionsJournal_Spring06.pdf.

5   Bigger power plants sending bigger bulk power flows through longer transmission lines tend to make the grid less stable (id. [66]). Leading engineering analysts of electric-grid theory are reaching similar conclusions, e.g. http://eceserv0.ece.wisc.edu/~dobson/PAPERS/carrerasHICSS03.pdf. FERC doesn't let resilient options compete.

6   See n. 2 for discreet details. Since the invasion of Iraq, private recommendations that its electricity infrastructure be rebuilt in decentralized form, virtually invulnerable to insurgent attack, have been repeatedly rejected.

7   A. B. & L. H. Lovins and L. Ross, "Nuclear Power and Nuclear Bombs," *For. Aff.* 58(5):1137–1177, Summer 1980. Had that article's recommendations been adopted, we would not today be worrying about Iran and North Korea. In brief, nuclear power makes widely and innocently available the key ingredients—fissile materials, equipment, technologies, skills—needed to make bombs by any of the ~20 known methods (other than stealing military bombs or parts). (New reactor types and the proposed reversal of the Ford–Cheney non-reprocessing policy greatly intensify these perilous links.) But in a world that took economics seriously, nuclear power would gracefully complete its demise, due to an incurable attack of market forces (n. 10), so these ingredients of do-it-yourself bomb kits would no longer be items of commerce. This would make them harder to get, more conspicuous to try to get, and politically far costlier to be caught trying to get, because for the first time the reason for wanting them would be *unambiguously* military. This would not make proliferation impossible, but would make it far more difficult and much easier to detect timely: intelligence resources could focus on needles, not haystacks. The US example is critical because if a country with such wealth, technical skill, and fuel resources claims it cannot meet its energy needs without nuclear energy and reprocessing, then it invites every other less fortunate country to make the same spurious claim. Yet

the US could still offer to meet the intent of the Non-Proliferation Treaty's Article IV bargain by sharing today's cheaper, faster, more effective energy technologies (n. 10) to boost global development. The NPT's specifically nuclear bargain was written by nuclear experts, in a nuclear context, around 1969–1970, when nuclear energy was widely believed to be cheap and indispensable. Now that the market has decided otherwise, Article IV should be reinterpreted to achieve the same electricity-for-development goal by more modern, speedy, and affordable means, starting immediately with US/Indian energy cooperation: improving the non-nuclear 97% of India's electricity system could produce enormously greater, wider, faster, and cheaper development benefits.

8   Low-carbon cogeneration plus decentralized no-carbon renewables surpassed nuclear power's global capacity in 2002 and its annual electricity output in 2005, and they are far outcompeting central stations despite typically lower subsidies and bigger obstacles. In 2004, micropower worldwide added ~2.9 times as much output and ~5.9 times as much capacity as nuclear power did (or at least ten times if electric efficiency were also included). Industry projects that in 2010, micropower will add ~160 times as much capacity as nuclear power adds. Micropower comprises cogeneration (combined-heat-and-power using 1–120 MWe gas turbines, 1–30 MWe engines, and steam turbines only if in China), plus renewables excluding big hydro (>10 MWe). Electricity *savings* are probably even bigger than micropower additions but are not being well tracked. See Chapter 13 of this book, and for details see "Nuclear Power: Economics and Climate-Protection Potential," 11 September 2005 / 6 January 2006, www.rmi.org/rmi/Library/E05-14_NuclearPowerEconomicsClimateProtection. Statistics at www.rmi.org/rmi/pid113 and www.ren21.net/globalstatusreport.

9   Choosing the best buys first could relieve climate concerns not at a cost but at a profit, because efficiency generally costs less than the energy it saves: A. B. Lovins, "More Profit With Less Carbon," *Sci. Amer.*, pp. 74–82, September 2005, www.sciam.com/media/pdf/Lovinsforweb.pdf, and its extended bibliography, www.rmi.org/images/other/Climate/C05-05a_MoreProfitBib.pdf. Reducing global energy intensity not by the normally assumed 1%/yr but by 2%/yr would eliminate CO_2 growth; slightly faster improvement would stabilize climate. Both the US and certain states have sustained intensity reductions well over 2%/yr, and attentive companies around 6%/yr, all at a handsome profit. Yet climate politics focus on cost, burden, and sacrifice rather than on profit, jobs, and competitive advantage. Fixing this sign error is the key to crafting a profitable climate solution. Of course, buying carbon-free resources judiciously, not indiscriminately, yields the most climate solution per dollar and per year. Expanding nuclear power would reduce and retard climate protection, simply because it's costlier and slower than its key competitors—cogeneration, certain renewables, and efficient end-use. See Lovins papers in n. 8.

10   Saving 1% of US electricity, including peak hours, can save 2% of total US natural gas consumption and cut the gas price by 3–4% (see n. 11, pp. 112–116, 219–220). In this decade, such straightforward efficiencies could cut $50 billion off the nation's annual gas and power bills and relieve many gas and electricity constraints without costly, controversial, and vulnerable supply-side investments. The main obstacles are that gas efficiency isn't on the federal policy agenda, and that 48 states reward utilities for selling more electricity and gas while penalizing them for cutting customers' bills. Scores of other barriers, too, block wider

purchases of energy efficiency in all sectors (see pp. 11–20 in www.rmi.org/images/other/Climate/C97-13_ClimateMSMM.pdf), but each obstacle can be turned into a business opportunity if policy focuses systematically on "barrier-busting."

11 *Winning the Oil Endgame: Innovation for Profits, Jobs, and Security*, RMI, 20 September 2004, by A. B. Lovins *et al.*; forewords by George Shultz and Sir Mark Moody-Stuart; PDF download free at www.oilendgame.com. That site also posts the Executive Summary, 24 Technical Annexes, lay summaries from *Ripon Forum* (www.rmi.org/rmi/Library/E05-02_OilOurFatalDependence) and many other articles and reviews, and offers the 331-page hard-copy book for $40.

12 Refiner's acquisition cost on the short-run margin, 2000 $, 5%/yr real discount rate.

13 Compared with EIA 1/04 Reference Case vehicle characteristics and fleet mix, fuel economy could be improved by 69% for cars at a levelized Cost of Saved Energy of 57¢/gal, by 65% for Class 8 trucks at 25¢/gal, and by ~65% for planes at ≤46¢/gal. The first 25% of truck and 20% of airplane fuel savings are free. Please see n. 11 and its Technical Annexes 4–6 and 12 for full analytic details and documentation.

14 Because the advanced composites' higher cost is offset by simpler automaking and smaller powertrains. See n. 11, pp. 44–73, Tech. Annex 5 (www.oilendgame.com/TechAnnex.html), and *Intl. J. Veh. Des.* 35(1/2):50–85 (2004), www.rmi.org/images/other/Trans/T04-01_HypercarH2AutoTrans.pdf. One cost-competitive carbon-composite structural manufacturing process, being commercialized by a small firm, Fiberforge®, of which (full disclosure) I'm Chairman and a small shareholder, is described at www.fiberforge.com/company/company.php and in trade press articles at www.fiberforge.com/news/news.php.

15 Such feebates (= fee + rebate) would broaden the price spread within each size class by charging fees on less efficient vehicles and using the revenue to pay rebates on more efficient vehicles. Whether you pay a fee or receive a rebate depends on your efficiency choice within the size class you prefer. A typical feebate slope—$1000 per 0.01 gallon/mile difference from the "pivot point" efficiency level set within each size class—would arbitrage the spread in discount rate between consumers and society, so a car buyer would consider full lifecycle fuel savings (nominally ~14 years) rather than just the first 2–3 years. DOE/ORNL modeling, closely matching RMI's, shows that such feebates yield both producer and consumer surplus. See n. 11, pp. 186–190.

16 See n. 11, pp. 169–190.

17 See n. 11, pp. 178–226. The 2005 Energy Policy Act's 3–5-year biofuel credits are too brief for investment horizons; any serious incentive, especially in an area fraught with investment uncertainties, should last at least a decade. However, I generally prefer abolishing energy subsidies to adding new ones, and I fear that the same broad policy conditions that created the energy market collapse of 1984–85 are now being repeated.

18 L. Scott, "Twenty First Century Leadership," 24 October 2005, http://walmartstores.com/sites/sustainabilityreport/2007/documents/21stCentury Leadership.pdf.

19 Fuel-efficient platforms offer huge benefits in force protection, tens of billions of dollars' annual savings in fuel logistics, and multi-divisional realignments from

tail to tooth: n. 11, pp. 84–93, 221, and 261–262; Defense Science Board, *More Capable Warfighting Through Reduced Fuel Burden*, 2001, www.dtic.mil/cgi-bin/GetTRDoc?AD=ADA392666&Location=U2&doc=GetTRDoc.pdf.

20 Both a quick low-budget experiment (www.nepinitiative.org) and the National Commission on Energy Policy (www.energycommission.org) revealed a broad ground for trans-ideological consensus on these general lines. The former effort found that a bipartisan group of private- and public-sector energy leaders could readily agree on a comprehensive, visionary, but practical framework for national energy policy by focusing on what they already agreed about—thus making what they disagreed about largely superfluous.

Chapter 25

DOD's Energy Challenge as Strategic Opportunity

2010

Editor's note: *This article was published in* Joint Force Quarterly, *issue 57, second quarter of 2010, and is reprinted with permission.*

Energy is the lifeblood of modern societies and a pillar of America's prowess and prosperity. Yet energy is also a major source of global instability, conflict, pollution, and risk. Many of the gravest threats to national security are intimately intertwined with energy, including oil supply interruptions, oil-funded terrorism, oil-fed conflict and instability, nuclear proliferation, domestic critical infrastructure vulnerabilities, and climate change (which changes everything).[1]

Every combatant command has significant and increasing energy-related missions. Energy has become such a "master key"—it is so pervasive in its tangled linkages to nearly every other security issue—that no national security strategy or doctrine can succeed without a broad and sharp focus on how the United States and the world get and use energy. For the first time, 37 years after the 1973 oil embargo, the 2010 Quadrennial Defense Review is expected to recognize energy's centrality to the mission of the Department of Defense (DOD), and to suggest how DOD can turn energy from a major risk into a source of breakthrough advantage.

DOD faces its own internal energy challenges. The heavy steel forces that defeated the Axis "floated to victory on a sea of oil," six-sevenths of which came from Texas. Today, Texas is a net importer of oil, and warfighting is about 16 times more energy-intensive: its oil intensity per warfighter rose 2.6 percent annually for the past 40 years and is projected to rise another 1.5 percent annually through 2017 due to greater mechanization, remote expeditionary conflict, rugged terrain, and irregular operations.[2] Fuel price volatility also buffets defense budgets: each $10 per barrel (bbl) rise in oil price costs DOD over $1.3 billion per year. But of immediate concern, DOD's mission is at risk (as recent wargaming confirms), and the Department is paying a huge cost in lives, dollars, and compromised warfighting capability, for two reasons:

1 pervasively inefficient use of energy in the battlespace;
2 ~99 percent dependence of fixed-facility critical missions on the vulnerable electricity grid.

This discussion of both issues draws heavily on the Defense Science Board's (DSB's) 2008 report *More Fight—Less Fuel*.[3] That analysis, building on and reinforcing its largely overlooked 2001 predecessor, found that solutions are available to turn these handicaps into revolutionary gains in warfighting capability, at comparable or lower capital cost and at far lower operating cost, without tradeoff or compromise. The prize is great. As the Logistics Management Institute stated, "Aggressively developing and applying energy-saving technologies to military applications would potentially do more to solve the most pressing long-term challenges facing DOD and our national security than any other single investment area."[4]

Fuel Logistics: DOD's Soft Underbelly

Fuel has long been peripheral to DOD's focus ("We don't *do* fuel—we *buy* fuel"), but turbulent oil markets and geopolitics have lately led some to question the Department's long-term access to mobility fuel. Echoing the International Energy Agency's chief economist, Fatih Birol—"We must leave oil before it leaves us"—some analysts assert world oil output capability has peaked or soon will.

They overlook recent evidence that "peak oil" is more clearly imminent in demand than in supply. US gasoline use—an eighth of world oil—is probably in permanent decline.[5] So may be Organisation for Economic Co-operation and Development countries' oil use, which has been falling since early 2005.[6] Deutsche Bank projects world oil use to peak in 2016, then be cut by electric cars to ~40 percent below the consensus forecast or ~8 percent below current levels by 2030.[7] This assumes China's new cars will be 26 percent electrified by 2020 (China's target is 80 percent), and omits lightweight and low-drag cars, superefficient trucks and planes, and other important oil savings well under way. Oil, as predicted for two decades, is becoming uncompetitive even at low prices before it becomes unavailable even at high prices.

Nobody knows how much oil is in the ground: governments, which often do not know or will not transparently reveal what they have, hold about 94 percent of reserves. But DOD, like the United States, has three compelling reasons to get off oil regardless: security, climate, and cost. Long-term oil availability concerns for DOD are misdirected; even more so, as we will see, are proposals to create a defense synthetic fuel industry. Indeed, DOD is probably the world's largest institutional oil buyer, consuming in fiscal year (FY) 2008 120 million barrels costing $16 billion—93 percent of all US government oil use (see Figure 25.1).

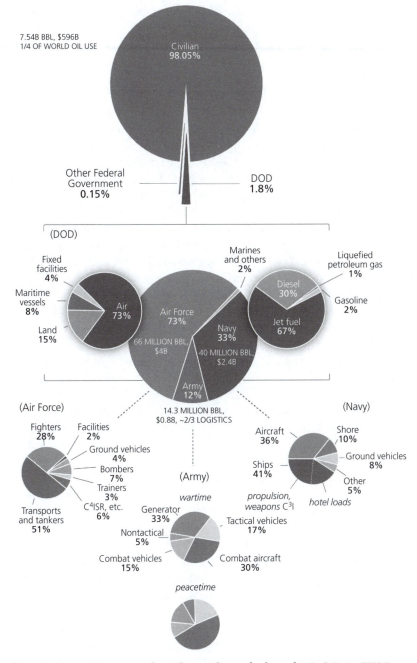

Figure 25.1 *Approximate liquid petroleum fuel use by DOD in FY05*

Notes: DOD's apparent fuel cost [FY05: $7.43B] is a modest fraction of true fully burdened delivered fuel cost; the added delivery costs are mainly for the 9% of Air Force fuel delivered aerially for >$49/gal, and for forward fuel to Army.

An unknown fraction of Air Force and Navy fuel transports Army materiel. Oil used by contractors that DOD has outsourced is unknown.

But oil is a largely fungible commodity in a global market; the Department uses only 0.4 percent of the world's oil output (about what two good-sized Gulf of Mexico platforms produce); and in a crisis, DOD has oil-buying priority. Rather, the issue is that DOD's unnecessarily inefficient use of oil makes it move huge quantities of fuel from purchase to use, imposing high costs in blood, treasure, and combat effectiveness.

Logistics uses roughly half the Department's personnel and a third of its budget. One-fifth of DOD's oil—at least 90 million gallons each month—supports Iraq and Afghanistan operations that have increased forward bases' oil use tenfold.[8] Of the tonnage moved when the Army deploys, roughly half is fuel.[9] A typical Marine combat brigade needs more than a half-million gallons per day. Desert Storm's flanking maneuver burned 70,000 tons of fuel in 5 days.[10] Delivering that quantity is a huge job for brigades of logistics personnel and for the personnel and assets needed to maintain and protect the logistics chain.

Despite extensive land and air forces trying to guard them—a "huge burden on the combat forces"[11]—fuel convoys are attractive and vulnerable targets, making them one of the Marine Corps Commandant's most pressing casualty risks in Afghanistan.[12] In FY07, attacks on fuel convoys cost the US Army 132 casualties in Iraq (0.026/convoy) and 38 in Afghanistan (0.034/convoy).[13] About 12 percent of *total* FY07 US casualties in Iraq and 35 percent in Afghanistan were Army losses—including contractors but not other Services or coalition partners—associated with convoys.[14] Their constrained routes expose them to improvised explosive devices (IEDs), which probably caused the majority of US fatalities in Afghanistan in 2009. Should that conflict follow an Iraq-like profile, its casualty rates could rise 17.5 percent annually.[15] Just the dollar cost of protecting fuel convoys can be "upward of 15 times the actual purchase cost of fuel, [increasing] exponentially as the delivery cost increases or when force protection is provided from air."[16]

Thus, attacks on fuel assets and other serious hazards to fuel convoys increase mission risk, while fuel logistics and protection divert combat effort and hammer oil-strained budgets. Yet the need for most of the fuel delivered at such high cost could have been avoided by far more efficient use. Efficiency lags because when requiring, designing, and acquiring the fuel-using devices, DOD has systematically assumed that fuel logistics is free and invulnerable—so much so that wargames did not and often could not model it. Instead of analyzing fuel logistics' burden on effectiveness and signaling it by price, DOD valued fuel at its wholesale price delivered in bulk to a secure major base (around $1–$3 per gallon), rather than at its fully burdened cost delivered to the platform in theater in wartime (usually tens and sometimes hundreds of dollars per gallon). Lacking requirements, instructions, shadow prices, rationales, or rewards for saving fuel, hardly anyone considered the military value of achieving, nor strove to achieve, high fuel-efficiency.

As consequences became obvious in theater and began to emerge in wargames, the Department in 2007 started changing its policy to value energy savings at the "Fully Burdened Cost of Fuel" (FBCF, in dollars per gallon), including force protection, *delivered* to its end-user in theater. The 2009 National Defense Authorization Act (NDAA) codified both FBCF and new energy Key Performance Parameters (KPPs, in gallons per day or mission). Those are to receive similar weight to traditional KPPs like lethality, protection, and reliability that encapsulate the Department's pursuit of capability. In principle, both FBCF and energy KPPs will guide requirements writing, analyses of alternatives, choices in the acquisition tradespace, and the focus of DOD's science and technology (S&T) investments. In practice, energy KPPs have not yet been applied (their "selective use" is allowed but not yet launched), and much work must be organized and resourced to get the FBCF numbers right and apply them systematically.[17]

The FBCFs initially in use are incomplete. Current guidance still appears to omit support pyramids, multipliers to rotational force strength, actual (not book) depreciation lives, full headcounts including borrowed and perhaps contractor personnel, theft and attrition adjustments,[18] and uncounted Air Force and Navy lift costs to and from theater.

All should be included: FBCF should count all assets and activities—at their end-to-end, lifecycle, fully burdened total cost of ownership—that will no longer be needed, or can be realigned, if a given gallon need no longer be delivered. Thus, if fielded fuel supply needs shrink, so do its garrison costs for related training and maintenance. Conversely, garrison costs should be additive to FBCF, not dilutive: some analysts average peacetime with wartime costs to water down FBCF, or even assume a peacetime operating tempo, but as the 2008 task force stated, "FBCF is a wartime capability planning factor, not a peacetime cost estimate."[19] Even before these conservatisms are made realistic, initial FBCF estimates value saved fuel often *one to two orders of magnitude* higher than previously. If these new metrics gain momentum and top-level focus, they could drive strategic shifts and innovations that could revolutionize military capability and effectiveness.

More Fight—Less Fuel mapped a detailed military energy reform agenda, broadly backed by DOD's 2008 Energy Security Task Force. DSB offered specific solutions for key findings: that DOD lacks the strategy, policies, metrics, information, and governance structure to properly manage its energy risks; that technologies are available to make DOD systems more energy-efficient, but are undervalued, slowing implementation and resulting in inadequate S&T investments; and that there are many opportunities to reduce energy demand by changing wasteful operational practices and procedures.

The 2009 NDAA [National Defense Authorization Act] codified reforms on the lines recommended by DSB, to be led by a new DOD Director of Operational Energy. As of 1 December 2009, that critical post remained vacant, but some encouraging Service adoption initiatives had begun, such as the Army Energy Security Implementation Strategy and Navy Secretary Ray Mabus's invigorating energy goals. But the DSB task force, not stopping with bureaucratic fixes, had added the even more incisive finding that "DOD's energy problems [are] sufficiently critical to add two new strategic vectors"—

an older term for "succinct descriptions of capabilities that would make a big difference in military operations"[20]—to complement the four historic ones: "speed, stealth, precision, and networking."

In today's more familiar language, Endurance and Resilience are new capabilities that both drive and apply new operational requirements. An Endurance capability will create transformational strategies and tactics that both tell the requirements writer to make a new platform fuel-efficient and inspire the force planner to exploit its increased range and agility. Today's DOD habits would instead tend to make it heavier with the same range—much as Detroit's engine improvements since the 1970s, rather than saving one-third of civilian cars' fuel, only made them more muscular.

The need to change entrenched habits in force planning and operational requirements makes big new capabilities both vital and hard. Driving them deeply into doctrine, strategy, organizational structures, cultures, training, reward systems, and behaviors requires strong, consistent, persistent senior leadership. But once so embedded, new capabilities disruptively and profoundly improve military effectiveness and cost-effectiveness.

The Endurance Capability

Endurance traditionally means "ability to sustain operations for an extended time without support or replenishment."[21] The DSB task force elaborated:

> *Endurance exploits improved energy efficiency and autonomous energy supply to extend range and dwell—recognizing the need for affordable dominance, requiring little or no fuel logistics, in persistent, dispersed, and remote operations, while enhancing overmatch in more traditional operations.*[22]

A lean or zero-fuel logistics tail also increases mobility, maneuver, tactical and operational flexibility, versatility, and reliability—all required to combat asymmetrical, adaptive, demassed, elusive, faraway adversaries. Endurance is needed in every "platform" using energy in the battlespace, from mobility platforms to expeditionary base power to battery-powered land-warrior electronics. Endurance is even more valuable in stability operations, which often need even more persistence, dispersion, and affordability than the combat operations with which they now enjoy comparable priority.[23]

The DSB report found "enormous technical potential to cost-effectively become more fuel-efficient and by so doing to significantly enhance operational effectiveness."[24] Current, near-term, and emerging efficiency technologies offer major fuel savings in land, sea, and air platforms,[25] with better warfighting capability (not one of 143 briefs disclosed a tradeoff), and with generally excellent economics and operational characteristics. Early adoption has begun at a modest scale. For example, field commanders in Iraq noticed that:

> *Fuel that is transported at great risk, great cost in lives and money, and substantial diversion of combat assets for convoy protection is burned in generator sets to produce electricity that is, in turn, used to air-condition un-insulated and even unoccupied tents. ... One recently analyzed FOB [forward operating base] used about 95% of its genset [engine-generator set] electricity for this purpose, and about one-third of the Army's total wartime fuel use is for running gensets.[26]*

A single typical 60-kilowatt genset burns 4 to 5 gallons per hour, or $0.7 million per year at a typical Afghanistan FBCF of $17.44/gal. Fueling one FOB's gensets might cost $34 million per year—plus, at the FY07 casualty rate, nearly one casualty.[27]

In response, DOD is spraying over 17 million square feet of insulating foam onto temporary structures in theater, saving over half their air-conditioning energy. This $146 million investment should repay its cost in 67 to 74 days at the estimated Iraq $13.80 per gallon FBCF—10 times faster than under the old assumption of undelivered and unprotected fuel. The first $22 million worth should save more than $65 million each year—and more than one convoy casualty.[28] Next steps include far more efficient gensets and air conditioners, encompassing emerging concepts for cooling without electricity.

Lieutenant General James Mattis's 2003 challenge to "unleash us from the tether of fuel" and Major General Richard Zilmer's 2006 operational request from Anbar Province for a "self-sustainable energy solution" stimulated the Army's Rapid Equipping Force to develop a portable renewable/hybrid energy supply system, demonstrated at the National Training Center but not yet fielded. In theater, at the fully burdened cost of fuel, it would probably have been paid back in months[29]—faster if credited for avoided casualties and enhanced combat capability. The Marines have pledged resources for such work.

Over several decades, concerted adoption of identified energy-efficiency technologies holds the estimated potential to cut total DOD mobility-fuel requirements by about two-thirds, perhaps even three-fourths. The fattest targets vary according to intent:

- The most gallons can be saved in aircraft, which use 73 percent of DOD fuel. Saving 35 percent of aircraft fuel would free up as much fuel as all DOD land and maritime vehicles plus facilities use. New heavy fixed-wing platforms can save at least 50 percent and new rotary-wing platforms 80 percent, since those fleets use designs that are, respectively, 50 to 60 and 30 to 50 years old.
- The biggest gains in combat effectiveness will come from fuel-efficient ground forces (land and vertical-lift platforms, land warriors, FOBs). For example, soldiers carry an average of 2 kilograms of batteries per mission-day.

- Savings downstream in a long logistics chain save more fuel: delivering 1 gallon to the Army speartip consumes about 1.4 extra gallons in logistics.
- Savings in aerially refueled aircraft and forward-deployed ground forces save the most delivery cost and thus realignable support assets.

Reset, such as the tens of billions of dollars slated for Humvee replacement, offers a ripe opportunity for leap-ahead performance if, for example, a break-through light tactical vehicle already substantially developed can get the "intensive development, design and competitive prototyping" recommended by the 2008 DSB task force. A vehicle as protective and lethal as a 23- to 29-ton mine-resistant ambush protected (MRAP) vehicle, but with acceleration, agility, and stability similar to a top-of-the-line pickup truck—and fuel economy, weight, and cost better than a 5- to 6-ton up-armored Humvee—sounds more promising than a Humvee or MRAP.

Yet the innovative competitor's prototyping remains stalled, and Office of the Secretary of Defense policy bars using reset funds for innovative platforms. Both DSB task forces recommended changes in DOD doctrine, structure, business processes, and other activities—emphasizing design and acquisition—to capture these opportunities aggressively and exploit five major military energy-efficiency benefits:

1 *Force protector*, with far fewer vulnerable fuel convoys.
2 *Force multiplier*, freeing up convoy guards for combat tasks—turning fuel-guarders into trigger-pullers.
3 *Force enabler*, equipping warfighters with the greatly enhanced dwell, reach, agility, and flexibility that can affordably dominate in both dispersed and focused combat.
4 *Key to transformational realignment* from tail to tooth—shifts totaling multidivisional size, worth many tens of billions of dollars per year.
5 *Catalyst for leap-ahead fuel savings in the civilian sector*, which uses more than 50 times as much fuel as DOD. Valuing saved military fuel at FBCF will drive astonishing innovations that accelerate civilian vehicle efficiency, much as past military S&T investment yielded the internet, Global Positioning System, and jet engine and microchip industries.

DSB's 2008 report summarized:

> *Unnecessarily high and growing battlespace fuel demand compromises operational capability and mission success; requires an excessive support force structure at the expense of operational forces; creates more risk for support operations than necessary; and increases life-cycle operations and support costs.*[30]

Yet radically boosting platforms' energy-efficiency and combat effectiveness at reasonable or reduced up-front cost can turn each of these energy risks into major warfighting gains. Requiring and exploiting Endurance can give DOD both more effective forces and a more stable world, at reduced cost and risk. This better-than-free opportunity must become a cornerstone of military doctrine.

This shift will not be easy. It requires fundamentally redesigning military energy flows to support fast-changing strategic, operational, and tactical requirements. It demands new DOD planning processes that recognize Endurance's operational value so it becomes a requirement in platforms now in development, and appreciate that delivering an operational effect within a fixed energy budget is itself an important capability. A new system's energy budget is an important requirement—as important as any other—and should be analytically based on the size of the logistics tail the system demands and the burden that assuring successful delivery of that logistics tail imposes on the force.

Severalfold greater platform fuel-efficiency comes from rapidly adopting and fielding advances in ultralight and ultra-strong materials, fluid dynamics, actuators, and propulsion, all synergistic with alternative fuel and power supplies. It also depends on transformational approaches, incentivized by FBCF and potentially required by energy KPPs but unfamiliar to most DOD contractors, that use integrative design to achieve expanding, not diminishing, returns to investments in energy efficiency—yielding major energy savings at lower capital cost without trading off nonenergy KPPs. Basic innovation in design and acquisition requires taking intelligent risks and rewarding those who do so. All this will require senior leadership to tackle head-on the issue that a previous DSB report described thus: "Often the very technology that can provide the United States with a disruptive advantage is itself disruptive to DOD's culture ... and antibodies rapidly and reflexively form to reject it."[31] Yet such disruptive concepts can be so clearly beneficial that masterful and resolute leadership breaks through hesitancy and resistance. This is the Department's imperative today.

Fuel and power autonomy

Very efficient energy use stretches fuel and power made in theater from wastes, opportunistically acquired feedstocks, or renewable energy flows. Fedex and Virgin Airways plan to fuel 30 percent and 100 percent of their respective fleets with biofuels by 2020. Domestically produced biofuels from centralized, specialized plants do little for DOD's expeditionary needs, but much cutting-edge research emphasizes portable biofuel converters akin to an "opportunistic foraging herbivore."[32] The 2008 DSB task force favored promising expeditionary biofuel and synfuel technologies, and the Services are examining some. In contrast, the DSB task force expressed "strong concerns" about the coal-to-liquids synfuels favored by the Air Force and Navy (but illegally carbon-intensive under a 2007 law), finding they "do not contribute to [solving] DOD's most critical fuel problem—delivering fuel to deployed forces," "do not appear to have a viable market future or contribute to reducing battlespace fuel demand," and do not appear to address a real problem. Fuel interdiction risk in theater is best countered by efficient use, diversified fuels and supply chains, and greater or more secure local stockpiling. If the concern is long-term fuel availability, military and civilian end-use efficiency is by far the cheapest choice. In 2005, Wal-Mart's giant Class 8 truck fleet launched gallon per ton-mile savings that reached 38 percent in 2008 and are targeted to reach 50 percent in 2015. General US adoption of those doubled-efficiency civilian trucks will save 6 percent of US oil—triple DOD's total use. The Secretary of Defense's JASON science advisers, whose energy report also pointedly failed to endorse coal-to-liquids, suggested saving oil by redesigning the Postal Service's delivery fleet.[33]

Nuclear power is sometimes suggested for land installations or even expeditionary forces,[34] typically without discussing cost (grossly uncompetitive), modern renewables (typically much cheaper), operational reliability (usually needing 100 percent backup), or security. For these and other reasons, the 2008 DSB and JASON task forces did not endorse this option. After vast investment in hardware and a unique technical culture, nuclear propulsion has proven its merit in submarines and aircraft carriers. In 2006–2009, congressional enthusiasts announced supposed Naval Sea Systems Command (NAVSEA) findings that nuclear propulsion in new medium surface combatants could beat $70/bbl oil. However, the 2008 DSB task force discovered that NAVSEA's actual finding ($75–$225/bbl) had improperly assumed a zero real discount rate. A 3 percent annual real discount rate yielded a $132–$345/bbl breakeven oil price; NAVSEA did not respond to requests to test the 7 percent annual real discount rate that the Office of Management and Budget probably mandates. Presumably, the Secretary of Defense will reject this option and focus resources on making ships optimally efficient. The 2008 DSB and JASON studies are redirecting military energy conversation from exotic, speculative, and often inappropriate supplies to efficient use, which makes autonomous in-theater supply important and often cost-effective. But all such choices depend on a further fundamental reform in DOD's metrics and procedures.

Gross versus net capability

A change that would boost operational capability by greatly increasing tooth-to-tail ratios was identified in a little-noticed but "important observation of the [2008 DSB] Task Force":

> What [the Joint Capabilities Integration and Development System][35] currently calls "capability" is actually the theoretical performance of a platform or system unconstrained by the logistics tail required for its operation. But tail takes money, people, and materiel that detract from tooth. True net capability,

constrained by sustainment, is thus the gross capability (perfor-mance) of a platform or system times its "effectiveness factor"—its ratio of effect to effort:

Effectiveness Factor = Tooth / (Tooth + Tail)

Also, in an actual budget, Tooth = (Resources – Tail), so

Effectiveness Factor = (Resources – Tail) / Resources.

Effectiveness factor ranges from zero (with infinite tail) to one (with zero tail). If tail >0, true net capability is always less than theoretical (tail-less) performance, but DOD consistently confuses these two metrics, and so misallocates resources. Buying more tooth that comes with more (but invisible) tail may achieve little, no, or negative net gain in true capability. While the Department recognizes the need to reduce tail, the analytical tools needed to inform decisions on how to do so are not in place. Focusing on reducing tail can create revolutionary capability gains and free up support personnel, equipment, and budget for realignment. The task force recommendations are intended to build the analytical and policy foundation to begin introducing this way of thinking into the requirements, acquisition and budget forecasting processes.[36]

To summarize, current force planning does not and cannot predict or compare competing options' needed tail size or their net capability, so after decades, the "tail is eating the tooth." Reversing this impairment needs five missing steps:

1 an Endurance capability to drive and exploit operational requirements for radical efficiency,
2 enforced by energy KPPs,
3 valued at FBCF,
4 competed on *net* capability, and
5 tested with wargaming and campaign-modeling tools revised so they "play fuel" and reveal the full operational value of lean fuel logistics.

All five together will help drive DOD toward ultimately breeding, where possible, a Manx force—one with no tail. Efficient and passively or renewably cooled tents in the desert can mean no gensets, no fuel convoys, no problem. Such a thrust toward efficiency in every use of fuel and electricity also strongly supports the second proposed new key capability—Resilience.

The Resilience Capability

Resilience "combines efficient energy use with more diverse, dispersed, renewable supply—turning the loss of critical missions from energy supply failures (by accident or malice) from inevitable to near-impossible."[37] This capability is vital because the

> *almost complete dependence of military installations on a fragile and vulnerable commercial power grid and other critical national infrastructure places critical military and Homeland defense missions at an unacceptably high risk of extended disruption. [Backup generators and their fuel supplies at military installations are generally sized] for only short-term commercial outages and seldom properly prioritized to critical loads because those are often not wired separately from non-essential loads. DOD's approach to providing power to installations is based on assumptions that commercial power is highly reliable, subject to infrequent and short term outages, and backups can meet demands. [These assumptions are] no longer valid and DOD must take a more rigorous risk-based approach to assuring adequate power to its critical missions.*[38]

The 2008 DSB task force found that the confluence of many risks to electric supply—grid overloads, natural disasters, sabotage or terrorism via physical or cyber attacks on the electric grid, and many kinds of interruptions to generating plants—hazards electricity-dependent hydrocarbon delivery, the national economy, social stability, and DOD's mission continuity.

The US electric grid was named by the National Academy of Engineering as the top engineering achievement of the 20th century. It is very capital-intensive, complex, technologically unforgiving, usually reliable, but inherently brittle. It is responsible for ~98–99 percent of US power failures, and occasionally blacking out large areas within seconds—because the grid requires exact synchrony across subcontinental areas and relies on components taking years to build in just a few factories or one (often abroad), and can be interrupted by a lightning bolt, rifle bullet, malicious computer program, untrimmed branch, or errant squirrel. Grid vulnerabilities are serious, inherent, and not amenable to quick fixes; current federal investments in the "smart grid" do not even require simple mitigations. Indeed, the policy reflex to add more and bigger power plants and power lines after each regional blackout may make the next blackout more likely and severe, much as suppressing forest fires can accumulate fuel loadings that turn the next unsuppressed fire into an uncontrollable conflagration.

Power-system vulnerabilities are even worse in-theater, where infrastructure and the capacity to repair it are often marginal: "attacks on the grid are one of the most common and effective tactics of insurgents in Iraq, and are increasingly seen in Afghanistan."[39] Thus *electric*, not oil, vulnerabilities now hazard national and theater energy security. Simple exploitation of domestic electric vulnerabilities could take down DOD's basic operating ability and the whole economy, while oil supply is only a gathering storm.

The DSB task force took electrical threats so seriously that it advised DOD—following prior but unimplemented DOD policy[40]—to replace grid reliance, for critical missions at US bases, with onsite (preferably renewable) power supplies in netted, islandable[41] microgrids. The Department of Energy's Pacific Northwest National Laboratory found ~90 percent of those bases could actually meet those critical power needs from onsite or nearby and mainly renewable sources, and often more cheaply. This could achieve zero daily net energy need for facilities, operations, and ground vehicles; full independence in hunker-down mode (no grid); and increased ability to help serve surrounding communities and nucleate blackstart of the failed commercial grid.

Implementing these sensible policies merits high priority: probably only DOD can move as decisively as the threat to national security warrants. And as with the Endurance capability, exploiting Resilience—building on DOD's position as the world's leading direct-or-indirect buyer of renewable energy—would provide leadership, market expansion, delivery refinement, and training that would accelerate civilian adoption. Already, the 2008 NDAA requires DOD to establish a goal to make or buy at least 25 percent of its electricity from renewables by 2020, and study solar and windpower feasibility for expeditionary forces. Under 2007 Executive Order 13423's government-wide mandate, DOD must also reduce energy intensity by FY15 to 30 percent below FY03. The Resilience capability would focus all these efforts on robust architectures and implementation paths, ensuring that bases' onsite renewables deliver reliable power to critical loads whether or not the commercial grid is working—a goal not achieved by today's focus on compliance with renewables quotas.

Resilience is even more vital and valuable abroad, in fixed installations and especially in FOBs (whose expeditionary character emphasizes the Endurance logic of Fully Burdened Cost of Electricity). Foreign grids are often less reliable and secure than US grids; protection and social stability may be worse; logistics are riskier and costlier in more remote and austere sites; and civilian populations may be more helped and influenced. Field commanders strongly correlate reliable electricity supplies with political stability. In Sadr City, Army Reserve Major General Jeffrey Talley's Task Force Gold proved in 2008–2009 that *making* electricity reliable, and thus underpinning systematic infrastructure-building, is an effective cornerstone of counterinsurgency.

Reconstruction in Iraq and Afghanistan is starting to define and capture this opportunity to build civic cohesion and dampen insurgency, while reducing attacks' disruption and attractiveness. A resilient, distributed electrical architecture can bring important economic and social side-benefits, as with Afghan microhydropower programs for rural development. Cuba lately showed, too, that aggressively integrating end-use efficiency with micropower can cut national blackouts—caused by decrepit infrastructure, not attacks—by one to two orders of magnitude in a year.

At home, DOD efficiency and micropower echo new domestic energy policy and startling developments in the marketplace. In 2006, micropower[42] delivered one-sixth of the world's electricity, one-third of its new electricity, and 16 to 52 percent of all electricity in a dozen industrialized countries (the United States lagged with 7 percent). In 2008, for the first time in about a century, the world invested more in renewable than in fossil-fueled power supplies; renewables (excluding big hydroelectric dams) added 40 billion watts of global capacity and got $100 billion of private investment. Their competitive and falling costs, short lead times, and low financial risks attract private capital. Shifting to these more resilient energy solutions goes with the market's flow.

Expanding DOD's Energy Voice

Endurance and Resilience offer synergistic national security benefits far beyond those internal to the Department's mission effectiveness. As a dozen retired flag officers concluded, "We can say, with certainty, that we need not exchange benefits in one dimension for harm in another; in fact, we have found that the best approaches to energy, climate change, and national security may be one and the same."[43] Moreover, whether we care most about national security, climate change, or jobs and competitiveness, we should do exactly the same things about energy. Thus, focusing on our energy actions' attributes and outcomes, not motives, could build broad consensus. The resulting benefits could be enlarged by bringing DOD's perspective and expertise more vigorously into national energy policymaking. A common critique holds that past federal energy policy has constituted the most comprehensive threat to national energy security by:

- perpetuating America's expanding oil dependence;
- strongly favoring overcentralized energy system architectures inherently vulnerable to disruption;
- creating attractive new terrorist targets;
- aiming to increase and prolong reliance on the most vulnerable domestic infrastructure; and
- promoting technologies that encourage proliferation.

Now that national energy policy is shifting—often for additional reasons such as economic recovery, competitive advantage, and climate protection—DOD's knowledge of energy-related security risks needs to inform the councils of government more systematically.

If past national security outcomes are not what DOD wants, it is the duty of military professionals to say so. Their guidance, and increasingly their achievements, can help the Department of Defense build a stronger America and a richer, fairer, cooler, and safer world. The United States can and must make oil obsolete as a strategic commodity—just as refrigeration did to salt (once so vital a preservative that countries fought over salt mines)[44]—and electric power a boon unshadowed by threat. DOD's leadership in adopting and exploiting the two new capabilities proposed here would dramatically speed that journey toward a world beyond oil—with "negamissions" in the Persian Gulf, Mission Unnecessary—and indeed beyond all energy vulnerabilities. Fighting for Endurance and Resilience in Pentagon decisions today can eliminate the need to fight for oil on the battlefield tomorrow.

Notes

1   Center for Naval Analyses, *National Security and the Threat of Climate Change*, 2007; Gwynne Dyer, *Climate Wars* (Toronto: Random House of Canada, 2008); Thomas Fingar, unclassified summary of *National Intelligence Assessment on the National Security Implications of Global Climate Change to 2030*, 25 June 2008, testimony to the US House of Representatives.

2   Deloitte, *Energy Security: America's Best Defense* (Washington, DC: Deloitte, 9 November 2009), available at www.deloitte.com/us/aerospacedefense/ energysecurity.

3   Defense Science Board (DSB), *"More Fight—Less Fuel": Report of the Defense Science Board Task Force on DOD Energy Strategy* (Washington, DC: Department of Defense, February 2008), available at www.acq.osd.mil/dsb/ reports/ADA477619.pdf.

4   D. Berkey *et al.*, *Energy Independence Assessment: Draft Final Briefing for Office of Force Transformation* (Washington, DC: Logistics Management Institute, 12 January 2005).

5   Russell Gold and Ana Campoy, "Oil Industry Braces for Drop in U.S. Thirst for Gasoline," *The Wall Street Journal*, 13 April 2009.

6   Cambridge Energy Research Associates, "Peak Oil Demand in the Developed World: It's Here," 29 September 2009, available at www.ihs.com/products/cera/energy-report.aspx?ID=106592338.

7   Paul Sankey *et al.*, "The Peak Oil Market," *Deutsche Bank Global Markets Research*, 4 October 2009, available at www.petrocapita.com/attachments/128_ Deutsche%20Bank%20-%20The%20Peak%20Oil%20Market.pdf

8   Deloitte, op. cit., 15.

9   Army Environmental Policy Institute, "Sustain the Mission Project: Energy and Water Costing Methodology and Decision Support Tool," July 2008, available at www.aepi.army.mil/docs/whatsnew/SMP_Casualty_Cost_Factors_Final1-09.pdf.

10   Marvin Baker Schaffer and Ike Chang, "Mobile Nuclear Power for Future Land Combat," *Joint Force Quarterly* 52 (1st Quarter 2009), 51.

11   Ashton Carter, 2009 congressional testimony, quoted in Deloitte, 15.

12   Ibid.

13   Army Environmental Policy Institute, "Sustain the Mission Project: Casualty Factors for Fuel and Water Resupply Convoys," September 2009, available at www.aepi.army.mil/docs/whatsnew/SMP_Casualty_Cost_Factors_Final1-09.pdf.

14   Deloitte; total US casualty data available at http://icasualties.org.

15   Ibid., 18.

16   Ibid., 19.

17   Andrew Bochman, "Measure, Manage, Win: The Case for Operational Energy Metrics," *Joint Force Quarterly* 55 (4th Quarter 2009), 113–119.

18   Deloitte also notes that attacks are far from the only hazard: bad weather, traffic accidents, and pilferage lost DOD some 44 trucks and 220,000 gallons of fuel in June 2008 alone (15).

19   DSB, op. cit., 31.

20   DSB, *Defense Science Board 2006 Summer Study Report on 21st Century Strategic Technology Vectors*, vol. 1, Main Report (February 2007), x–xi, available at www.acq.osd.mil/dsb/reports/ADA463361.pdf.

21   DSB, "*More Fight—Less Fuel*," 25.

22   Ibid., 35.

23   DOD Instruction 3000.05, "Stability Operations," 16 September 2009, §4.1, available at www.dtic.mil/whs/directives/corres/pdf/300005p.pdf.

24   Ibid., 37.

25   DSB, "*More Fight—Less Fuel*." Innovation was encouraging on the supply side in the recent Wearable Power Prize Competition but seems to lag in efficient use.

26   Ibid., 29–30.

27   Army Environmental Policy Institute, "Sustain the Mission Project: Casualty Factors for Fuel and Water Resupply Convoys," September 2009; for genset and FOB data, G. D. Kuntz, "Renewable Energy Systems: Viable Options for Contingency Operations," *Environmental Practice* 9 (2007), 157–161.

28   Troy Wilke and Bradley Frounfelker, "Tent Foam Insulation Cost Benefit Analysis," 48th Annual Army Operations Research Symposium, Fort Lee, VA, 14–15 October 2009; personal communications from Troy Wilke and John Spiller (29 November–1 December 2009).

29   DSB, "*More Fight—Less Fuel*," 45.

30   Ibid., 3.

31   DSB, *Defense Science Board 2006 Summer Study Report*, xviii.

32   Amory B. Lovins and James Newcomb, "Bioconversion: What's the Right Size?" 20 February 2008, brief to National Research Council Panel on Alternative Liquid Transportation Fuels.

33   JASON, The MITRE Corporation, *Reducing DoD Fossil-Fuel Dependence*, JSR–06–135 (McLean, VA: MITRE, September 2006), available at www.fas.org/irp/agency/dod/jason/fossil.pdf.

34   Schaffer and Chang, op. cit.

35   *Manual for the Operation of the Joint Capabilities Integration and Development System*, 31 July 2009, available at https://acc.dau.mil/adl/en-US/267116/file/ 41245/JCIDS%20Manual%20-%2031%20July2009.pdf.

36   DSB, "*More Fight—Less Fuel*," 28–29.

37   As of FY97, Defense Science Board Summer 1998 Study Task Force, *DOD Logistics Transformation*, Annotated Briefing Slides, slide 7, which also shows that "Active duty combat forces [were then] half [the] size of active logistics forces." One estimate of DOD's FY09 logistics and sustainment cost is $270 billion—over half the base budget (35).

38   Ibid., 3 and 53.

39   DSB, "*More Fight—Less Fuel*," 55.

40   Ibid., 59–60; DOD Instruction 1470.11 §5.2.3.

41   *Islandable* describes onsite supplies that can continuously serve the base and neighboring communities whether or not the commercial grid is operating.

42   Defined here as cogeneration plus renewables minus big (>10 megawatt electrical) hydro. RMI maintains a global database [www.rmi.org/rmi/Library/2010-06_MicropowerDatabase].

43   Center for Naval Analyses Military Advisory Board, *Powering America's Defense: Energy and the Risks to National Security*, May 2009, available at www.cna.org/sites/default/files/Powering%20Americas%20Defense.pdf.

44   R. James Woolsey and Anne Korin, "Turning Oil into Salt," *National Review*, 25 September 2007. (An unabridged version of this article, slightly expanding the text, providing more references, and restoring deleted text and graphics about technological options for platform efficiency, is at www.rmi.org/rmi/Library/2010-05_DODsEnergy Challenge.)

Chapter 26

On Proliferation, Climate, and Oil: Solving for Pattern

2010

Editor's note: *This article was published in* Foreign Policy *online, 21 January 2010, and is reprinted with permission. A data-rich unabridged version is at www.rmi.org/rmi/Library/2010-02_ProliferationOilClimate Pattern.*

The problems of proliferation, climate change, and oil dependence share both a nuclear non-solution that confounds US policy goals and a non-nuclear solution that achieves them.

The first four months of 2010 offer a unique opportunity to align the United States' foreign-policy goals with domestic energy policy and new market developments, and thereby to stem what the Pentagon's Nuclear Posture Review will reportedly rank equally with great-power threats—the spread of nuclear weapons.

Epistemologist Gregory Bateson and farmer-poet Wendell Berry counseled "solving for pattern"—harnessing hidden commonalities to resolve complex challenges without making more. President Obama's speech at the recent UN climate summit in Copenhagen hinted at such an approach by linking an efficient, clean-energy, climate-safe economy with three other key issues: prosperity, oil displacement, and national security. Keeping proliferation, climate, and oil in separate policy boxes has in the past stalled progress on the first two issues over North/South splits that the third issue intensifies. Yet these three problems share profitable solutions, and seem tough only because of a wrong economic assumption.

One false assumption can distort and defeat policies vital to paramount national interests. The Copenhagen climate conference proved again how pricing carbon and winning international collaboration are hard if policymakers assume climate protection is costly, focusing debate on cost, burden, and sacrifice.

That assumption is backwards: business experience proves climate protection is *not costly but profitable*, because saving fuel costs less than buying fuel. Changing the conversation to profits, jobs, and competitive advantage sweetens the politics, melting resistance faster than glaciers. Whether you care most about security, prosperity, or environment, and whatever you think about climate science, you'll favor exactly the same energy choices: focusing on outcomes, not motives, can forge broad consensus.

For instance, a January 2009 study by McKinsey & Company demonstrated how it was possible to cut projected 2030 global greenhouse-gas emissions by 70 percent at a trivial average cost: $6 per metric ton of CO_2. Newer technologies and integrative design, which often makes very large energy savings cost *less* than small or no savings, turning diminishing into expanding returns, could make even bigger abatements cost less than zero dollars.

This can be done fast enough. Consider that from 1977 through 1985, US oil intensity (barrels per real GDP dollar) fell 5.2 percent per year. Today, cutting global energy intensity at an annual rate of about 3–4 percent, vs. the historic 1 percent, could abate further climate damage. The United States has long achieved 2–4 percent cuts each year without paying attention; China achieved more than 5 percent reductions from 1976 through 2001 and is on track for 4 percent reductions from 2005 through 2010. Individual firms have been able to achieve 6–16 percent reductions. So why should 3–4 percent be hard, especially with most of the global economic growth in China and India, where making new infrastructure efficient is easier than fixing it later?

Since energy efficiency consistently makes money (billions for many firms), why should this be costly? And why should climate negotiators adopt economists' assumptions about cost rather than business leaders' experiences of profit? The climate conversation gets vastly easier and less necessary when it's shifted from shared sacrifice to informed self-interest.

Many policymakers likewise assume US oil dependence and imports must be permanent. Yet a 2004 Pentagon-cosponsored independent study showed how it was possible to *eliminate* US oil use by the 2040s at an average cost of about $15 per barrel, led by business for profit. Implementation was launched in 2005 by "institutional acupuncture," then spurred by the 2008 price shock, 2009 policy shifts, and military innovation.

That effort now looks to be on or ahead of schedule: in 2009, "peak oil" emerged, but on the *demand* side. US gasoline demand reached its apex in 2007. Cambridge Energy Research Associates doubts OECD oil demand will regain its 2005 peak. Deutsche Bank forecasts light-vehicle electrification (at one-third China's planned rate, and without counting the other revolutionary innovations under way) will turn *world* oil demand downward from 2016—reaching, by 2030, 8 percent below 2009. Suburban sprawl is reversing. Just in 2008, government-mandated "feebates" cut inefficient cars' sales in France 42 percent and raised efficient cars' sales 50 percent. Thus oil is becoming uncompetitive even at low prices before it becomes unavailable even at high prices.

Yet with oil as with climate, official assessments ignore these solutions as too detailed, disruptive, novel, or integrative to contemplate. When offered cramped old choices, policymakers all too often perpetuate largely incremental policies. Private firms are more likely to innovate, while governments play catch-up. And intergovernmental negotiations learn slowest of all.

Similar adherence to outmoded orthodoxies now cripples nonproliferation. Policy still rests on the fatally contradictory assumption that nuclear *power* is economical, necessary, and experiencing a revival. This makes the proliferation problem insoluble. Fortunately, that assumption is counterfactual—and correcting it can make the proliferation problem largely soluble. Here's how.

In a 1980 *Foreign Affairs* article, I first set out with two coauthors an economically based, logically consistent approach to nonproliferation. Eerily presaging today's conditions, the article said:

> For fundamental reasons ... nuclear power is not commercially viable, and questions of how to regulate an inexorably expanding world nuclear regime are moot. ...
>
> The collapse of nuclear power in response to the discipline of the marketplace is to be welcomed, for nuclear power is both the main driving force behind proliferation and the least effective known way to displace oil: indeed, it retards oil displacement by the faster, cheaper and more attractive means which new developments in energy policy now make available to all countries. So far, nonproliferation policy has gotten the wrong answer by persistently asking the wrong questions, creating "a nuclear armed crowd" by assuming its inevitability. We shall argue instead that acknowledging and taking advantage of the nuclear collapse, as part of a pragmatic alternative program, can offer an internally consistent approach to nonproliferation, as well as a resolution to the bitter dispute over Article IV of the Non-Proliferation Treaty (NPT).
>
> On the eve of the second NPT Review Conference, to be held in Geneva in August 1980, fatalism is becoming fashionable as the headlines show proliferation slipping rapidly out of control. Yet ... an effective nonproliferation policy, though impossible with continued commitments to nuclear power, may become possible without it—if only we ask the right questions.

Thirty years later, as the eighth NPT Review Conference prepares to convene in Vienna on 30 April 2010, just one word needs updating: now that *oil* generates less than 6 percent of the world's electricity, today's nuclear expansion is meant instead to displace *coal* to protect climate.

That rationale is identically unsound. In principle, quadrupling today's global nuclear power capacity—to replace, then triple, retiring units—could provide up to one-tenth of needed carbon reductions. But nuclear power is the least effective method: using it does save carbon, but about 2–20 times less per dollar and 20–40 times less per year than buying its winning competitors (mentioned below). Nuclear expansion would thus reduce and retard climate protection. We must invest judiciously, not indiscriminately, to get the most climate solution per dollar and per year. Expanding nuclear power does the opposite.

The 1980 article's logic remains sound:

- We can have proliferation with nuclear power, via either end of any fuel cycle: "*every* form of every fissionable material in *every* nuclear fuel cycle can be used to make military bombs, either on its own or in combination with other ingredients made widely available by nuclear power."
- We can't have nuclear power without proliferation, because its vast flows of materials, equipment, skills, knowledge, and skilled people create do-it-yourself bomb kits wrapped in innocent-looking civilian disguise. Safeguards to prevent that misuse "cannot succeed either in principle or in practice," because national rivalries, subnational instabilities, and human frailties trump treaties and policing.
- We can have proliferation without nuclear power—but needn't if we do it right: with unimportant exceptions, "*every* known civilian route to bombs involves *either* nuclear power *or* materials and technologies whose possession, indeed whose existence in commerce, is a direct and essential consequence of nuclear fission power."
- Crucially, in a *world* without nuclear power, the ingredients needed to make bombs by any known method would no longer be ordinary items of commerce. They'd become harder to get, more conspicuous to try to get, and politically costlier to be caught trying to get (or supply), because their purpose would be *unambiguously military*. This disambiguation would make proliferation not impossible but far harder—and easier to detect timely, because intelligence resources could focus on needles, not haystacks. Thus phasing out nuclear power is a necessary and nearly sufficient condition for nonproliferation.

The American Academy of Arts and Sciences' 2009 nuclear study, confident of nuclear power's necessity and viability, ignored its decades-long collapse in market economies due to unsupportable economic costs and financial risks. That study simply overlooked the data: shrinking global nuclear output, less than 5 percent nuclear share of capacity under construction, retirements outpacing additions for decades to come, every plant under construction bought by central planners (none by conventional free-market transactions), and zero equity investment despite extremely generous new subsidies in the United States, roughly equivalent to or greater than construction cost.

The fact is, nuclear investment has no business case: With or without a price on carbon, nuclear power and big fossil-fueled power plants simply cost far more than "micropower" generation (renewables except big hydropower, plus cogenerating electricity with useful heat) or saving electricity through efficient use. Micropower has surpassed nuclear output since 2006, when it produced one-sixth of global electricity, one-third of new electricity, and 16–52 percent of all electricity in a dozen industrial countries.

In 2007 alone, the United States added more megawatts of windpower than it added in coal generation from 2003 through 2007, or than the world added nuclear power in 2007. And in 2008, renewables attracted more global investment than fossil-fueled generation; distributed renewables added 40 billion watts and got $100 billion of private investment while nuclear added and got zero. In each year since 2005, nuclear power has added only a few percent as much output as micropower, and since 2008, less than photovoltaics.

No policy can change this: even France's uniquely dirigiste 1970–2000 nuclear program suffered 3.5-fold capital escalation, nearly doubled construction time, and acute strains. "New" reactor types aren't materially different, though they often pose more proliferation danger. Even more today than when I wrote in 1980, nuclear power's "risks, including proliferation, are ... not a minor counterweight to enormous advantages but rather a gratuitous supplement to enormous disadvantages."

Today, these market realities present a brief opportunity to align US nonproliferation policy with the Obama Administration's emphasis on efficient energy use and renewable, distributed sources.

If a country with America's wealth, infrastructure, skills, and fuels claims it needs more nuclear power, all countries gain a strong excuse to follow suit. But US acknowledgment of the market verdict favoring non-nuclear alternatives would encourage less richly endowed countries to seek profit and prestige from similar modernity. Aligning America's energy words, deeds, and offers would transform her journey beyond fossil fuels from a seeming plot to choke global development into routine, rational, replicable pursuit of least cost, green jobs, and industrial renewal.

Nobody need be antinuclear. The issue, just as I framed it in 1980, "is not whether to maintain a thriving [nuclear] enterprise, but rather whether to accept the verdict of the very calculations on which free-market economies rely." Making nuclear power compete on a level playing field, after 56 years of enormous subsidies, would be a good start. De-subsidizing all energy across the board would be an even sounder approach.

Since Washington proposes nuclear fuel security initiatives, why not broader energy security initiatives? What if the Obama Administration announced it would help spread the best buys it's adopting—efficiency, renewables, distributed energy systems—to all desirous developing countries, unconditionally and nondiscriminatorily? Most such countries are renewable-rich, but infrastructure-poor. They could welcome "Sunbeams for Peace" for the same hard-nosed reasons that made China the world leader in five renewable technologies, with energy efficiency its top strategic priority—not forced by treaty, but informed by Premier Wen Jiabao's and his fellow-leaders' understanding that otherwise Beijing can't afford to develop. Perhaps the United States, which invented many of these technologies, could even try to reclaim part of the burgeoning market it abandoned to China, Japan, and Europe.

Attendees at the upcoming NPT Review Conference are expected to clash on implementation of two main points in the original treaty: weapons states' underfulfilled obligation under Article VI to pursue nuclear disarmament, and developing-country signatories' right under Article IV to access nuclear technology for exclusively peaceful purposes.

Progress in and beyond the new round of Strategic Arms Reduction Treaty talks between the United States and Russia should help on Article VI; policy shifts building on Obama's Nobel Peace Prize speech can help too. But progress on Article IV depends on recognizing one simple yet unnoticed fact. When the NPT was drafted in 1958–1968, nuclear power was widely expected to be cheap, easy, abundant, and indispensable.

Non-weapons states' reward for forgoing nuclear weapons was therefore *framed* as access to nuclear power but only, as I explained in 1980:

> because of the nuclear context and background of the negotia-tors, not as an expression of the essential purpose of Article IV. … The time is therefore ripe to reformulate the bargain in the light of new knowledge. Instead of denying or hedging their obligations, the exporting nations should fulfill it—in a wider sense based on a pragmatic reassessment of what recipients say their real interests are.

Having abjured bombs, recipients want reliable and affordable energy for development. The past half-century has revealed manifestly cheaper, faster, surer, more flexible methods than nuclear power, so now, just as I put it in 1980, "recipients should insist on aid in meeting their declared central need: not nuclear power *per se* but rather *oil* [and now coal] *displacement and energy security.*"

Reinterpreting Article IV in light of a half century of energy experience can isolate legitimate from illegitimate motives and help smoke out proliferators, advancing the treaty's central goal. Let countries that still want specifically *nuclear* energy, rather than cheaper and more suitable options, explain why.

Now let's solve for pattern. The help developing countries expect under NPT Article IV is exactly the same help they sought in Copenhagen to get off fossil fuels, and the same help many also want to escape oil dependence. President Obama's Copenhagen pledge of climate mitigation aid must now echo in Vienna's NPT context. That linkage would attain many big policy goals for the price of one, and remove the contradiction undermining the NPT.

Launching this new energy conversation in Vienna is America's best opportunity to inhibit the spread of nuclear bombs and start breaking the Copenhagen political logjam on climate justice.

At home, proposals to expand nuclear subsidies—whether to buy Senate climate-bill votes, or motivated by a sincere but mistaken belief that nuclear expansion will help protect climate—will amount to lose–lose scenarios; that approach will only prop up a failed climate non-solution that also makes proliferation unstoppable and weakens American values of free markets and a free society.

Yet applying internationally the sound *non*-nuclear elements of current domestic energy policy could profitably and simultaneously help solve the proliferation, climate, and oil problems. It would reinforce global development, transparency, democracy, women's advancement, energy resilience, and economic and political stability. It makes sense. It makes money. It would expose and discomfit only those who lack competitive offerings or harbor ulterior motives.

The surest path to a richer, fairer, cooler, safer world—where energy insecurity, oil, climate change, most proliferation, and many development problems fade away—would be a US energy policy that takes economics seriously. It would let all ways to save or produce energy compete fairly, at honest prices, regardless of their type, technology, location, size, or ownership. Who's not in favor of that? Why don't we find out? And why can't such a least-cost domestic energy strategy inform, integrate, and inspire foreign policy too?

Section 7

Business and Climate: Making Sense, Making Cash, Making Good

The way a business operates, the way we account for things we extract, shape, manufacture, package, and sell (and then usually throw away), typically seems to ignore the planet, from whence all value actually comes. Somehow, as humans have created economic and industrial models and systems, they've failed to factor in the Earth, without which there are no people and no economic transactions. In 1999, this material omission prompted Amory to team up with entrepreneur and green guru Paul Hawken and his regular writing partner Hunter Lovins to compose the influential business book *Natural Capitalism: Creating the Next Industrial Revolution*. It shows how business can profitably adopt four major principles (radically increasing the productivity of natural resources; shifting to biologically inspired production models and materials; adopting a "solutions economy" business model that rewards those shifts; and reinvesting in natural capital)—which seem so common-sense today that we'd be hard-pressed to argue why we wouldn't. Yet not all companies do. The CliffsNotes version, "A Roadmap for Natural Capitalism," included here, appeared in *Harvard Business Review* in May–June 1999 and has become among its most-reprinted articles. Like *Natural Capitalism*, it has also been widely misinterpreted as depending on internalizing hidden (external) costs. On the contrary, since there's little agreement on what those costs are worth, and no time to reach agreement, both the book and the article describe how to do business *as if* nature and people were properly valued—thereby achieving the same benefits internalization would.

Natural Capitalism and its HBR précis are something of a next step up from Amory and Hunter's 1997 white paper *Climate: Making Sense* and *Making Money*. Written for business leaders at the Kyoto Climate Conference, at first glance it appears more esoteric, less going for the jugular, than *Natural Capitalism*. Yet it's not. The climate is simply part of our home, and the argument for protecting it (besides the obvious) follows economic lines. As they noted in the introduction:

> *Arguments that protecting the Earth's climate will cost a lot rest on theoretical economic assumptions flatly contradicted by business experience. Most climate/economics models assume that almost all energy-efficiency investments cost-effective at present prices have already been made. Actually, huge opportunities to save money by saving energy exist, but are being blocked by dozens of specific obstacles at the level of the firm, locality, or society. Even if climate change were not a concern, it would be worth clearing these barriers in order to capture energy-efficiency investments with rates of return that often approach and can even exceed 100 percent per year. Focusing private and public policy on barrier-busting can permit businesses to buy energy savings that are large enough to protect the climate, intelligent enough to improve living standards, and profitable enough to strengthen economic vitality, employment, and competitiveness.*

At 44 pages, the original *Climate: Making Sense* and *Making Money* is excerpted here, where you'll find a "prospector's guide" to the solutions. For the original piece, plus extensive references, see www.rmi.org/rmi/Library/C97-13_ClimateSenseMoney.

Chapter 27

A Roadmap for Natural Capitalism

1999

Editor's note: *This essay is reprinted with permission from* Harvard Business Review, *May–June 1999.*

On 16 September 1991, a small group of scientists was sealed inside Biosphere II, a glittering 3.2-acre glass and metal dome in Oracle, Arizona. Two years earlier, when a radical attempt to replicate the Earth's main ecosystems in miniature ended, the engineered environment was dying. The gaunt researchers had survived only because fresh air had been pumped in. Despite $200 million worth of elaborate equipment, Biosphere II had failed to generate breathable air, drinkable water, and adequate food for just eight people. Yet, Biosphere I, the planet we all inhabit, effortlessly performs those tasks for 6 billion of us every day.

Disturbingly, Biosphere I is now itself at risk. The Earth's ability to sustain life, and therefore economic activity, is threatened by the way we extract, process, transport, and dispose of a vast flow of resources—some 220 billion tons a year, or more than 20 times the body weight of the entire US population every day. With dangerously narrow focus, our industries look only at the exploitable resources of the Earth's ecosystems—its oceans, forests, and plains—and not at the larger services that those systems provide for free. Resources and ecosystem services both come from the Earth—even from the same biological systems—but they're two different things. Forests, for instance, not only produce the resource of wood fiber but also provide such ecosystem services as water storage, habitat, and regulation of the atmosphere and climate. Yet companies that earn income from harvesting the wood fiber resource often do so in ways that damage the forest's ability to carry out its other vital tasks.

Unfortunately, the cost of destroying ecosystem services becomes apparent only when the services start to break down. In China's Yangtze basin in 1998, for example, deforestation triggered flooding that killed 3700 people, dislocated 223 million, and inundated 60 million acres of cropland. That $30 billion disaster forced a logging moratorium and a $12 billion crash program of reforestation.

The reason companies (and governments) are so prodigal with ecosystem services is that the value of those services doesn't appear on the business balance sheet. But that's a staggering omission. The economy, after all, is embedded in the environment. Recent calculations published in the journal *Nature* conservatively estimate the value of all the Earth's ecosystem services to be at least $33 trillion a year. That's close to the gross world product, and it implies a capitalized book value on the order of half a quadrillion dollars. What's more, for most of these services, there is no known substitute at any price, and we can't live without them.

This article puts forward a new approach not only for protecting the biosphere but also for improving profits and competitiveness. Some very simple changes to the way we run our businesses, built on advanced techniques for making resources more productive, can yield startling benefits for today's shareholders and for future generations.

This approach is called *natural capitalism* because it's what capitalism might become if its largest category of capital—the "natural capital" of ecosystem services—were properly valued. The journey to natural capitalism involves four major shifts in business practices, all vitally interlinked:

1   **Dramatically increase the productivity of natural resources.** Reducing the wasteful and destructive flow of resources from depletion to pollution represents a major business opportunity. Through fundamental changes in both production design and technology, farsighted companies are developing ways to make natural resources—energy, minerals, water, forests—stretch 5, 10, even 100 times further than they do today. These major resource savings often yield higher profits than small resource savings do—or even saving no resources at all would—and not only pay for themselves over time but in many cases reduce initial capital investments.

2   **Shift to biologically inspired production models.** Natural capitalism seeks not merely to reduce waste but to eliminate the very concept of waste. In closed-loop production systems, modeled on nature's designs, every output is either returned harmlessly to the ecosystem as a nutrient, like compost, or becomes an input for manufacturing another product. Such systems can often be designed to eliminate the use of toxic materials, which can hamper nature's ability to reprocess materials.

3   **Move to a solutions-based business model.** The business model of traditional manufacturing rests on the sale of goods. In the new model, value is instead delivered as a flow of services—providing illumination, for example, rather than selling light bulbs. This model entails an entirely new perception of value, a move from the acquisition of goods as a measure of affluence to one where well-being is measured by the continuous satisfaction of changing expectations for quality, utility, and performance. The

new relationship aligns the interests of providers and customers in ways that reward them for implementing the first two innovations of natural capitalism—resource productivity and closed-loop manufacturing.

4 **Reinvest in natural capital.** Ultimately business must restore, sustain, and expand the planet's ecosystems so that they can produce their vital services and biological resources even more abundantly. Pressures to do so are mounting as human needs expand, the costs engendered by deteriorating ecosystems rise, and the environmental awareness of consumers increases. Fortunately, these pressures all create business value.

Natural capitalism is not motivated by a current scarcity of natural resources. Indeed, although many biological resources, like fish, are becoming scarce, most mined resources, such as copper and oil, seem ever more abundant. Indices of average commodity prices are at 28-year lows, thanks partly to powerful extractive technologies, which are often subsidized and whose damage to natural capital remains unaccounted for. Yet, even despite these artificially low prices, using resources manyfold more productively can now be so profitable that pioneering companies—large and small—have already embarked on the journey toward natural capitalism.[1]

Still the question arises—if large resource savings are available and profitable, why haven't they all been captured already? The answer is simple: scores of common practices in both the public and private sectors systematically reward companies for wasting natural resources and penalize them for boosting resource productivity. For example, most companies expense their consumption of raw materials through the income statement but pass resource-saving investment through the balance sheet. That distortion makes it more tax-efficient to waste fuel than to invest in improving fuel efficiency. In short, even though the road seems clear, the compass that companies use to direct their journey is broken. Later, we'll look in more detail at some of the obstacles to resource productivity—and some of the important business opportunities they reveal. But first, let's map the route toward natural capitalism.

Dramatically Increase the Productivity of Natural Resources

In the first stage of a company's journey toward natural capitalism, it strives to wring out the waste of energy, water, minerals, and other resources throughout its production system and other operations. There are two main ways companies can do this at a profit. First, they can adopt a fresh approach to design that considers industrial systems as a whole rather part by part. Second, companies can replace old industrial technologies with new ones, particularly with those based on natural processes and materials.

Implementing whole-system design

Inventor Edwin Land once remarked that "people who seem to have a new idea have often simply stopped having an old idea." This is particularly true when designing for resource savings. The old idea is one of diminishing returns—the greater the resource saving, the higher the cost. But that old idea is giving way to the new idea that bigger savings can cost less—that saving a large fraction of resources can actually cost less than saving a small fraction of resources. This is the concept of expanding returns, and it governs much of the revolutionary thinking behind whole-system design. Lean manufacturing is an example of whole-system thinking that has helped many companies dramatically reduce such forms of waste as lead times, defect rates, and inventory. Applying whole-system thinking to the productivity of natural resources can achieve even more.

Consider Interface Corporation, a leading maker of materials for commercial interiors. In its new Shanghai carpet factory, a liquid had to be circulated through a standard pumping loop similar to those used in nearly all industries. A top European company designed the system to use pumps requiring a total of 95 horsepower. But before construction began, Interface's engineer, Jan Schilham, realized that two embarrassingly simple design changes would cut that power requirement to only 7 horsepower—a 92 percent reduction. His redesigned system cost less to build, involved no new technology, and worked better in all respects.

What two design changes achieved this 12-fold saving in pumping power? First, Schilham chose fatter-than-usual pipes, which create much less friction than thin pipes do and therefore need far less pumping energy. The original designer had chosen thin pipes because, according to the textbook method, the extra cost of fatter ones wouldn't be justified by the pumping energy that they would save. This standard design tradeoff optimizes the pipes themselves but "pessimizes" the larger system. Schilham optimized the *whole* system by counting not only the higher capital cost of the fatter pipes but also the *lower* capital cost of the smaller pumping equipment that would be needed. The pumps, motors, motor controls, and electrical components could all be much smaller because there'd be less friction to overcome. Capital cost would fall far more for the smaller equipment than it would rise for the fatter pipe. Choosing big pipes and small pumps—rather than small pipes and big pumps—would therefore make the whole system cost less to build, even before counting its future energy savings.

Schilham's second innovation was to reduce the friction even more by making the pipes short and straight rather than long and crooked. He did this by laying the pipes out first, then positioning the various tanks, boilers, and other equipment that they connected. Designers normally locate the production equipment in arbitrary positions and then have a pipefitter connect everything. Awkward placement forces the pipes to make numerous bends that greatly increase friction. The pipe fitters don't mind: they're paid by the hour, they profit from the extra pipes and fittings, and they don't pay for oversized pumps or inflated electric bills. In addition to reducing those four kinds of costs, Schilham's short, straight pipes were easier to insulate, saving an extra 70 kilowatts of heat loss and repaying the installation's cost in three months.

This small example has big implications for two reasons. First, pumping is the largest application of motors, and motors use three-quarters of all industrial electricity. Second, the lessons are very widely relevant. Interface's pumping loop shows how simple changes in design mentality can yield huge resource savings and returns on investment. This isn't rocket science; often it's just a rediscovery of Victorian engineering principles that have been lost because of specialization.

Whole-system thinking can help managers find small changes that lead to big savings that are cheap, free, or even better than free (because they make the whole system cheaper to build). They can do this because often the right investment in one part of the system can produce multiple benefits throughout the system. For example, companies would gain 18 distinct economic benefits—of which direct energy savings are only one—if they switched from ordinary motors to premium-efficiency motors or from ordinary lighting ballasts (the transformer-like boxes that control fluorescent lamps) to electronic ballasts that automatically dim the lamps to match available daylight. If everyone in America integrated these and other selected technologies into all existing motor and lighting systems in an optimal way, the nation's $220-billion-a-year electric bill would be cut in half. The after-tax return on investing in these changes would in most cases exceed 100 percent per year.

The profits from saving electricity could be increased even further if companies also incorporated the best off-the-shelf improvements into their building structure and their office, heating, cooling, and other equipment. Overall, such changes could cut national electricity consumption by at least 75% and produce returns of around 100 percent a year on the investments made. More important, because workers would be more comfortable, better able to see, and less fatigued by noise, their productivity and the quality of their work output would rise. Eight recent case studies of people working in well-designed, energy-efficient buildings measured labor productivity gains of 6 percent to 16 percent. Since a typical office pays about 100 times as much for people as it does for energy, this increased productivity in people is worth about 6 to 16 times as much as eliminating the entire energy bill.

Energy-saving, productivity-enhancing improvements can often be achieved at even lower cost by piggybacking them onto the periodic renovations that all buildings and factories need. A recent proposal for reallocating the normal 20-year renovation budget for a standard 200,000-square-foot glass-clad office tower near Chicago, Illinois, shows the potential for whole-system design. The proposal suggested replacing the aging glazing system with a new kind of window that lets in nearly six times more daylight than the old sun-blocking glass units. The new windows would reduce the flow of heat and noise four times better than traditional windows do. So even though the glass costs slightly more, the overall cost of the renovation would be reduced because the windows would let in cool, glare-free daylight that, when combined with more efficient lighting and office equipment, would reduce the need for air-conditioning by 75 percent. Installing a fourfold more efficient, but fourfold smaller, air-conditioning system would cost $200,000 less than giving the old system its normal 20-year renovation. The $200,000 saved would, in turn, pay for the extra cost of the new windows and other improvements. This whole-system approach to renovation would not only save 75 percent of the building's total energy use, it would also greatly improve the building's comfort and marketability. Yet it would cost essentially the same as the normal renovation. There are about 100,000 twenty-year-old glass office towers in the United States that are ripe for such improvement.

Major gains in resource productivity require that the right steps be taken in the right order. Small changes made at the downstream end of a process often create far larger savings further upstream. In almost any industry that uses a pumping system, for example, saving one unit of liquid flow or friction in an exit pipe saves about ten units of fuel, cost, and pollution at the power station.

Of course, the original reduction in flow itself can bring direct benefits, which are often the reason the changes are made in the first place. In the 1980s, while California's industry grew 30 percent, for example, its water use was cut by 30 percent, largely to avoid increased wastewater fees. But the resulting reduction in pumping energy (and the roughly tenfold larger saving on power-plant fuel and pollution) delivered bonus savings that were at the time largely unanticipated.

To see how downstream cuts in resource consumption can create huge savings upstream, consider how reducing the use of wood fiber disproportionately reduces the pressure to cut down forests. In round numbers, half of all harvested wood fiber is used for such structural products as lumber; the other half is used for paper and cardboard. In both cases, the biggest leverage comes from reducing the amount of the retail product used. If it takes, for example, three pounds of harvested trees to produce one pound of product, then saving one pound of product will save three pounds of trees—plus all the environmental damage avoided by not having to cut them down in the first place.

The easiest savings come from not using paper that's unwanted or unneeded. In an experiment at its Swiss headquarters, for example, Dow Europe cut flow by about 30 percent in six weeks simply by discouraging unneeded information. For instance, mailing lists were eliminated and senders of memos got back receipts indicating whether each recipient had wanted the information. Taking those and other small steps, Dow was also able to increase labor productivity by a similar proportion because people could focus on what they really needed to read. Similarly, Danish hearing-aid maker Oticon saved upwards of 30 percent of its paper as a by-product of redesigning its business processes to produce better decisions faster. Setting the default on office printers and copiers to double-sided mode reduced AT&T's paper costs by about 15 percent. Recently developed copiers and printers can even strip off toner and old printer ink, permitting each sheet to be reused about ten times.

Further savings can come from using thinner but stronger and more opaque paper, and from designing packaging more thoughtfully. In a 30-month effort at reducing such waste, Johnson & Johnson saved 2750 tons of packaging, 1600 tons of paper, $2.8 million, and at least 300 acres of forest annually. The downstream savings in paper use are multiplied by the savings further upstream, as less need for paper products (or less need for fiber to make each product) translates into less raw paper, less raw paper means less pulp, and less pulp requires fewer trees to be harvested from the forest. Recycling paper and substituting alternative fibers such as wheat straw will save even more.

Comparable savings can be achieved for the wood fiber used in structural products. Pacific Gas and Electric, for example, sponsored an innovative design developed by Davis Energy Group that used engineered wood products to reduce the amount of wood needed in a stud wall for a typical tract house by more than 70 percent. These walls were stronger, cheaper, more stable, and insulated twice as well. Using them enabled the designers to eliminate heating and cooling equipment in a climate where temperatures range from freezing to 113°F. Eliminating the equipment made the whole house much less expensive both to build and to run while still maintaining high levels of comfort. Taken together, these and many other savings in the paper and construction industries could make our use of wood fiber so much more productive that, in principle, the entire world's present wood fiber needs could probably be met by an intensive tree farm about the size of Iowa.

Adopting innovative technologies

Implementing whole-system design goes hand in hand with introducing alternative, environmentally friendly technologies. Many of these are already available and profitable but not widely known. Some, like the "designer catalysts" that are transforming the chemical industry, are already runaway successes. Others are still making their way to market, delayed by cultural rather than economic or technical barriers.

The automobile industry is particularly ripe for technological change. After a century of development, motorcar technology is showing signs of age. Only 1 percent of the energy consumed by today's cars is actually used to move the driver: only 15 percent to 20 percent of the power generated by burning gasoline reaches the wheels (the rest is lost in the engine and drivetrain) and 95 percent of the resulting propulsion moves the car, not the driver. The industry's infrastructure is hugely expensive and inefficient. Its convergent products compete for narrow niches in saturated core markets at commodity-like prices. Automaking is capital intensive, and product cycles are long. It is profitable in good years but subject to large losses in bad years. Like the typewriter industry just before the advent of personal computers, it is vulnerable to displacement by something completely different.

Enter the Hypercar®. Since 1993, when Rocky Mountain Institute placed this automotive concept in the public domain, several dozen current and potential auto manufacturers have committed billions of dollars to its development and commercialization. The Hypercar integrates the best existing technologies to reduce the consumption of fuel as much as 85 percent and the amount of materials used up to 90 percent by introducing four main innovations.

First, making the vehicle out of advanced polymer composites, chiefly carbon fiber, reduces its weight by two-thirds while maintaining crashworthiness. Second, aerodynamic design and better tires reduce air resistance by as much as 70 percent and rolling resistance by up to 80 percent. Together, these innovations save about two-thirds of the fuel. Third, 30 percent to 50 percent of the remaining fuel is saved by using a "hybrid-electric" drive. In such a system, the wheels are turned by electric motors whose power is made onboard by a small engine or turbine, or even more efficiently by a fuel cell. The fuel cell generates electricity directly by chemically combining stored hydrogen with oxygen, producing pure hot water as its only by-product. Interactions between the small, clean, efficient power source and the ultralight, low-drag auto body then further reduce the weight, cost, and complexity of both. Fourth, much of the traditional hardware—from transmissions and differentials to gauges and certain parts of the suspension—can be replaced by electronics controlled with highly integrated, customizable, and upgradeable software.

These technologies make it feasible to manufacture pollution-free, high-performance cars, sport utilities, pickup trucks, and vans that get 80 to 200 miles per gallon (or its energy equivalent in other fuels). These improvements will not require any compromise in quality or utility. Fuel savings will not come from making the vehicles small, sluggish, unsafe, or unaffordable, nor will they depend on government fuel taxes, mandates, or subsidies. Rather, Hypercars will succeed for the same reason that people buy compact discs instead of phonograph records: the CD is a superior product that redefines market expectations. From the manufacturers' perspective, Hypercars will cut cycle times, capital needs, body part counts, and assembly effort and space by as much as tenfold. Early adopters will have a huge competitive advantage—which is why dozens of corporations, including most automakers, are now racing to bring Hypercar-like products to market.[2]

In the long term, the Hypercar will transform industries other than automobiles. It will displace about an eighth of the steel market directly and most of the rest eventually, as carbon fiber becomes far cheaper. Hypercars and their cousins could ultimately save as much oil as OPEC now sells. Indeed, oil may well become uncompetitive as a fuel long before it becomes scarce and costly. Similar challenges face the coal and electricity industries because the development of the Hypercar is likely to accelerate greatly the commercialization of inexpensive fuel cells. These fuel cells will help shift power production from centralized coal-fired and nuclear power stations to networks of decentralized, small-scale generators. In fact, fuel-cell-powered Hypercars could themselves be part of these networks. They'd be, in effect, 20-kilowatt power plants on wheels. Given that cars are left parked—that is, unused—more than 95 percent of the time, these Hypercars could be plugged into a grid and could then sell back enough electricity to repay as much as half the predicted cost of leasing them. A national Hypercar fleet could ultimately have five to ten times the generating capacity of the national electric grid.

As radical as it sounds, the Hypercar is not an isolated case. Similar ideas are emerging in such industries as chemicals, semiconductors, general manufacturing, transportation, water and wastewater treatment, forestry, energy, real estate, and urban design. For example, the amount of carbon dioxide released for each microchip manufactured can be reduced almost 100-fold through improvements that are now profitable or soon will be.

Some of the most striking developments come from emulating nature's techniques. In her book *Biomimicry*, Janine Benyus points out that spiders convert digested crickets into silk that's as strong as Kevlar without the need for boiling sulfuric acid and high-temperature extruders. Using no furnaces, abalone can convert seawater into an inner shell twice as tough as our best ceramics. Trees turn sunlight, water, soil, and air into cellulose, a sugar stronger than nylon but one-fourth as dense. They then bind it into wood, a natural composite with a higher bending strength than concrete, aluminum alloy or steel. We may never become as skillful as spiders, abalone, or trees, but smart designers are already realizing that nature's environmentally benign chemistry offers attractive alternatives to industrial brute force.

Whether through better design or through new technologies, reducing waste represents a vast business opportunity. The US economy is not even 10 percent as energy efficient as the laws of physics allow. Just the energy thrown off as waste heat by US power stations equals the total energy use of Japan. Materials efficiency is even worse: only 1 percent of all the materials mobilized to serve America is actually made into products still in use six months after sale. In every sector, there are opportunities for reducing the amount of resources that go into a production process, the steps required to run that process, and the amount of pollution generated and by-products discarded at the end. These all represent avoidable costs and hence profits to be won.

Redesign Production According to Biological Models

In the second stage on the journey to natural capitalism, companies use closed-loop manufacturing to create new products and processes that can totally prevent waste. This plus more efficient production processes could cut companies' long-term materials requirements by more than 90 percent in most sectors.

The central principle of closed-loop manufacturing, as architect Paul Bierman-Lytle of the engineering firm CM2 Hill puts it, is "waste equals food." Every output of manufacturing should be either composted into natural nutrients or remanufactured into technical nutrients—that is, should be returned to the ecosystem or recycled for further production. Closed-loop production systems are designed to eliminate any materials that incur disposal costs, especially toxic ones, because the alternative—isolating them to prevent harm to natural systems—tends to be costly and risky. Indeed, meeting EPA and OSHA standards by eliminating harmful materials often makes a manufacturing process cost less than the hazardous process it replaced. Motorola, for example, formerly used chlorofluorocarbons for cleaning printed circuit boards after soldering. When CFCs were outlawed because they destroy stratospheric ozone, Motorola at first explored such alternatives as orange-peel terpenes. But it turned out to be even cheaper—and to produce a better product—to redesign the whole soldering process so that it needed no cleaning operations or cleaning materials at all.

Closed-loop manufacturing is more than just a theory. The US remanufacturing industry in 1996 reported revenues of $53 billion—more than consumer-durables manufacturing (appliances, furniture, audio, video, farm and garden equipment). Xerox, whose bottom line swelled by more than $700 million from remanufacturing, expects to save another $1 billion just by remanufacturing its new, entirely reusable or recyclable line of "green" photocopiers. What's more, policymakers in some countries are already taking steps to encourage industry to think along these lines. German law, for example, makes many manufacturers responsible for their products forever, and Japan is following suit.

Combining closed-loop manufacturing with resource efficiency is especially powerful. DuPont, for example, gets much of its polyester industrial film back from customers after they use it and recycles it into new film. DuPont also makes its film ever stronger and thinner so it uses less material and costs less to make. Yet, because the film performs better, customers are willing to pay more for it. As DuPont chairman Jack Krol noted in 1997, "Our ability to continually improve the inherent properties [of our films] enables this process [of developing more productive materials, at lower cost, and higher profits] to go on indefinitely."

Interface is leading the way to this next frontier of industrial ecology. While its competitors are "down cycling" nylon-and-PVC-based carpet into less valuable carpet backing, Interface has invented a new floor-covering material called Solenium, which can be completely remanufactured into identical new product. This fundamental innovation emerged from a clean-sheet redesign. Executives at Interface didn't ask how they could sell more carpet of the familiar kind; they asked how they could create a dream product that would best meet their customers' needs while protecting and nourishing natural capital.

Solenium lasts four times longer and uses 40 percent less material than ordinary carpets—an 86 percent reduction in materials intensity. What's more, Solenium is free of chlorine and other toxic materials, is virtually stainproof, doesn't grow mildew, can easily be cleaned with water, and offers aesthetic advantages over traditional carpets. It's so superior in every respect that Interface doesn't market it as an environmental product—just a better one.

Solenium is only one part of Interface's drive to eliminate every form of waste. Chairman Ray C. Anderson defines waste as "any measurable input that does not produce customer value," and he considers all input to be waste until shown otherwise. Between 1994 and 1998, this zero-waste approach led to a systematic treasure hunt that helped to keep resource inputs constant while revenues rose by $200 million. Indeed, $67 million of the revenue increase can be directly attributed to the company's 60 percent reduction in landfill waste.

Subsequently, President Charlie Eitel expanded the definition of waste to include all fossil-fuel inputs, and now many customers are eager to buy products from the company's recently opened solar-powered carpet factory. Interface's green strategy has not only won plaudits from environmentalists, it has also proved a remarkably successful business strategy. Between 1993 and 1998, revenue has more than doubled, profits have more than tripled, and the number of employees has increased by 73 percent.

Change the Business Model

In addition to its drive to eliminate waste, Interface has made a fundamental shift in its business model—the third stage on the journey toward natural capitalism. The company has realized that clients want to walk on and look at carpets—but not necessarily to own them. Traditionally, broadloom carpets in office buildings are replaced every decade because some portions look worn out. When that happens, companies suffer the disruption of shutting down their offices and removing their furniture. Billions of pounds of carpets are removed each year and sent to landfills, where they will last up to 20,000 years. To escape this unproductive and wasteful cycle, Interface is transforming itself from a company that sells and fits carpets into one that provides floor-covering services.

Under its Evergreen Lease, Interface no longer sells carpets but rather leases a floor-covering service for a monthly fee, accepting responsibility for keeping the carpet fresh and clean. Monthly inspections detect and replace worn carpet tiles. Since at most 20 percent of an area typically shows 80 percent of the wear, replacing only the worn parts reduces the consumption of carpeting material by about 80 percent. It also minimizes the disruption that customers experience—worn tiles are seldom found under furniture. Finally, for the customer, leasing carpets can provide a tax advantage by turning a capital expenditure into a tax-deductible expense. The result: the customer gets cheaper and better services that cost the supplier far less to produce. Indeed, the energy saved from not producing a whole new carpet is in itself enough to produce all the carpeting that the new business model requires. Taken together, the 5-fold savings in carpeting material that Interface achieves through the Evergreen Lease and the 7-fold materials savings achieved through the use of Solenium deliver a stunning 35-fold reduction in the flow of materials needed to sustain a superior floor-covering service. Remanufacturing, and even making carpet initially from renewable materials, can then reduce the extraction of virgin resources essentially to the company's goal of zero.

Interface's shift to a service-leasing business reflects a fundamental change from the basic model of most manufacturing companies, which still look on their businesses as machines for producing and selling products. The more products sold, the better—at least for the company, if not always for the customer or the Earth. But any model that wastes natural resources also wastes money. Ultimately, that model will be unable to compete with a service model that emphasizes solving problems and building long-term relationships with customers rather than making and selling products. The shift to what James Womack of the Lean Enterprise Institute calls a "solutions economy" will almost always improve customer value *and* providers' bottom lines because it aligns both parties' interests, offering rewards for doing more and better with less.

Interface is not alone. Elevator giant Schindler, for example, prefers leasing vertical transportation services to selling elevators because leasing lets it capture the savings from its elevators' lower energy and maintenance costs. Dow Chemical and Safety Kleen prefer leasing dissolving services to selling solvents because they can reuse the same solvent scores of times, reducing costs. United Technologies' Carrier division, the world's largest manufacturer of air conditioners, is shifting its mission from selling air conditioners to leasing comfort. Making its air conditioners more durable and efficient may compromise future equipment sales, but it provides what customers want and will pay for—better comfort at lower cost. But Carrier is going even further. It's starting to team up with other companies to make buildings more efficient so that they need less air conditioning, or even none at all, to yield the same level of comfort. Carrier will get paid to provide the agreed-upon level of comfort, however that's delivered. Higher profits will come from providing better solutions rather than from selling more equipment. Since comfort with little or no air conditioning (via better building design) works better and costs less than comfort with copious air conditioning, Carrier is smart to capture this opportunity itself before its competitors do. As they say at 3M: "We'd rather eat our *own* lunch, thank you."

The shift to a service business model promises benefits not just to participating businesses but to the entire economy as well. Womack points out that by helping customers reduce their need for capital goods such as carpets or elevators, and by rewarding suppliers for extending and maximizing asset values rather than for churning them, adoption of the service model will reduce the volatility in the turnover of capital goods that lies at the heart of the business cycle. That would significantly reduce the overall volatility of the world's economy. At present, the producers of capital goods face feast or famine because the buying decisions of households and corporations are extremely sensitive to fluctuating income. But in a continuous-flow-of-services economy, those swings would be greatly reduced, bringing a welcome stability to businesses. Excess capacity—another form of waste and source of risk—need no longer be retained for meeting peak demand. The result of adopting the new model would be an economy in which we grow and get richer by using less and become stronger by being leaner and more stable.

Reinvest in Natural Capital

The foundation of textbook capitalism is the prudent reinvestment of earnings in productive capital. Natural capitalists who have dramatically raised their resource productivity, closed their loops, and shifted to a solutions-based business model have one key task remaining. They must reinvest in restoring, sustaining, and expanding the most important form of capital—their own natural habitat and biological resource base.

This was not always so important. Until recently, business could ignore damage to the ecosystem because it didn't affect production and didn't increase costs. But that situation is changing. In 1998 alone, violent weather displaced 300 million people and caused upwards of $90 billion worth of damage, representing more weather-related destruction than was reported through the entire decade of the 1980s. The increase in damage is strongly linked to deforestation and climate change, factors that accelerate the frequency and severity of natural disasters and are the consequences of inefficient industrialization. If the flow of services from industrial systems is to be sustained or increased in the future for a growing population, the vital flow of services from living systems will have to be maintained or increased as well. Without reinvestment in natural capital, shortages of ecosystem services are likely to become the limiting factor to prosperity in the next century. When a manufacturer realizes that a supplier of key components is overextended and running behind on deliveries, it takes immediate action lest its own production lines come to a halt. The ecosystem is a supplier of key components for the life of the planet, and it is now falling behind on its orders.

Failure to protect and reinvest in natural capital can also hit a company's revenues indirectly. Many companies are discovering that public perceptions of environmental responsibility, or its lack thereof, affect sales. MacMillan Bloedel, targeted by environmental activists as an emblematic clear-cutter and chlorine user, lost 5 percent of its sales almost overnight when dropped as a UK supplier by Scott Paper and Kimberly-Clark. Numerous case studies show that companies leading the way in implementing changes that help protect the environment tend to gain disproportionate advantage, while companies perceived as irresponsible lose their franchise, their legitimacy, and their shirts. Even businesses that claim to be committed to the concept of sustainable development but whose strategy is seen as mistaken, like Monsanto, are encountering stiffening public resistance to their products. Not surprisingly, University of Oregon business professor Michael Russo, along with many other analysts, has found that a strong environmental rating is "a consistent predictor of profitability."

The pioneering corporations that have made reinvestments in natural capital are starting to see some interesting paybacks. The independent power producer AES, for example, has long pursued a policy of planting trees to offset the carbon emissions of its power plants. That ethical stance, once thought quixotic, now looks like a smart investment because a dozen brokers are now starting to create markets in carbon reduction. Similarly, certification by the Forest Stewardship Council of certain sustainably grown and harvested products has given Collins Pine the extra profit margins that enabled its US manufacturing operation to survive brutal competition. Taking an even longer view, Swiss Re and other European reinsurers are seeking to cut their storm-damage losses by pressing for international public policy to protect the climate and by investing in climate-safe technologies that also promise good profits. Yet most companies still do not realize that a vibrant ecological web underpins their survival and their business success. Enriching natural capital is not just a public good—it is vital to every company's longevity.

It turns out that changing industrial processes so that they actually replenish and magnify the stock of natural capital can prove especially profitable because nature does the production; people need just step back and let life flourish. Industries that directly harvest living resources, such as forestry, farming, and fishing, offer the most suggestive examples. Here are three:

1   Allan Savory of the Center for Holistic Management in Albuquerque, New Mexico, has redesigned cattle ranching to raise the carrying capacity of rangelands which have been degraded, not by overgrazing, but by under-grazing and grazing the wrong way. Savory's solution is to keep the cattle moving from place to place, grazing intensely but briefly at each site, so that they mimic the dense but constantly moving herds of native grazing animals that coevolved with grasslands. Thousands of ranchers are estimated to be applying this approach, improving both their range and their profits. This "management-intensive rotational grazing" method, long standard in New Zealand, yields such clearly superior returns that over 15 percent of Wisconsin's dairy farms have adopted it in the past few years.

2   The California Rice Industry Association has discovered that letting nature's diversity flourish can be more profitable than forcing it to produce a single product. By flooding 150,000 to 200,000 acres of Sacramento Valley rice fields—about 30 percent of California's rice-growing area—after harvest, farmers are able to create seasonal wetlands that support millions of wildfowl, replenish groundwater, improve fertility, and yield other valuable benefits. In addition, the farmers bale and sell the rice straw, whose high silica content—formerly an air pollution hazard when the straw was burned—adds insect resistance and hence value as a construction material when it's resold instead.

3   John Todd of Living Technologies in Burlington, Vermont, has used biological Living Machines—linked tanks of bacteria, algae, plants, and other organisms—to turn sewage into clean water. That not only yields cleaner water at a reduced cost, with no toxicity or odor, but also produces commercially valuable flowers and makes the plant compatible with its

residential neighborhood. A similar plant at the Ethel M Chocolates factory in Las Vegas, Nevada, not only handles difficult industrial wastes effectively but is showcased in its public tours.

Although such practices are still evolving, the broad lessons they teach are clear. In almost all climates, soils, and societies, working with nature is more productive than working against it. Reinvesting in nature allows farmers, fishermen, and forest managers to match or exceed the high yields and profits sustained by traditional input-intensive, chemically driven practices. Although much of our mainstream business is still headed the other way, the profitability of sustainable, nature-emulating practices is already being proved. In the future, many industries that don't now consider themselves dependent on a biological resource base will become more so as they shift their raw materials and production processes more to biological ones. There is evidence that many business leaders are starting to think this way. The consulting firm Arthur D. Little surveyed a group of North American and European business leaders and found that 83 percent of them already believe that they can derive "real business value [from implementing a] sustainable-development approach to strategy and operations."

A Broken Compass?

If the road ahead is clear, why are so many companies straying or falling by the wayside? We believe the reason is that the instruments companies use to set their targets, measure their performance, and hand out rewards are faulty. In other words, the markets are full of distortions and perverse incentives. Of the more than 60 specific forms of misdirection that we have identified,[3] the most obvious involve the ways companies allocate capital and the way governments set policy and impose taxes. Merely correcting these defective practices would uncover huge opportunities for profit.

Consider how companies make purchasing decisions. Decisions to buy small items are typically based on their initial cost rather than their full life-cycle cost, a practice that can add up to major wastage. Distribution transformers that supply electricity to buildings and factories, for example, are a minor item at just $320 apiece, and most companies try to save a quick buck by buying the lowest-price models. Yet nearly all the nation's electricity must flow through transformers, and using the cheaper but less efficient models wastes $1 billion a year. Such examples are legion. Equipping standard new office-lighting circuits with fatter wire that reduces electrical resistance could generate after-tax returns of 193 percent a year. Instead, wire as thin as the National Electrical Code permits is usually selected because it costs less up-front. But the code is meant only to prevent fires from overheated wiring, not to save money. Ironically, an electrician who chooses fatter wire—thereby reducing long-term electricity bills—doesn't get the job. After paying for the extra copper, he's no longer the low bidder.

Some companies do consider more than just the initial price in their purchasing decision but still don't go far enough. Most of them use a crude payback estimate rather than more accurate metrics like discounted cash flow. A few years ago, the median simple payback these companies were demanding from energy efficiency was 1.9 years. That's equivalent to requiring an after-tax return of around 71 percent per year—about six times the marginal cost of capital.

Most companies also miss major opportunities by treating their facilities costs as an overhead to be minimized, typically by laying off engineers, rather than as a profit center to be optimized—by using those engineers to save resources. Deficient measurement and accounting practices also prevent companies from allocating costs—and waste—with any accuracy. For example, only a few semiconductor plants worldwide regularly and accurately measure how much energy they're using to produce a unit of chilled water or clean air for their cleanroom production facilities. That makes it hard for them to improve efficiency. In fact, in an effort to save time, semiconductor makers frequently build new plants as exact copies of previous ones—a design method nicknamed "infectious repetitis."

Many executives pay too little attention to saving resources because they are often a small percentage of total costs (energy costs run about 2 percent in most industries). But those resource savings drop straight to the bottom line and so represent a far greater percentage of profits. Many executives also think they already "did" efficiency in the 1970s, when the oil shock forced them to rethink old habits. They're forgetting that with today's far better technologies, it's profitable to start all over again. Malden Mills, the Massachusetts maker of such products as Polartec, was already using "efficient" metal-halide lamps in the mid-1990s. But a recent warehouse retrofit reduced the energy used for lighting by another 93 percent, improved visibility, and paid for itself in 18 months.

The way people are rewarded often creates perverse incentives. Architects and engineers, for example, are traditionally compensated for what they spend, not for what they save. Even the striking economics of the retrofit design for the Chicago office tower described earlier wasn't incentive enough to actually implement it. The property was controlled by a leasing agent who earned a commission every time she leased space, so she didn't want to wait the few extra months needed to retrofit the building. Her decision to reject the efficiency-quadrupling renovation proved costly for both her and her client. The building was so uncomfortable and expensive to occupy that it didn't lease, so ultimately the owner had to unload it at a firesale price. Moreover, the new owner will for the next 20 years be deprived of the opportunity to save capital cost.

If corporate practices obscure the benefits of natural capitalism, government policy positively undermines it. In nearly every country on the planet, tax laws penalize what we want more of—jobs and income—while subsidizing what we want less of—resource depletion and pollution. In every state but

Oregon, regulated utilities are rewarded for selling more energy, water, and other resources, and penalized for selling less even if increased production would cost more than improved customer efficiency. In most of America's arid western states, use-it-or-lose-it water laws encourage inefficient water consumption. Additionally, in many towns, inefficient use of land is enforced through outdated regulations, such as guidelines for ultrawide suburban streets recommended by 1950s civil-defense planners to accommodate the heavy equipment needed to clear up rubble after a nuclear attack.

The costs of the perverse incentives are staggering: $300 billion in annual energy wasted in the United States, and $1 trillion already misallocated to unnecessary air-conditioning equipment and the power supplies to run it (about 40 percent of the nation's peak electric load). Across the entire economy, unneeded expenditures to subsidize, encourage, and try to remedy inefficiency and damage that should not have occurred in the first place probably account for most, if not all, of the GDP growth of the past two decades. Indeed, according to former World Bank economist Herman Daly and his colleague John Cobb (along with many other analysts), Americans are hardly any better off than they were in 1980. But if the US government and private industry could redirect dollars currently earmarked for remedial costs toward reinvestment in natural and human capital, they could bring about a genuine improvement in the nation's welfare. Companies, too, are finding that wasting resources also means wasting money and people. These intertwined forms of waste have equally intertwined solutions. Firing the unproductive tons, gallons, and kilowatt-hours often makes it possible to keep the people, who will have more and better work to do.

Recognizing the Scarcity Shift

In the end, the real trouble with our economic compass is that it points in exactly the wrong direction. Most businesses are behaving as if people were still scarce and nature still abundant—the conditions that helped fuel the first Industrial Revolution. At that time, people were relatively scarce compared with the present-day population. The rapid mechanization of the textile industries caused explosive economic growth that created labor shortages in the factory and on the field. The Industrial Revolution, responding to those shortages and mechanizing one industry after another, made people a hundred times more productive than they had ever been.

The logic of economizing on the scarcest resource, because it limits progress, remains correct. But the pattern of scarcity is shifting: now people aren't scarce but nature is. This shows up first in industries that depend directly on ecological health. Here, production is increasingly constrained by fish rather than by boats, by forests rather than by chainsaws, by fertile topsoil rather than by plows. Moreover, unlike the traditional factors of industrial production—capital and labor—the biological limiting factors cannot be substituted for one another. In the industrial system, we can easily exchange machinery for labor. But no technology or amount of money can substitute for a stable climate and a productive biosphere. Even proper pricing can't replace the priceless. Natural capitalism addresses those problems by reintegrating ecological with economic goals. Because it is both necessary and profitable, it will subsume traditional industrialism with a new economy and a new paradigm of production, just as industrialism subsumed agrarianism.

The companies that first make the changes we have described will have a competitive edge. Those that don't make that effort won't be a problem because ultimately they won't be around. In making that choice, as Henry Ford said, "Whether you believe you can, or whether you believe you can't, you're absolutely right."

Notes

1   Our book *Natural Capitalism* provides hundreds of examples of how companies of almost every type and size, often through modest shifts in business logic and practice, have dramatically improved their bottom lines.
2   For more information, see www.rmi.org/rmi/Library/T04-01_HypercarsHydrogenAutomotiveTransition.
3   Summarized in the report "Climate: Making Sense *and* Making Money" at www.rmi.org/rmi/Library/C97-13_ClimateSenseMoney. See also the next selection, "Marketplace Energy Savings," which reproduces its "prospector's guide" from pp. 11–20, without its extensive notes and references.

Chapter 28

Marketplace Energy Savings: Turning Obstacles into Opportunities

1997

Editor's note: *This essay was published in* Climate: Making Sense *and* Making Money *by Amory B. Lovins and L. Hunter Lovins, Rocky Mountain Institute, 13 November 1997.*

If big savings in fuel, cost, and emissions are both feasible and profitable, why haven't they all been done? Because the free market, effective though it is, is burdened by subtle imperfections that inhibit the efficient allocation and use of resources. It is necessary at the outset, writes Professor Stephen DeCanio, Senior Staff Economist for President Reagan's Council of Economic Advisers,

> to discard the baggage carried by most economists (the author confesses membership of that much-maligned group) that immersion in a market environment guarantees efficient behavior by the market participants. Much of modern economic theory practically defines efficiency as the outcome of competitive market exchanges. But the bloodless "competition" of mathematical general equilibrium models bears only a partial relationship to the actual experience of real firms.

This is tacitly conceded whenever market economists, as a senior government official recently wrote, "are unpersuaded that just because an act seems to make good economic sense it will happen." Many economically rational things don't happen—precisely because of real-world obstacles and complexities that aren't reflected in the perfect-market economic models relied upon for the conventional conclusion that saving much energy will require much higher energy prices. In fact, those barriers block economically optimal investment in efficient use of energy in at least eight main ways. The good news is that *each of these obstacles represents a business opportunity*. Consider some examples of how they match up.

Capital Misallocation

Energy is only 1–2 percent of most industries' costs, and most managers pay little attention to seemingly small line-items, even though small savings can look big when added to the bottom line. Surprisingly many executives focus on the top line and forget where saved overheads go; and without managerial attention, nothing happens. In addition, manufacturing firms tend to be biased toward investments that increase output or market share and away from those that cut operating costs.

OBSTACLES

About four-fifths of firms don't assess potential energy savings using discounted-cash-flow criteria, as sound business practice dictates; instead, they require a simple payback whose median is 1.9 years. At (say) a 36 percent total marginal tax rate, a 1.9-year payback means a *71 percent* real after-tax rate of return, or around six times the marginal cost of capital. (For example, before state and then federal standards prohibited worse options, high-efficiency magnetic ballasts, with a 60% real internal rate of return, won only a 9 percent market share.) Many capital-constrained industries use even more absurd hurdle rates: in some, the energy managers can't buy anything beyond a six-month payback.

Many supposedly sophisticated firms count lifecycle cost only for big items and make routine "small" purchases based on first cost alone. Thus 90 percent of the 1.5 million electric distribution transformers bought every year, including the ones on utility poles, are bought for lowest first cost—passing up an after-tax return on investment (ROI) of at least 14 percent a year and many operational advantages, and misallocating $1 billion a year.

If you invest to save energy in your business or home, you probably want your money back within a couple of years, whereas utilities are content to recover their power-plant investments in 20–30 years—about ten times as long. Thus householders (and many corporate managers) typically require tenfold higher returns for saving energy than for producing it, equivalent to a tenfold price distortion. This practice makes us buy far too much energy and too little efficiency. Not fairly comparing ways to save with ways to supply energy means not choosing the best buys first, hence misallocating capital. Until the late 1980s, the US wasted on uneconomic power plants and their subsidies (each roughly $30 billion a year) about as much as it invested in all durable goods manufacturing industries, badly crimping the nation's competitiveness.

Capital Misallocation

A few years ago, the CEO of a *Fortune* 100 company heard that one of his sites had an outstanding energy manager who was saving $3.50 per square foot per year. He said, "That's nice—it's a million-square-foot facility, isn't it? So that guy must be adding $3.5 million a year to our bottom line." Then in the next breath, he added: "I can't really get excited about energy, though—it's only a few percent of my cost of doing business." He had to be shown the arithmetic to realize that similar results, if achieved in his 90-odd million square feet of facilities worldwide, would boost his corporation's net earnings that year by 56 percent. The energy manager was quickly promoted so he could spread his practices across the company.

Top finance firms have joined the US Department of Energy to create the International Performance Measurement and Verification Protocol now adopted in more than 20 countries, including Brazil, China, India, Mexico, Russia, and Ukraine. This voluntary industry-consensus approach, like Federal Housing Administration (FHA) mortgage rules, standardizes streams of energy-cost savings in buildings so they can be aggregated and securitized. Only a year old, the Protocol is creating a booming market in which loans to finance energy (and water) savings can be originated as fast as they can be sold into the new secondary market. Achieving the savings therefore no longer requires one's own capital, can be afford-ably financed, and needn't compete with other internal investment needs.

A new generation of buildings is overcoming the psychological barrier of suppos-edly higher capital cost. A hundred case-studies demonstrate that large energy savings, often of 75 percent or more, can come with superior comfort, amenity, and real-estate market and financial performance—yet *identical or lower* capital cost, because integrated design creates synergies that help displace equipment and infra-structure.

Arbitrageurs make fortunes from spreads of a tenth of a percentage point. The spread between the discount rates used in buying energy savings and supply are often hundreds of times bigger than that—surely big enough to overcome the trans-action costs of marketing and delivering lots of small savings. (Scores of utilities proved this in well-designed 1980s and early 1990s programs that delivered efficiency improvements at total costs far cheaper than just *operating* existing thermal power stations.) This is the basis of the energy service company (ESCO) concept, where entrepreneurs offer to help cut your energy bills for nothing up front—just a share of the savings. Skilled firms of this type are flourishing world-wide, although the American ESCO industry is still in its shakeout phase, and many federal agencies don't yet hire ESCOs because of rigid procurement habits.

OPPORTUNITIES

Capital Misallocation

High consumer discount rates are especially tough: people used to paying 50¢ for an incandescent light bulb are often unwilling or unable to pay $15–20 for a compact fluorescent lamp which, over its 13-fold-longer life, keeps nearly a ton of CO_2 out of the air and saves tens of dollars more in power-plant fuel, replacement lamps, and installation labor than it costs. It's a good deal, but sounds like too much up-front money out of pocket.

Most international vehicles for investing in national or utility-level electric power systems consider only supply-side, not demand-side, options and have no way to compare them. The resulting misallocation is like the recipe for elephant and rabbit stew—one elephant, one rabbit.

O B S T A C L E S

Organizational Failures

Old habits die hard. A famous company that hasn't needed steam for years still runs a big boiler plant, with round-the-clock licensed operators, simply to heat distribution pipes (many uninsulated and leaking) lest they fail from thermal cycling; nobody has gotten around to shutting the system down. Why rock the boat to make someone else look good? Why stick your neck out when the status quo seems to work and nobody's squawking?

Schedules conquer sensible design. One of us called the chief engineer of a huge firm to introduce opportunities like a cleanroom that uses a small fraction of the energy he was used to, performs better, costs less, and builds faster. His reply: "Sounds great, but I pay a $100,000-an-hour penalty if I don't have the drawings for our next plant done by Wednesday noon, so I can't talk to you. Sorry. Bye." The result is "infectious repetitis"—like the semiconductor plant where a pipe took an inexplicable jog in mid-air as if it were going around some invisible obstacle. The piping design had been copied from another plant that had a structural pillar in that location. In short, intense schedule pressures combine with design professionals' poor compensation and prestige, overspecialized training, and utterly dis-integrated processes to yield commoditized, lowest-common-denominator technical design.

Capital Misallocation

Southern California Edison Company gave away more than a million compact fluorescent lamps because doing that saved energy more cheaply than running power stations could produce it. SCE then cut the lamps' retail price by about 70 percent via a temporary subsidy paid not to buyers but to lamp manufacturers, thus leveraging all the markups. Some other utilities *lease* the lamps for, say, 20¢ per lamp per month, with free replacements; customers can thus pay over time, just as they now pay for power stations, but the lamps are cheaper.

Rapidly growing new investment funds, partly funded by the climate-risk-averse insurance industry, are bypassing utilities altogether and investing directly in developing countries' house-level "leapfrog" efficiency-plus-solar power systems. Those often cost less than villagers are now paying for lighting kerosene and radio batteries, and represent a new market of two billion people.

Organizational Failures

Columbia University had entrenched practices too. A tough new energy manager, Lindsay Audin, was told to cut 10 percent off its $10-million-a-year energy bill, with uncompromised service and no up-front capital. Authorizations were painfully slow—until Audin showed the delays were costing $3000 a day in lost savings, more than the delayers' monthly paychecks. Five years later he was saving $2.8 million a year, 60 percent of it just in lighting; had won 9 awards and $3 million in grants and rebates; and had brought 16 new efficiency products to market.

Both such designers and their clients can get away with poor design, and probably won't notice it, so long as their competitors use the same methods, consultants, and vendors. But once such striking improvements are introduced to a given market segment, the laggards must adopt them or lose market share. Thus competitive forces can do automatically much of the marketing and outreach normally required. Rocky Mountain Institute, having successfully promoted superefficient buildings and cars by this method, is now helping with a new initiative to overhaul the semiconductor industry, which has $100 billion worth of fabrication plants on the drawing boards worldwide, all very inefficient. The opportunity for clean-sheet redesign is intriguing industry leaders who now understand that they can't compete internationally without leapfrogging over old methods. For example, energy cost per East Asian-made hard-disk drive now differs by as much as 54-fold—many times the margin critical to market share.

OPPORTUNITIES

Organizational Failures

Few firms carefully measure how their buildings and processes actually work. Their design assumptions are therefore untested and often incorrect. Their design process is linear—require, design, build, repeat—rather than cyclic—require, design, build, *measure, analyze, improve*, repeat. No measurement, no improvement. And no discoveries—like the plant that for decades had been unwittingly running a 40-kilowatt electric heater year-round under its parking lot to melt snow. Nobody remembered or noticed until measurement found the books didn't balance, and the wiring was traced to track down the discrepancy.

Departments often don't or can't cooperate. A noted firm calculated that its proposed new office building should get all-new, superefficient office equipment, because the extra cost of buying it early (rather than waiting for normal turnover) would be less than the up-front savings from smaller cooling equipment. No deal: the chiller was in one budget, office equipment in another. Similarly, Federal buildings are bought from one budget, then operated from another; they may even be forbidden to share investments so as to reduce taxpayers' total costs.

If you save, the bean-counters simply cut your budget some more. Institutional or personal rewards for cutting energy costs are rare, even in the private sector. It's equally hard to prime the investment pump so savings from one project can help pay for the next.

Corporate turmoil spoils continuity. Many firms, assuming they'd already done all the worthwhile energy savings, have downsized their energy managers right out of a job, stuffed the task onto other overloaded agendas, and watched it slip to an invisible priority. How many economists does it take to screw in a compact fluorescent lamp? None, goes the joke—the free market will do it. But we all know that somebody actually has to get the lamp from shelf to socket; otherwise the wealth isn't created. In many firms, that somebody doesn't exist.

Companies full of smart, competent, rational, and profit-oriented people often fail to optimize because of even deeper kinds of inherent organizational failures well described in the economic literature.

Organizational Failures

The late economist Kenneth Boulding said hierarchies are "an ordered arrangement of wastebaskets, designed to prevent information from reaching the executive." But letting viscous information flow freely to those who need it stimulates intelligence, curiosity, and profits. At a large hard-disk-drive factory, the cleanroom operator started saving lots of money once the gauge that showed when to change dirty filters was marked not just in green and red zones but in "cents per drive" and "thousand dollars' profit per year." In another plant, just labeling the light-switches, so everyone could see which switches controlled which lights, saved $30,000 in the first year.

Electric utilities traditionally dis-integrate their operations too. But Canada's giant Ontario Hydro inverted its culture to make end-use efficiency and distribution planning its primary focus and generation an afterthought. Its first three experiments in meeting customers' needs by the cheapest means—typically demand-side investments plus better wires management—rather than reflexively building transmission and generating capacity cut its investment needs by up to 90 percent, saving US$600 million. Such achievements can motivate deep structural and cultural reforms.

Washington State routinely shares the savings between their achievers, the General Fund, and an account reserved for reinvestment in more savings. The 1997 Federal Energy Bank Bill, modeled on Texas's LoanSTAR, would set up a revolving fund for such savings.

After Ken Nelson,[1] the sparkplug of the remarkable Dow/Louisiana savings, retired in 1993, a reorganization disbanded his organizing committee, tracking ceased, and it became impossible to evaluate how much progress, if any, continued without him. (Lacking a champion, the neighboring Texas division reportedly never undertook a comparable effort in the first place.) But now Mr Nelson, like Southwire's[2] Mr Clarkson and some of their ablest peers, is an independent consultant, sharing his skills with more firms.

Proper measurement and incentives help: a utility that started paying its efficiency marketing staff a dollar for every measured kilowatt they saved quickly found that verified savings got bigger and cheaper—both by an order of magnitude.

OPPORTUNITIES

Regulatory Failures

All but a handful of states and nations reward regulated utilities for selling more energy and penalize them for cutting your bill, so shareholders and customers have opposite goals—with predictable results. Many proposed restructuring efforts would enshrine the same perverse incentive in new commodity-based market rules—rewarding the sale of as many kilowatt-hours as possible at the lowest possible price, rather than rewarding better service at lower cost. Similarly, New York State just cut ConEd's efficiency investments by 95 percent and is bringing back declining-block rates that make savings unprofitable.

In some (though increasingly rare) cases, obsolete codes, standards (as for cement composition), specifications (including those for corporate and military procurement), and laws actually prohibit sound and efficient practices. Far more often, standards meant to set a floor—like "meets code" (euphemism for "the worst building you can put up without being sent to jail"), or the British expression "catnap" (Cheapest Available Technology Narrowly Avoiding Prosecution)—are misinterpreted as a ceiling or as an economic optimum. For example, almost all US buildings use wire sizes equal to National Electrical Code minimum requirements, because the wire size is selected and its cost passed through by the low-bid electrician. But in a typical lighting circuit, the next larger wire size yields about a 169 percent per year after-tax return. Few electricians know this; even fewer care, since their reward for lower-loss wires is typically a lost bid.

The transportation sector is the fastest-growing and seemingly most intractable source of carbon emissions precisely because it is the most socialized, subsidized, and centrally planned sector of the US economy—at least for favored modes like road transport and aviation. It has the least true competition among modes, and the most untruthful prices, with hidden costs of hundreds of billions of dollars per year for US road vehicles alone. These distortions leverage more billions into otherwise uneconomic infrastructural and locational decisions. In particular, the dispersion of uses that causes so much excessive driving is mandated by obsolete single-use zoning rules meant to segregate noxious industries that scarcely exist today. Congestion is specifically caused by non-pricing or underpricing of the road resource: most roads are supported by taxes, not users, so they look free to drivers who behave much as Soviet customers did in demanding a great deal of energy when it looked free. Congestion is not only unpriced, but is further exacerbated by building more subsidized roads that elicit even more traffic, and by requiring developers to provide as much parking as people use when they pay nothing for it. Future generations will marvel that the incredible social costs of these policies—costs intertwined with many inner-city ills—went so long uncorrected: all ways to get around, or not to need to, were never made to compete fairly against each other, and drivers neither got what they paid for nor paid for what they got.

Thailand loses a sixth of its GDP to Bangkok traffic jams, so it's building Los Angeles-style freeways that will create more traffic.

OBSTACLES

Regulatory Failures

Simple accounting innovations in a few states decouple utilities' profits from their sales volumes, and let utilities keep as extra profit part of whatever they save off their customers' bills. The nation's largest investor-owned utility, PG&E, thus added over $40 million of riskless return to its 1992 bottom line while saving customers nine times that much. In California alone, Governor Wilson's PUC found that efficiency investments rewarded and motivated by this incentive system's emulation of efficient market outcomes, just during 1990–1993, had saved customers a net present value of nearly $2 billion. Thoughtful utility restructuring can do the same.

To encourage developers to exceed the minimal energy-saving requirements of building codes, Santa Barbara County entitled overcompliers (by 15–45+ percent) to jump the queue for approvals—a valuable reward at no cost. Elsewhere, some builders of superinsulated homes that leapfrogged far beyond code requirements have won credibility, and dominant market share, by offering to pay any heating bills over, say, $100 a year, or all utility bills for the first five years' ownership.

The private sector is also starting to highlight profit opportunities from exceeding code minima. The Copper Development Association, for example, publishes wire-size tables optimized to save money, not just to prevent fires. However, these will do little good unless winning bidders are chosen for minimizing lifecycle cost, not just first cost.

Strong evidence is emerging that co-locating where people live, play, shop, and work creates such desirable, friendly, low-crime, walking-and-biking-dominated neighborhoods that they yield exceptional market performance. Such co-location, and land-use policies that integrate housing and jobs with transit, can be further encouraged by "locationally efficient mortgages"—the subject of a $1-billion Fannie Mae experiment—that effectively let homebuyers capitalize the avoided costs of the car they no longer need in order to get to work.

Under a 1997 legal innovation, employers can profit from "cashing out" employee parking spaces—charging fair market value for each space, and paying each employee a "commuting allowance" of equal after-tax value. By monetizing competition between all means of getting to work (or, through sensible land-use or telecommuting, of not needing to), this will typically reduce demand for parking spaces—which often cost $10,000–$30,000 apiece—by enough to make employees, employers, and the Treasury all better off.

Real-estate developers can profit from annuitizing perpetual transit passes rather than providing a $25,000 parking place with each housing unit (which yields less but costlier housing). Allowing residents to rent out their daytime parking spaces can yield enough income to pay their home property tax.

Singapore is almost congestion-free because it charges drivers their true social cost and invests the proceeds in effective public transit and coordinated land use.

Informational Failures

The extremely high returns implicitly demanded for buying efficiency often reflect a paucity of accurate and up-to-date information. Do you know where to get everything you would need to optimize your own energy use, how to shop for it, how to get it properly installed, who would stand behind it? If any of the preceding examples of big, cheap savings surprised you, you've just observed a market barrier: if you don't know something is possible, you can't choose to do it.

Misinformation is also a problem. The United States, for example, uses about 1000 megawatts continuously (the output of one Chernobyl-sized power station) to run television sets that are turned off. Adding VCRs' and other household devices' standby loads roughly quintuples this waste. It's typically described as a convenience feature (no warm-up delay, TV turns on at previously selected channel, etc.). But few customers or manufacturers realize that exactly the same convenience is available with 80–95 percent less standby power. Similarly, few customers, vendors, or plumbers know that the best high-performance shower-heads can deliver just as wet, strong, and satisfying a shower as poorly designed models that use 2–6 times as much hot water.

"Hassle factor" and transaction costs prevent efficient micro-decisions in day-to-day life. For example, how much do you pay at home for a kilowatt-hour of electricity, and how many kilowatt-hours does your refrigerator—typically the biggest single user in the household unless you have electric space or water heating—use each year? If you don't know, because you're too busy living to delve into such minutiae, then you're part of another market barrier. And if you do know, then there's probably another barrier, because for the same price, you could have bought a seemingly identical refrigerator 2–3-fold more efficient, or nearer 20-fold with advanced techniques not yet brought to the mass market.

Informational Failures

Labeling tells buyers how competing models compare. Some voluntary labeling systems (as of a quarter-million San Francisco houses in 1978–1980) have swept the market because buyers quickly became suspicious of any house that wasn't labeled. EPA's voluntary Energy Star standard for office equipment did the same, now embracing over 2000 products by more than 400 manufacturers, because the efficient machines worked better, cost the same or less, and were therefore mandated for federal purchasing. They're saving a half-billion dollars a year, could nearly double that by 2000, and promise a profitable ten-million-ton-a-year carbon saving by 2005. Other voluntary programs that provide informational, technical, and trade-ally support, like EPA's Green Lights, are succeeding because they create competitive advantage. Involving more than 2300 organizations, Green Lights' retrofits save over half the lighting energy with 30 percent ROI and unchanged or improved lighting quality. The national potential for this effort alone is a $16-billion annual saving, plus a 12 percent reduction in utilities' carbon and other emissions. Just the new EPA voluntary standard to reduce unnecessary standby energy in TVs, VCRs, etc. can save, at zero cost, about eight million tons of carbon per year—as much as eight million cars now emit.

It's precisely to make such decisions hassle-free—and because most appliances are bought not by billpayers but by landlords, homebuilders, and public housing authorities—that Congress almost unanimously approved mandatory efficiency standards for household appliances. They merit extension to some commercial and industrial devices too. Such standards knock the worst equipment off the market and reward manufacturers for continuous improvement. That's largely why careless shopping for a same-priced refrigerator can sacrifice only 2–3-fold efficiency gains in America, vs. 6-fold in Europe. Smart utilities also reinforce standards by rewarding customers for beating them.

O
P
P
O
R
T
U
N
I
T
I
E
S

Risks to Manufacturers and Distributors

Industry lacks information too—about what customers really want and whether they'll put their money where their mouths are. Manufacturers often hesitate to take the risk of developing and making new energy-saving products, because of limited confidence that customers will buy them in the face of all the obstacles listed here. For example, the Idaho National Engineering Laboratory has developed a very promising and affordable ultralight elevated train called CyberTran, but it's so different from conventional trains that manufacturers aren't sure it will sell, so nobody is yet making it, so nobody can buy it—even though it appears able to relieve many communities' road congestion at far lower cost than building more roads, and without needing land.

Efficient equipment often isn't available when and where it's needed—as anyone knows who's tried to replace a burned-out water-heater, furnace, refrigerator, etc. on short notice. Yet distributors, aware of the slow uptake of efficient devices, don't want to take the risk of carrying inventory that may sell slowly or not at all. Thus British Columbia Hydro found that the huge motors in that Province's mining and pulp-and-paper mills were virtually all inefficient, simply because that's what local vendors customarily stocked; anything else took too long to order, and the mills couldn't afford to wait.

Corporations may think they won't be liable for their products once sold to someone else—then be unpleasantly surprised by laws and litigation that pursue deep pockets back through the value chain. This uncertainty leads to inefficient defensive behavior and discourages choices that minimize societal cost.

Risks to Manufacturers and Distributors

Swedish official Hans Nilsson pioneered contests for bringing efficient devices into the mass market. A major public-sector purchasing office issues a Request for Proposal guaranteeing to buy a large number of devices, bid from certain prices, if they meet certain technical specifications, including energy savings highly cost-effective to the user. This explicit expression of market demand has already elicited many important innovations giving a strong advantage to Swedish industry in both home and export markets. A "golden carrot" devised by Dr David Goldstein of the Natural Resources Defense Council followed suit, improving US refrigerators. Pioneer customers could also be encouraged to try such technologies as CyberTran by a system analogous to one EPA formerly used: the first adopters of an innovative wastewater treatment system would get a free replacement with a conventional alternative if the novel one didn't work.

BC Hydro paid a small, temporary subsidy to stock only efficient models, covering vendors' extra carrying cost. In three years, premium-efficiency motors' market share soared from 3 percent to 60 percent. The subsidy was then phased out, supported by a modest backup standard. Similarly, PG&E found in the 1980s that rather than paying customers a rebate for buying efficient refrigerators, it could improve refrigerator efficiencies faster, at less than a third the cost, by paying retailers a small bonus for each efficient model stocked, but nothing for stocking inefficient ones. The inefficient models quickly vanished from the shops, so when you wanted the next unit the dealer could put on the truck, it'd be efficient, because that's all they'd have.

Under the "cycle principle" pioneered in Germany, manufacturers own their products forever. This leads to design for minimum lifecycle (cradle-to-cradle) costs and maximum lifecycle efficiency. Both then become new sources of profit, as illustrated by the remanufacturing and service-leasing examples given elsewhere.

OPPORTUNITIES

Perverse Incentives

Compensation to architects and engineers worldwide is based directly or indirectly on a percentage of the *cost* of the building or equipment specified. Designers who work harder to eliminate costly equipment therefore end up with lower fees, or at best with the same fees for more work. Such backwards incentives have led the US to misallocate about $1 trillion to air-conditioning equipment (and utility systems to power them) that wouldn't have been bought if the same buildings had been optimally designed to produce the same or better comfort at least cost.

The real-estate value chain is full of incentives so perverse that each of the 25 or so parties in a typical large deal is systematically rewarded for inefficiency and penalized for efficiency. The 75 percent energy saving designed for the Chicago office tower mentioned earlier (see Chapters 19 and 27), with instant payback, wasn't bought: the property was controlled by a local leasing office, incentivized on dealflow, that didn't want to delay its commissions a few months by retrofitting before leasing up the building. The building then proved unmarketably costly and uncomfortable, so it had to be sold off to a bottom-feeder. Yet the owner wasn't unsophisticated: it was one of the world's largest fiduciaries.

Split incentives—one party selecting the technology, another paying its energy costs—limit ultimate consumers' choices by substituting intermediaries who don't bear the cost of their poor decisions. This issue is ubiquitous. Why should you fix up your rented premises if you don't own them? Why should the landlord do it if you pay the energy bills? Alternatively, if you *don't* pay the bills, why use energy thoughtfully (for example, why maintain or efficiently drive a company car whose costs are paid for you)? In the Shanghai pumping example described in Chapter 27, the pipefitters don't mind putting in lots of extra bends, because they're paid by the hour and they won't pay the equipment or electricity bills. Efficiency measures used in owned space often aren't in rented.

Similar split incentives apply to the makers and users of all kinds of equipment used in buildings and factories. Such equipment is almost always inefficient and designed for low first cost alone, since those who sell it won't pay the operating costs and most buyers won't shop carefully. (Indeed, for most kinds of equipment, efficient equipment simply isn't available—until a big customer demands better, as Wal-Mart successfully did for daylighting and air-conditioning equipment.)

In one respect the market works all too well: wasteful old equipment often gets salvaged for resale in the secondary market—mainly to poor people who can least afford the high running costs that motivated the scrappage in the first place. Such "negative technology transfer" can cripple development efforts.

OBSTACLES

Perverse Incentives

Pilot projects launched by RMI are now testing how much more efficient buildings become if their designers are rewarded for what they save, not what they spend, by letting them keep several years' measured energy savings as extra profit. Early results are encouraging. The German and Swiss architectural associations are pursuing similar reforms.

Careful case studies are revealing that in well-designed, highly efficient buildings, the better visual, acoustic, and thermal comfort enables people to do about 6–16 percent more and better work. In a typical office, where people cost 100 times as much as energy, that boost in labor productivity is about 6–16 times as valuable to the bottom line as eliminating the entire energy bill (see "A Roadmap for Natural Capitalism"). Analogous benefits, big enough to create decisive competitive advantage, are also being found in retail sales and manufacturing. These results may help to explain why firms participating in EPA's voluntary Green Lights lighting-efficiency programs showed stronger earnings growth than nonparticipants. Increasingly educated tenants will not long tolerate buildings that don't contribute to their success.

Lease riders can fairly share savings between landlords and tenants so both have an incentive to achieve them. Energy utilities could also (as some water/wastewater utilities already do) apply "feebates" to new building hookups: you pay a fee or get a rebate to connect to the system, but which and how big depends on how efficient you are, and each year the fees pay for the rebates. Unlike building codes and appliance standards—which are better than nothing, but become instantly obsolete and offer no incentive to beat the standard—such a revenue-neutral economic instrument drives continuous improvement. It also signals lifecycle costs up front, when the long-term investment decisions are being made.

The world's largest maker of air conditioners, Carrier Corporation, is leasing comfort services—much as elevator-maker Schindler leases vertical transportation services and Dow leases solvent services. This improves not only resource efficiency but also incentives: the *more* efficient, durable, and flexible Carrier's air-conditioning systems become, the *greater* its profits, and the better the service it provides at lower cost to more customers. Service leasing aligns the providers' incentive with their customers' objective.

Some big California utilities buy up inefficient old motors, refrigerators, and other devices in order to scrap them before they enter the second-hand market: they're worth far more dead than alive. Unocal even bought and scrapped numerous polluting old cars in order to gain pollution credits for its refinery near Los Angeles.

OPPORTUNITIES

False or Absent Price Signals

Energy prices are often badly distorted by subsidies and by uncounted external (larcenous) costs not internalized by the Clean Air Act's laudable trading system. The US in 1989 still subsidized energy supply by about $21–36 billion per year, mostly for the least competitive options and essentially all for supply. Significant costless (or better) reductions in carbon emissions are therefore available just by removing subsidies, a process already under way. And that doesn't count even bigger subsidies to security of supply that make the true cost over $100 per barrel for Persian Gulf oil (though more was at stake in the Gulf War than just oil).

OBSTACLES

Energy price signals are diluted by other costs. For example, US gasoline, cheaper than bottled water, is only an eighth of the total cost of driving, even though the car is cheaper per pound than a Big Mac. Why buy a 50- instead of a 20-mpg car when both cost about the same per mile to own and run?

Few firms track energy costs as a line-item for which profit centers are accountable. Firms in rented space may have energy bills prorated rather than submetered. Most billing systems give no end-use information that lets customers link costs to specific devices. Many firms, especially chains and franchises, never even see their energy bills, which are sent directly to a remote accounting department for payment. Some large firms still assume that utility bills are a fixed cost not worth examining.

Appraisers rarely credit efficient buildings for their actual energy savings, so efficiency's value isn't capitalized. Most leasing brokers base pro forma financials on average assumed operating costs, not actual ones. Few buildings have efficiency labels. Few renters have access to past energy bills.

Tax asymmetries further distort energy choices. For example, energy purchases are deductible business expenses, but investments to save energy get capitalized.

Market prices don't include many environmental costs and risks: [enforcement of] the Clean Air Act, for example, created a cap-and-trade regime for sulfur but not for carbon emissions.[3]

False or Absent Price Signals

Subsidies are under increasing pressure by a more skeptical Congress, a better-informed public, and more transparent prices. Utility regulators in about 30 of the United States also take account of some externalities in considering utilities' proposed resource acquisition decisions. Some proposals for industry restructuring would worsen but others would help to correct these longstanding distortions, improving economic efficiency.

Global annual energy subsidies are estimated to have fallen from about $350–400 billion in the early 1990s to about $250–300 billion in the mid-1990s. Their further transparency and reduction will reduce the risk of making investments not justified by fundamentals.

Feebates (see above) can reward turning over big capital stocks like car fleets more quickly, getting the worst ones off the road soonest. This offers a huge new market opportunity—especially if the rebate for your efficient new car depends on the *difference* in efficiency between the new one you buy and the old one you scrap.

New bill-paying and -minimizing service companies are springing up to meet exactly this need. Many provide submetering and two-way communications to pinpoint opportunities for improvement. Such simple efforts as ensuring that each meter generating a bill is actually on the customer's premises often generate big savings.

Some jurisdictions have right-to-know laws; others get similar results by training renters and buyers to be assertively inquisitive. Smart leasing brokers are distinguishing their services by offering valuable advice on minimizing occupancy costs. Home and commercial-building energy rating systems are rapidly emerging.

Some countries do better. When the Japanese government wanted to clean up sulfur emissions from power plants, it allowed scrubbers to be expensed in one year.

The Natural Resources Defense Council published an index of relative exposure to carbon-tax risks for all U.S. utilities, and let capital markets adjust ratings accordingly.

OPPORTUNITIES

Incomplete Markets and Property Rights[4]

There is no market in saved energy: "negawatts" aren't yet a fungible commodity subject to competitive bidding, arbitrage, secondary markets, derivatives, and all the other mechanisms that make efficient markets in copper, wheat, and sowbellies. You can't yet go bounty-hunting for wasted energy, trade negawatt futures and options (or bid them in a spot market against megawatts), or even, in general, bid them fairly against expansions of energy supply.[5] You can seldom sell reduced demand or reduced uncertainty of demand; yet both are valuable resources that deserve markets. Property rights in most forms of depletion-and-pollution avoidance are incomplete or absent and hence cannot be traded.

OBSTACLES

Incomplete Markets and Property Rights

When Morro Bay, California, ran short of water, it simply required any developer wanting a building permit to save, somewhere else in town, twice as much water as the new building would use. Many creative transactions occurred as developers discovered what saved water is worth. Two-fifths of the houses were retrofitted with efficient fixtures in the first four years. A more comprehensive market transformation effort enabled Goleta, California, to cut per-capita residential water use by over 50 percent, and total water use by over 30 percent, in one year and with no loss of service quality—thereby deferring indefinitely a multi-million-dollar wastewater-treatment-plant expansion.

O
P
P
O
R
T
U
N
I
T
I
E
S

Notes

1   In 1981, Dow Chemical's 2400-worker Louisiana division started prospecting for overlooked savings. Engineer Ken Nelson set up a shop-floor-level contest for energy-saving ideas. Proposals had to offer at least 50% annual return on investment. The first year's 27 projects averaged 173% ROI. Startled at this unexpected bounty, though expecting it to peter out quickly, Nelson persevered. The next year, 32 projects averaged 340% ROI. Twelve years and almost 900 implemented projects later, the workers had averaged (in the 575 projects subjected to *ex post* audit) 202% predicted and 204% audited ROI. In the later years, the returns and the savings were both getting *bigger*, because the engineers were learning faster than they were exhausting the "negawatt" resource. In only one year did returns dip into double digits (97% annual ROI). By 1993, the whole suite of projects was paying Dow's shareholders $110 million every year.

2   Southwire is the top independent US maker of rod, wire, and cable, a very energy-intensive business, and has nearly 50 acres of industrial facilities under roof. During 1981–1987, the firm cut its electricity use per pound of product by 40%, gas by 60%—then kept on saving even more energy and money, still within two-year paybacks. The resulting savings created nearly all the company's profits in a tough period when competitors were going under. The two engineers responsible may have saved 4000 jobs at ten plants in six (now nine) states. The lead engineer, Jim Clarkson, says the technologies were all simple and available; their effective use took only "an act of management will and design mentality, consistently applied." Indeed, Southwire found that such dramatic energy savings both require and facilitate better management and production systems that are vital anyhow for competitiveness. America's energy-saving potential—sufficient "to cut industrial energy use in half," as Southwire did—tags along almost for free.

3   Actually the Supreme Court recently found the Act did authorize carbon trading, so it's only EPA that hasn't yet used that authority, though it is now starting to.

4   Compare the "actually existing market" in the left column above with the requirements of a *theoretical* free market: perfect information about the future, perfectly accurate and complete price signals, perfect competition, no monopoly or monopsony (sole buyer), no unemployment or underemployment of any resource, no unmarketed resources, no transaction costs, no subsidies, no barriers to market entry or exit, and so forth. It's a whole different universe. But under *actual* market conditions, can energy efficiency be implemented rapidly without the high energy prices that many economists and businesspeople fear?

5   By 2010, 13 states (in the New England and PJM power pools) were finally allowing negawatts to be bid into supply-side auctions, with excellent results.

Section 8

Miscellany: A Poem, a Letter, and an Opinion

Amory's intellectual pursuits are far too wide-ranging to be limited to essays and commentaries on (just!) the natural world, resources, energy, buildings, vehicles, and security. Amory grew up studying everything from Latin and Greek to mathematics and music. Indeed, by 12 or 13, he'd abandoned a potential career as a pianist and composer. Amory's parents kept a good library and he and his sister, Julie, who'd later become a computer linguist, had access to many books on the physical and social sciences, the arts, and culture ("although there were many holes in my background—I never learned much history, for example," Amory says).

This Miscellany reflects the "intense curiosity" about the world that the Lovinses shared and which Amory has encouraged in the thousands of students, interns, colleagues, and mentors he's worked with in the past five decades. For example, as a "recovering physicist" he's long striven to learn some biology, and his "A Tale of Two Botanies" essay has proven all too prophetic. The preceding letter about US constitutional law, which he later read at Harvard, echoes important English legal history that he then read at Oxford, as in Katherine Drinker Bowen's book about Lord Coke, *The Lion and the Throne*. "Vaguely We Walk," the oldest piece of writing in this collection (written when Amory was a high-school sophomore—fourth form in the UK), stands out for a special reason. The poem was influenced by poet Robert Frost, who said there are only two natural forms of English verse—loose iambic and strict iambic. Frost sometimes lived near, and endowed teaching chairs at, Amory's school in Amherst, Massachusetts. Frost's famous poem, it should be noted, also influenced the title of "Energy

Strategy: The Road Not Taken?" Life is, after all, about finding the right path. And Amory's 1962 poem makes the same point that Spanish poet Antonio Machado (1875–1939) made:

> *Wanderer, there is no path*
> *The path is made by walking.*

Chapter 29

Vaguely We Walk

1962

Trudging over the common
through the bleak bones of trees
set off sharp by the dirty white
winter-killed snow, I noticed the path
trod by those who had gone
the same way before:
doubtless the first,
grinning rue at the then bare
expanse of snow,
crossed the common, and left
the mincing cat's-pathway of footprints
winding around the oak in the middle,
finally reaching the plowed place;
then the others, seeking to save
the trouble of snow in their shoes,
meandered on off-balance, awkwardly
setting their strides
to those of the first comer, and soon found it
needless to follow exactly, as the
clumsy treading enlarged the path,
gouged out craters,
then iced them over,
slick as though
they were glazed for skating.

I tramped to the pathway,
not thinking the ice would have
followed from so many feet,
and started along it, slipping a bit
at each hesitant step
on the wavy glass,

and wondering: as for that pioneer
of this path, sober as he
may have been,
my stride fitted not quite to him
nor his to me,
and it were useless
to say otherwise
for the sake of a doubtful convenience;
so I diverged and
impressed a new trail beside the slipperiness,
pushing through powder and loud-crunching crust,
stepping the way to the clear place.

I stayed not to think
of those who would come after.
They would perhaps ask
how one could have lost himself
from a path so broad—
I, who in so doing
have found what they themselves have lost;
and they would be faced
with the ancient quandary of the fork in the way:
to study between the common path;
mine;
and the third choice.

Chapter 30

Gulf of Tonkin Resolution a Red Herring

1964

Editor's note: *This unpublished letter, penned for* The New York Times *while Amory was en route from high school to Harvard (the Resolution passed in August), was about the "attack" (later exposed as a fake) that justified the vast expansion of President Johnson's power to prosecute the war in Viet-Nam.*

The question whether the US Congress may have waived its war-making powers in favor of Presidential discretion seems wide of the mark.

The Constitution, whether or not glossed by such documents as the Federalist Papers, clearly provides that only the Congress can declare war and only the President can thereafter command such troops as the Congress is willing to pay for. Nothing short of Constitutional Amendment can alter this separation of powers. Therefore it would be just as contrary to the Constitution for the Congress to abrogate or delegate its sole power of declaring war as for the President to usurp that power.

If this view is correct, then it is irrelevant exactly how we construe the Gulf of Tonkin Resolution, for insofar as it is interpreted as ceding to the President the authority laid on the Congress as a Constitutional duty, to that extent it must be void and without effect, and the President cannot rely on it to justify encroachment on Legislative prerogative.

In short, it is idle for the President to say that the Congress has authorized him to act on its behalf, for it is impossible for the Congress to have done this; he can only claim that what he has done has not infringed the lawful powers of the Congress. Whether invading a neutral country is "declaring war" is a question of construction for the courts to decide. The Gulf of Tonkin Resolution is a red herring.

The Senators and Representatives, like the President, have sworn to uphold the Constitution. The President seems to think that his oath enjoins him to "preserve, protect and defend" not the Constitution but his own Administration. Let us hope that all Members of Congress will instead see in their oath an obligation to defend the Constitutional basis of the Republic against the most dangerous attack in its history. The President must learn that he is under the law.

Chapter 31

A Tale of Two Botanies

1999

Editor's note: *In 1999, the International Union of Biological Sciences held its 16th International Botanical Congress in St. Louis. Appalled by the state of agriculture and the rapidly spreading use of biologically uninformed genetic modification in our agriculture, Amory coauthored a piece with Hunter Lovins for the* St Louis Post-Dispatch, *from which this version is slightly adapted.*

We all owe a debt to the subject matter of the [International] Botanical Congress now meeting at the Missouri Botanical Garden. Plants, shaped into incredible diversity by 3.8 billion years of evolution, make possible all life, underpin every ecosystem, and are resilient against almost any threat—except human destructiveness. From botany came the genetics of Mendel and Lamarck, formalizing the patient plant-breeding that's created 10,000 years of agriculture.

Now, however, in the name of feeding a growing human population, the process of biological evolution is being transformed. A St Louis firm is practicing a completely different kind of botany which, in the Cartesian tradition of reducing complex wholes to simple parts, strives to alter isolated genes while disregarding the interactive totality of ecosystems. Seeking what Sir Francis Bacon called "the enlarging of the bounds of Human Empire, to the effecting of all things possible," its ambition is to replace nature's wisdom with people's cleverness; to treat nature not as model and mentor but as a set of limits to be evaded when inconvenient; not to study nature but to restructure it.

As biophysicist Dr Donella Meadows notes, the new botany aims to align the development of plants not with their evolutionary success but with their economic success: survival not of the fittest but of the fattest, those best able to profit from wide sales of monopolized products. (High-yield, open-pollinated seeds abound; the new crops were created not because they're productive but because they're patentable.) Their economic value is mainly oriented not toward helping subsistence farmers to feed themselves but toward feeding more livestock for the already overfed rich. Most worryingly, the transformation of plant genetics is being accelerated from the measured pace of biological evolution to the speed of next quarter's earnings report. Such haste makes it impossible to foresee and forestall: unintended consequences appear only later, when they may not be fixable, because novel lifeforms aren't recallable.

In nature, all experiments are rigorously tested over eons. Single mutations venture into an unforgiving ecosystem and test their mettle. Whatever doesn't work gets recalled by the Manufacturer. What's alive today is what worked; only successes yield progeny. But in the brave new world of artifice, organisms are briefly tested by their creators in laboratory and field (no government agency systematically tests for nor certifies their long-term safety), then mass-marketed worldwide. The United States Department of Agriculture (USDA) has already approved about 50 genetically engineered crops for unlimited release; US researchers have tested about 4500 more. Just during 1995–1999, the non-Chinese farmland planted to such new crops expanded from zero to an eighth of a billion acres, about the size of Germany. Over half the world's soybeans and a third of the corn now contain genes spliced in from other forms of life. You've probably eaten some lately—unwittingly, since our government prohibits their labeling. The official assumption is that they're different enough to patent but similar enough to make identical food, so Europe's insistence on labeling, to let people choose what they're eating, is considered an irrational barrier to free trade.

Traditional agronomy transfers genes between plants whose kinship lets them interbreed. The new botany mechanically transfers genes between organisms that can never mate naturally: an antifreeze gene from a fish (Arctic flounder) rides a virus host to become part of a potato or a strawberry. Such patchwork, done by people who've seldom studied evolutionary biology and ecology, uses so-called "genetic engineering"—a double misnomer. It moves genes but is not about genetics. "Engineering" implies understanding of the causal mechanisms that link actions to effects, but nobody understands the mechanisms by which genes, interacting with each other and the environment, express traits. Transgenic manipulation inserts foreign genes into random locations in a plant's DNA to see what happens. That's not engineering; it's the industrialization of life by people with a narrow understanding of it.

The results, too, are more worrisome than those of mere mechanical tinkering, because unlike mechanical contrivances, genetically modified organisms reproduce, genes spread, and mistakes literally take on a life of their own, extending like Africanized bees. Herbicide-resistance genes may escape to make "superweeds." Insecticide-making genes may kill beyond their intended targets. Both these problems have already occurred; their ecological effects are not yet known. Among other recent unpleasant surprises, spliced genes seem unusually likely to spread to other organisms. Canola pollen can waft spliced genes more than a mile, and common crops can hybridize with completely unrelated weeds. Gene-spliced Bt insecticide in corn pollen kills Monarch butterflies; that insecticide, unlike its natural forbear, can build up in soil; and corn borers' resistance to it is apparently a dominant trait, so planned anti-resistance procedures won't work.

It could get worse. Division into species seems to be nature's way of keeping pathogens in a box where they behave properly (they learn that it's a bad strategy to kill your host). Transgenics may let pathogens vault the species barrier and enter new realms where they have no idea how to behave. It's so hard to eradicate an unwanted wild gene that we've intentionally done it only once—with the smallpox virus.

Since evolution is a fundamental process, it must occur at every scale at which it's physically possible, down to and including the nanoecosystem of the genome. Shotgunning alien genes into the genome is thus like introducing exotic species into an ecosystem. (Such invasives are among the top threats to global biodiversity today.) It's unwise to assume, as "genetic engineers" generally do, that 90+ percent of the genome is "garbage" or "junk" because they don't know its function. That mysterious, messy, ancient stuff is the context that influences how genes express traits. It's the genetic version of biodiversity, which in larger ecosystems is the source of resilience and endurance.

Transgenics is showing disturbing historical parallels to another problematic invention, nuclear fission—"a fit technology," said Nobel laureate Robert Sinsheimer, "for a wise, farseeing, and incorruptible people." In both enterprises, technical ability has evolved faster than social institutions; skill has outrun wisdom. Both have overlooked fundamentals, often from other disciplines wrongly deemed irrelevant. Both have overreached—too far, too fast, too uncritical. Both have failed to take their values from their customers and their discipline from the market. The rise and fall of such technologies seems to go something like this:

1 Promoters promise public benefits. Gifted scientists relish the "sweet" technology. Commercial enthusiasm and pride, bolstered by government promotion, draw huge investments. Advocates shield the promoters from political and market accountability, suppress dissent, and reject independent assessment. Rapid growth speeds industrial capture of the regulatory apparatus. (The combination of greedy firms, sleepy watchdogs, and sparse disinterested scrutiny is a recipe for trouble, since systems without feedback are by definition stupid.)

2 Initial technical stumbles and troublesome questions elicit public concern, deflected by PR. Public concern increases because the more people find out about the innovation, the less they like it. The PR grows stronger but less persuasive. Emergent whistleblowers raise awkward questions. Many bad surprises dwarf the few benefits.

3 Operational disappointments abound as it becomes clear that the problems with the innovation are fundamental. Simultaneously, many people realize that the alternatives, often long known, actually work better and cost less.

4 Smart money and insurance coverage exit; practitioners stop having fun; some have nightmares without a safe place to discuss them. The product can be sold only by concealing its identity—a mockery of economic principles. Almost everyone realizes the business is dying of an incurable attack of market forces.
5 With insubstantial benefits, mediocre performance, real risks, and unrewarding economics finally undeniable, the technology fades away, leaving behind socialized hazards, failed firms, disappointed investors, delegitimized institutions, and a cynical public.

Where's the "You Are Here" sign for transgenic crops? Europe is already at Stage 4. The US is around Stage 2, and moving rapidly in the same direction.

With transgenic crops, as with nuclear fission, the key choices are not between unwelcome alternatives—nuclear warheads or subjugation, nuclear power or freezing in the dark, transgenic crops or starvation—but between those bad choices and attractive ones outside the orthodoxy. For crops, the best choice would be fairer distribution of food grown by a respectful and biologically informed agriculture that stops treating soil like dirt. But sound choices tend to emerge and get adopted in time only if we take seriously the discipline of mindful markets and the wisdom of informed democracy. The botanists now being welcomed to St Louis can help us see beyond molecules and genes to plants, and beyond plants to ecosystems. Botanists have a professional duty to help us all understand the vital differences between biology and biotechnology—between the foundations of their traditional science and the smartaleck, scientifically immature, but commercially hell-for-leather enterprise, a billion times younger, that aims to replace it.

Section 9

Final Thoughts: Finding the Path

Perhaps Amory's greatest legacy is his outlook, often mistakenly considered optimistic, if not Panglossian. His focus on "applied hope" undoubtedly started with his childhood and his parents, and then was nurtured by various individuals throughout the journey. In an interview with a Japanese media executive after he won the important Blue Planet Prize, Amory observed, "It's like the old argument about 'Is the glass half empty? Is it half full?' ... Dave Brower taught me an interesting thing about this. He said, 'Optimism and pessimism are different sides ... different aspects of the same simplistic and irresponsible surrender to fatalism—treating the future as fate, not choice, and therefore not taking responsibility for creating the future we want.'" These last three essays—visions, really—capture the essence of Amory Lovins, and as such they are worth nurturing, sharing, and amplifying, with all and for all. "Imagine a World" (2007) was a keynote address by Lovins celebrating Rocky Mountain Institute's 25th anniversary; "Remarks on Acceptance of the Blue Planet Prize" (2007) challenges Japan's energy strategy to rise to the occasion (as it may have begun to do after the 2011 Fukushima disaster); and "Applied Hope" (2008) was written for RMI's 2008 Annual Report.

Chapter 32

Imagine a World ...

2007

Let me tell you a story.[1] In the early 1950s, the Dayak people in Borneo had malaria. The World Health Organization had a solution: spray DDT. They did; mosquitoes died; malaria declined; so far, so good. But there were side-effects. House roofs started falling down on people's heads, because the DDT also killed tiny parasitic wasps that had previously controlled thatch-eating caterpillars. The colonial government gave people sheet-metal roofs, but the noise of the tropical rain on the tin roofs kept people awake. Meanwhile, the DDT-poisoned bugs were eaten by geckoes, which were eaten by cats. The DDT built up in the food chain and killed the cats. Without the cats, the rats flourished and multiplied. Soon the World Health Organization was threatened with potential outbreaks of typhus and [sylvatic] plague, and had to call in RAF Singapore to conduct Operation Cat Drop—parachuting a great many live cats into Borneo.

This story—our guiding parable at Rocky Mountain Institute—shows that if you don't understand how things are connected, often the cause of problems is solutions. Most of today's problems are like that. But at RMI we harness hidden connections so the cause of *solutions* is solutions: we solve, or better still avoid, not just one problem but many, without making new ones, before someone has to go parachuting more cats. So join me in envisioning where these linked, multiplying solutions can lead if we take responsibility for creating the world we want.

Imagine a world, a few short generations hence, where spacious, peppy, ultrasafe, 120-to-200-mpg cars whisper through revitalized cities and towns, convivial suburbs, and fertile, prosperous countryside, burning *no* oil and emitting pure drinking water—or nothing; where sprawl is no longer mandated or subsidized, so stronger families eat better food on front porches and more kids play in thriving neighborhoods; where new buildings and plugged-in parked cars produce enough surplus energy to power the now-efficient old buildings; and where buildings make people healthier, happier, and more productive, creating delight when entered, serenity when occupied, and regret when departed.

Imagine a world where oil and coal are nearly phased out and nuclear energy has disappeared, all vanquished by the competitors whose lower costs, risks, and delays have *already* enabled them to capture most of the world's market for new electrical services—energy efficiency, distributed renewables, combined-heat-and-power—and by advanced biofuels that use no cropland and move carbon from air to topsoil; where resilient, right-sized energy systems make major failures impossible, not inevitable; where collapsing oil demand and price have defunded enemies, undermined dictatorship and corruption, and doused the Mideast tinderbox; where energy policy is no longer a gloomy multiple-choice test—do you prefer to die from (a) climate change, (b) oil wars, or (c) nuclear holocaust? We choose (d) *none* of the above.

Imagine, therefore, a world where carbon emissions have long been steadily *declining*—at a handsome profit, because saving fuel costs less than buying fuel; where global climate has stabilized and repair has begun; and where this planetary near-death experience has finally made antisocial and unacceptable the arrogance that let cleverness imperil the whole human prospect by outrunning wisdom.

Imagine a world where the successful industries, rather than wasting 99.98 percent of their materials, follow Ray Anderson's lead: they take nothing, waste nothing, and do no harm; where the cost of waste is driving *un*natural capitalism extinct; where service providers *and* their customers prosper by doing more and better with less for longer, so products become ever more efficient to make and to use; where integrative engineering and biomimicry create abundance by design; and where elegant frugality turns scarcities and conflicts in energy, water, land, and minerals into enough, for all, for ever.

Imagine a world where the war against the Earth is over; where forests are expanding, farms emulate natural systems, rivers run clean, oceans are starting to recover, fish and wildlife are returning, and a stabilizing, radically resource-efficient human population needs ever less of the world's land and metabolism, leaving more for all the relatives who give us life.

Imagine a world where we don't just know more—we also know better; where overspecialization and reductionism have gone from vitally fashionable to unaffordably foolish; where vision across boundaries triumphs, simply because it works better and costs less.

Imagine a world secure, free from fear of privation or attack; where conflict prevention is as normal as fire prevention; where conflicts not avoided are peacefully resolved through strengthened international laws, norms, and institutions; where threatened aggression is reliably deterred or defeated by non-provocative defense that makes others feel and be more secure, not less; where all people can be nourished, healthy, and educated; and where all know Dr King's truth that "Peace is not the absence of war; it is the presence of justice."

Imagine a world where reason, diversity, tolerance, and democracy are once more ascendant; where economic and religious fundamentalism are obsolete; where tyranny is odious, rare, failing, and dwindling; and where global consciousness has transcended fear to live and strive in hope.

This is the astonishing world we are all gradually creating together. It's being built before our eyes by a myriad world-weavers, half of whom are women, many poor, many linked via millions of grass-roots groups. In their many ways, they're mobilizing society's most potent forces—businesses in mindful markets and citizens in vibrant civil society—to do what is necessary at this pivotal moment, the most important moment since we walked out of Africa: the moment when humanity has *exactly enough time, starting now*.

Running through this emerging tapestry is a bright thread: a small group of unusual people who—with humor and fearlessness, *chutzpah* and humility, eager enthusiasm and relentless patience—are composing their lives and combining their efforts to make it so.

Here we are. And now imagine the power of *all of us together* to make it so.

Note

1   The story is true in all essential features, and Alan AtKisson wrote a song about it, though some minor details remain obscure. My efforts to confirm the logistics of Operation Cat Drop in the Kelabit Highlands of Borneo haven't yet succeeded because the RAF keeps its files by date, not operation name; perhaps some RAF veteran reading this book can provide the clue. But though original source Tom Harrison has died, some key elements are confirmed by scientific colleagues and literature, others by the 2010 Borneo interview of a Dayak elder by Dr Willie Smits, and the whole story by the longest-serving WHO employee when the US television program *60 Minutes* rang that agency to check the story after I told it on camera in 1986 and a caller dismissed it as an urban myth. She retorted, said CBS, "Yes, that's exactly what we did. Do you mean those fools have forgotten already? We should have put it up on every office wall in the building so we'd remember that mistake and not make it again." She was then just about to retire, so three-odd decades after the event, the institutional memory was about to have been lost.

Chapter 33

Remarks on Acceptance of the Blue Planet Prize

2007

Your Royal Highnesses, Excellencies, distinguished guests, ladies and gentlemen:

This precious award honors decades of collaboration with my colleagues at Rocky Mountain Institute and around the world, including Japan. Special meaning comes from this prize's roots in Japan—the world leader in eliminating waste (*muda*), in beautifully simple design that harmoniously integrates people within nature, and in social ability to form and quickly adopt a new consensus. These attributes can uniquely equip Japan—*if* the Japanese people choose to accept this mission—to lead the world on the historic shift to benign, secure, and affordable energy, for all, for ever.

This leadership will challenge the Japanese people to make four changes:

1 Japan's extraordinary gains in energy efficiency after the 1970s oil shocks have faded into complacency. Japan, once the pioneer of energy efficiency, now has passenger vehicles scarcely more efficient, and buildings less efficient, than in America. The average person uses more electricity in Japan than in California or New York, and that use is growing as fast as in Texas. Some Japanese firms do keep getting more efficient, but few Japanese people pay much attention. Too many think climate protection means cost, burden, and sacrifice—not profit, competitive advantage, and higher quality of life. Today's techniques can profitably at least *triple* Japanese energy efficiency, enhance security, and help protect our blue planet—*if* Japanese people realize this is possible and insist that it happen.

2 Japan is poor in *fuels*, but is the richest of all major industrial countries in renewable *energy* that can meet the entire long-term energy needs of an energy-efficient Japan, at lower cost and risk than current plans. Japanese industry can do it faster than anyone—*if* Japanese policymakers acknowledge and allow it.

3 The old idea that a big industrial economy requires giant, vulnerable power

plants is now obsolete. The revolutions in miniaturization and information make millions of smart distributed electric generators cheaper, faster to build, and more reliable than a few big plants—*if* old institutions and habits stop favoring central plants.

4 Today's fast-moving energy technologies and markets make old bureaucratic and monopolistic habits no longer in the national interest. Japanese energy policy needs to become more diverse, agile, and open. Japan's technological and commercial genius will best flower *if* all ways to save or produce energy can compete fairly, at honest prices—no matter which kind they are, what technology they use, where they are, how big they are, or who owns them.

Japan's energy future and the world's depend on these four big *if*s. Each is a big challenge—and a huge opportunity. So let me end with a fifth *if*:

If the nation with the sacred sun on its flag now turns these potentials into reality, and by its example, leadership, and investment quickly shares them with its neighbors, then Japan's highest purpose in history will be achieved; your country will have led the whole world to be healthier, safer, richer, fairer, and cooler; and all beings everywhere will be as happy and grateful as I am today.

Chapter 34

Applied Hope

2008

The early bioneer Bill McLarney was stirring a vat of algae in his Costa Rica research center when a brassy North American lady strode in. What, she demanded, was he doing stirring a vat of green goo when what the world really needs is *love*? "There's *theoretical* love," Bill replied, "and then there's *applied* love"—and kept on stirring.

At Rocky Mountain Institute, we stir and strive in the spirit of applied hope. Our 90 people work hard to make the world better, not from some airy theoretical hope, but in the practical and grounded conviction that starting with hope and acting out of hope can cultivate a different kind of world worth being hopeful about, reinforcing itself in a virtuous spiral. Applied hope is not about some vague, far-off future but is expressed and created moment by moment through our choices.

Applied hope is not mere optimism. The optimist treats the future as fate, not choice, and thus fails to take responsibility for making the world we want. Applied hope is a deliberate choice of heart and head. The optimist, says RMI Trustee David Orr, has his feet up on the desk and a satisfied smirk knowing the deck is stacked. The person living in hope has her sleeves rolled up and is fighting hard to change or beat the odds. Optimism can easily mask cowardice. Hope requires fearlessness.

Fear of specific and avoidable dangers has evolutionary value. Nobody has ancestors who weren't mindful of saber-toothed tigers. But pervasive dread, currently in fashion and sometimes purposely promoted, is numbing and demotivating. When I give a talk, sometimes a questioner details the many bad things happening in the world and asks how dare I propose solutions: isn't resistance futile?

The only response I've found is to ask, as gently as I can, "Does feeling that way make you more effective?"

To be sure, mood does matter. The last three decades of the 20th century reportedly saw 46,000 new psychological papers on despair and grief, but only 400 on joy and happiness. If psychologists want to help people find joy and happiness, they're looking in the wrong places. Empathy, humor, and reversing both inner and outer poverty are all vital. But the most solid foundation we know for feeling better about the future is to improve it—tangibly, durably, reproducibly, and scalably.

At RMI we're practitioners, not theorists. We do solutions, not problems. We do transformation, not incrementalism. In a world short of both hope and time, we seek to practice Raymond Williams's truth that "To be truly radical is to make hope possible, not despair convincing." Hope becomes possible, practical—even profitable—when advanced resource efficiency turns scarcity into abundance. The glass, then, is neither half empty nor half full; rather, it has a 100 percent design margin, expandable by efficiency.

In this [RMI's 2008] *Annual Report*, my colleagues outline the latest steps in RMI's long journey of applied hope. As signs of RMI's effectiveness proliferate, our challenges are chiefly those of success—of needing ever more discriminating focus as the world moves our way, demanding that our limited resources be rapidly scaled to serve nearly infinite needs. We can't do everything; doing just anything may miss the mark; doing nothing is unacceptable; but doing the right things at the right time can make all the difference. We are intently engaged in discerning and reaching those goals.

In a world so finely balanced between fear and hope, with the outcome in suspense and a whiff of imminent shift in the air, we choose to add the small stubborn ounces of our weight on the side of applied hope.

Index